An Introduction to Pollution Science

ISBN-10: 0-85404-829-4
ISBN-13: 978-0-85404-829-8

A catalogue record for this book is available from the British Library

Published by The Royal Society of Chemistry,
Thomas Graham House, Science Park, Milton Road,
Cambridge CB4 0WF, UK

Registered Charity Number 207890

For further information see our web site at www.rsc.org

Typeset by Macmillan India Ltd, Bangalore, India
Printed by Henry Ling Ltd, Dorchester, Dorset, UK

Preface

While this book is in its first edition, it nonetheless has a lengthy pedigree, which derives from a book entitled *Understanding Our Environment: An Introduction to Environmental Chemistry and Pollution*, which ran to three editions, the last of which was published in 1999. *Understanding Our Environment* has proved very popular as a student textbook but changes in the way that the subject is taught had necessitated its splitting into two separate books.

When *Understanding Our Environment* was first published, neither environmental chemistry nor pollution science was taught in many Universities, and most of those courses which existed were relatively rudimentary. In many cases, no clear distinction was drawn between environmental chemistry and pollution science and the two were taught largely hand-in-hand. Nowadays, the subjects are taught in far more institutions and in a far more sophisticated way. There is consequently a need to reflect these changes in what would have been the fourth edition of *Understanding Our Environment*, and after discussion with contributors to the Third Edition and with the Royal Society of Chemistry, it was decided to divide the former book into two and create new books under the titles respectively of *Pollution Science* and *Principles of Environmental Chemistry*. Because of the authoritative status of the authors of *Understanding Our Environment* and very positive feedback which we had received on the book, where possible it was decided to retain the existing chapters with updating in the new structure and enhance them through the inclusion of further chapters.

This division of the earlier book into two new titles is designed to accommodate the needs of what are now two rather separate markets. *Pollution Science* is designed for courses within degrees in environmental sciences, environmental studies and related areas including taught postgraduate courses which are not embedded in a specific physical science or life science discipline such as chemistry, physics or biology. The level of basic scientific knowledge assumed of the reader is therefore only that of the generalist and the book should be accessible to a very wide readership including those outside of the academic world wishing to acquire a broadly based knowledge of pollution phenomena. The second title, *The Chemistry of the Environment* assumes a significant knowledge of chemistry and is aimed far more at courses on environmental chemistry which are embedded within chemistry degree courses. The book will therefore be suitable for students taking second or third year option courses in environmental chemistry or those taking specialised Masters courses, having studied the chemical sciences at first degree level.

In this volume I have been fortunate to retain the services of a number of authors from *Understanding Our Environment*. The approach has been where possible to update chapters from that book, although some of the new authors have decided to take a completely different approach. The book deals with the atmosphere, the aquatic environment and the solid earth as separate components, which is followed by chapters on investigating the environment and on ecological and health effects of chemical pollution. The final chapter deals with the technical and institutional basis of environmental management in all compartments.

I am grateful to the authors for making available their great depth and breadth of experience to the production of this book and for tolerating my many editorial quibbles. I believe that their contributions have created a book of widespread appeal, which will find many eager readers both on taught courses and in professional practice.

Roy M. Harrison
Birmingham, UK

Contents

CHAPTER 1

Introduction

ROY M. HARRISON

School of Geography, Earth and Environmental Sciences,
University of Birmingham, Birmingham, UK

1.1   WHAT IS POLLUTION SCIENCE?

There are various definitions for environmental pollution, all of which contain two key components. The first is that pollution involves some kind of change to the environment. The most obvious kinds of changes are the addition of man-made chemicals that do not occur naturally. Equally important, however, can be the additions of chemicals which do occur naturally in the environment, provided the resultant concentration meets the second criterion. This criterion is that for the phenomenon to be described as pollution, then the perturbation suffered by the environment must in some way be harmful. Not all pollution phenomena are the results of chemicals in the environment. Other important forms of pollution include thermal pollution, an example of which is the discharge of relatively warm power station cooling waters into coastal seas where they can lead to a significant change in the ecology of aquatic organisms. A further example is that of light pollution, caused by the massive amount of urban street lighting, which has a deleterious effect on the environment through obscuring our view of stars in the nighttime sky. Noise pollution has important aesthetic impacts through causing widespread annoyance, but is increasingly suspected of causing adverse effects on health.

This book concerns itself almost exclusively with chemical pollution. It considers the environment as a set of compartments that is familiar to us from our everyday existence. Therefore, separate chapters deal with the atmosphere, the world's waters, and soils and the solid earth. While such a subdivision is convenient in that pollutants behave very differently in each of these media, it is of course not the full story. The atmosphere is very mobile and many pollutants have lifetimes in the atmosphere of only hours or days, although some remain for much longer. Pollutants in the aquatic environment, unless rapidly biodegraded, will often be present for days or weeks, and persistent pollutants for many years. In soils and sediments, however, pollutants can remain relatively immobile for tens or hundreds of years. The rates of mixing are also very different. An atmospheric pollutant with a lifetime of more than a year will become globally mixed, whereas the same degree of mixing throughout the oceans will take centuries and throughout the solid earth will never occur unless there are pathways through water and air. Exchange processes between these major environmental compartments can be very important. Thus, for example, the largest inputs of some pollutants to the North Sea arise through deposition from the atmosphere. There are other examples where pollutants discharged to the sea can become suspended in sea spray and lead to contamination of both the air and the land.

There are two main ways of monitoring chemical pollution. The first and the most obvious is through chemical analysis and that is the main focus of Chapter 5. However, there are also techniques of biological monitoring, which depend upon evaluating the effects of chemical pollution on certain sensitive organisms. Both methods have their advantages and disadvantages. Biological monitoring is rarely specific to a single substance whereas chemical monitoring is. On the other hand, chemical monitoring cannot establish an adverse effect, only a concentration, whereas biological monitoring establishes the effect rather than the concentration. The two are therefore essentially complementary. Chapter 6 examines the ecological and health effects of the chemical pollutants and Chapter 7 the institutional framework for managing the environment.

Pollution science is a relatively new discipline that brings together the various areas of traditional science, mainly from within chemistry, physics and biology, necessary to understand the behaviour of pollutants in the environment, to appreciate their effects on the environment and humans, and to monitor and manage those pollutants.

1.2   THE CHEMICALS OF INTEREST

A very wide range of chemical substances are considered in this book. They fall into three main categories:

   (a) *Chemicals of concern because of their human toxicity.* Some metals such as lead, cadmium and mercury are well known for their adverse effects on human health at high levels of exposure. These metals have no known essential role in the human body and therefore exposures can be divided into two categories (see Figure 1). For these non-essential

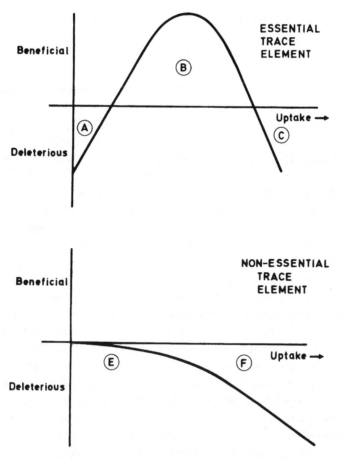

Figure 1 *Comparison of the consequences of exposure to essential and non-essential trace elements. For the essential trace elements, Region A represents the deficiency syndrome when intakes are insufficient, Area B is the optimum exposure window and in Area C, excessive intake leads to toxic consequences. In the case of the non-essential trace elements at low exposures (Zone E) the element is tolerated and little if any adverse effect occurs. In Zone F toxic symptoms are developed*

elements, at very low exposures the metals are tolerated with little, if any, adverse effects, but at higher exposures their toxicity is exerted and health consequences are seen. In the case of the so-called essential trace elements (see Figure 1) the human body requires a certain level of the element, and if intakes are too low then deficiency syndrome diseases will result. These can have consequences as severe as those which result from excessive intakes. In between, there is an acceptable range of exposures within which the body is able to regulate an optimum level of the element. Fluoride is an example of a chemical with a very narrow window of optimal exposure. Fluoridation of water supplies is typically at a level of about $1 \, mg \, L^{-1}$. Half of this concentration may still result in deficiency syndrome and weakened teeth, while double this concentration can lead to the start of adverse effects on teeth and bones.

Environmental exposure to chemical carcinogens is very topical despite the minuscule risks associated with many such exposures at typical environmental concentrations. Examples of chemical carcinogens are benzene (largely from vehicle emissions), and polynuclear aromatic hydrocarbons (generated by combustion of fossil fuels). Figure 2 shows the structures of benzene, benzo(a)pyrene (the best known of the carcinogenic polycyclic aromatic hydrocarbons), and 2, 3, 7, 8-tetrachlorodibenzodioxin (the most toxic of the chlorinated dioxin group of compounds). Despite great public concern over the emissions of the last compound, the evidence for carcinogenicity in humans is quite limited.

(b) *Chemicals, which cause damage to non-human biota but are not believed to harm humans at current levels of exposure.* Many elements and compounds come into this category. For example, copper and zinc are essential trace elements for humans and their environmental exposures very rarely present risk to health. These elements are, however, toxic to growing plants and there are regulations limiting their addition to soil in materials such as sewage sludge which are disposed of to the land. Another category of substance for which there is ample evidence of harm to biota, but as yet little, if any hard evidence of impacts on human populations, are the endocrine-disrupting chemicals. These synthetic chemicals mimic natural hormones and can disrupt the reproduction and growth of wildlife species. Thus, for example, bis-tributyl tin oxide (TBTO) interferes with the sexual development of oysters and its use as an anti-fouling paint for inshore vessels is now banned in most parts of the world. A wide range of other chemicals including polychlorinated biphenyls (PCBs), dioxins and many chlorinated species are also believed to have oestrogenic or androgenic potential, although the level of evidence for adverse effects is variable.

(c) *Chemicals not directly toxic to humans or other biota at current environmental concentrations, but capable of causing environmental damage.*

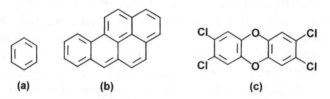

(a)　　　　**(b)**　　　　　　　　**(c)**

Figure 2 *Some molecules believed to have human carcinogenic potential: (a) benzene; (b) benzo(a)pyrene; and (c) 2,3,7,8-tetrachlorodibenzodioxin*

The prime example is the CFCs, which found widespread use precisely because of their stability and low toxicity to humans, but which at parts per trillion levels of concentration are capable of causing major disruption to the chemistry of the stratosphere.

1.3 UNITS OF CONCENTRATION

The concentration units used in environmental pollution are often confusing to the newcomer. Concentrations of pollutants in soils are most usually expressed in mass per unit mass, for example, milligrams of lead per kilogram of soil. Similarly, the concentrations in vegetation are also expressed in $mg\,kg^{-1}$ or $\mu g\,kg^{-1}$. In the case of vegetation and soils, it is important to distinguish between wet weight and dry weight concentrations, in other words, whether the kilogram of vegetation or soil is determined before or after drying. Since the moisture content of vegetation can easily exceed 50%, the data can be very sensitive to this correction.

In aquatic systems, concentrations can also be expressed as mass per unit mass and in the oceans some trace constituents are present at concentrations of $ng\,kg^{-1}$ or $\mu g\,kg^{-1}$. More often, however, the sample sizes are measured by volume and concentrations expressed as $mg\,L^{-1}$ are expressed as parts per million (ppm), $\mu g\,L^{-1}$ as parts per billion (ppb) and $ng\,L^{-1}$ as parts per trillion (ppt). This is unfortunate as it leads to confusion with the same units used in atmospheric chemistry with a quite different meaning.

Concentrations of trace gases and particles in the atmosphere can be expressed also as mass per unit volume, typically $\mu g\,m^{-3}$. The difficulty with this unit is that it is not independent of temperature and pressure. Thus, as an airmass becomes warmer or colder or changes in pressure, its volume will change, but the mass of the trace gas will not. Therefore, air containing $1\,\mu g\,m^{-3}$ of sulfur dioxide at 0°C will contain less than $1\,\mu g\,m^{-3}$ of sulfur dioxide if heated to 25°C. For gases (but not particles) this difficulty is overcome by expressing the concentration of a trace gas as a volume-mixing ratio. Thus, $1\,cm^3$ of pure sulfur dioxide dispersed in $1\,m^3$ of polluted air would be described as a concentration of one part per million (ppm). Reference to the gas laws tells us that not only is this one part per 10^6 by volume, it is also one molecule in 10^6 molecules and one mole in 10^6 moles, as well as a partial pressure of 10^{-6} atmospheres. Additionally, if the temperature and pressure of the airmass change, this affects the trace gas in the same way as the air in which it is contained and the volume-mixing ratio does not change. Thus, ozone in the stratosphere is present in the air at considerably higher mixing ratios than in the lower atmosphere (troposphere), but if the concentrations are expressed in $\mu g\,m^{-3}$ they are little different because of the much lower density of air at stratospheric attitudes. Chemical kineticists often express atmospheric concentrations in molecules per cubic centimetre (molec cm^{-3}), which has the same problem as the mass per unit volume units.

1.3.1 Worked Example

The concentration of nitrogen dioxide in polluted air is 85 ppb. Express this concentration in units of μgm^{-3} and molec cm^{-3} if the air temperature is 20°C and the pressure 1005mb (1.005×10^5Pa). Relative molecular mass of NO_2 is 46; Avogadro number is 6.022×10^{23}.

The concentration of NO_2 is 85 μl m^{-3}, at 20°C and 1005 mb,

$$85 \,\mu l \; NO_2 \; \text{weigh} \; 46 \times \frac{85 \times 10^{-6}}{22.41} \times \frac{273}{293} \times \frac{1005}{1013}$$

$$= 161 \times 10^{-6} \, g$$

(Since 46 g (1 mol) of NO_2 occupy 22.41 L at 273K and 1013 mb)

$$NO_2 \; \text{concentration} = 161 \; \mu g \, m^{-3}$$

This is equivalent to 161 pg cm^{-3}, and

$$161 \; \text{pg} \; NO_2 \; \text{contain} \; 6.022 \times 10^{23} \times \frac{161 \times 10^{-12}}{46} = 2.1 \times 10^{12} \; \text{molecules}$$

and

$$NO_2 \; \text{concentration} = 2.1 \times 10^{12} \; \text{molec cm}^{-3}$$

Aquatic concentrations (including rainwater) are often expressed in chemical equivalents (*e.g.* μeq L^{-1}) and airborne concentrations can be expressed in μeq m^{-3}. The useful aspect of this convention is that for charge neutrality:

$$\sum \text{anions in equivalents} = \sum \text{cations in equivalents}$$

The amount in chemical equivalents of anion is calculated from:

$$\text{equivalents} = \frac{\text{mass (g)}}{\text{relative molecular mass}} \times \text{charge}$$

and

$$\mu eq = \frac{\text{mass} \; (\mu g)}{\text{relative molecular mass}} \times \text{charge}$$

REFERENCES

For readers requiring knowledge of basic chemical principles:
R.M. Harrison and S.J. de Mora, *Introductory Chemistry for the Environmental Sciences*, 2nd edn, Cambridge University Press, Cambridge, 1996.

For more detailed information upon pollution phenomena:
R.M. Harrison (ed), *Pollution: Causes, Effects and Control*, 4th edn, Royal Society of Chemistry, Cambridge, 2001.

CHAPTER 2

The Atmosphere

J.A. SALMOND[1], A.G. CLARKE[2] AND A.S. TOMLIN[2]

[1] Division of Environmental Health and Risk Management, School of Geography, Earth and Environmental Sciences, University of Birmingham, Birmingham B15 2TT, UK
[2] Energy and Resources Research Institute, University of Leeds, Leeds LS2 9JT UK

2.1 THE GLOBAL ATMOSPHERE

2.1.1 The Structure of the Atmosphere

2.1.1.1 Troposphere and Stratosphere. The atmosphere consists of a mixture of different gases and particles (both liquid and solid). While there is no fixed boundary between the limit of the Earth's atmosphere and space, most of the atmosphere is found in a zone within 100 km above the Earth's surface. In this zone, which is known as the *homosphere*, the atmosphere consists of a mixture of gases. Nitrogen (78%) and oxygen (21%) are the two dominant gases, and account for 99% of the total mixture. Other important gases are argon, neon, helium, methane, krypton and hydrogen.

The atmosphere also consists of a large number of trace gases. Depending on their toxicity and proximity to the Earth surface, these gases, which make up only a tiny percentage of the volume of the Earth's atmosphere, can have a disproportionate impact on human health and the environment.

The concentration of trace gases is variable in time and space. This means that the concentrations of trace gases in a sample of air taken from any given point in the atmosphere (in time or space) are likely to be different to those measured in a second sample. This is caused by variations in the processes that release the gases (characteristics of the source), chemical processes operating within the atmosphere and changes in the rate at which they are removed from the atmosphere. For example, in the tropics the concentration of water vapour is about 4% while near the South Pole the concentration is likely to be less than 0.00001%. The length of time trace gases spend within the atmosphere is known as the residence time. For example, the mean residence time for water vapour is about 11 days. After this, condensation and precipitation processes will typically result in its removal from the atmospheric system.

The atmosphere can be divided into 5 different categories or layers which each have different properties. These layers are

- the troposphere 0–10 km;
- the stratosphere 10–50 km;
- the mesosphere 50–90 km;
- the thermosphere 90–500 km; and
- the exosphere >500 km.

The mesosphere, thermosphere and exosphere account for less than 1% of the total mass of the atmosphere, so emphasis in this chapter is placed on the troposphere (90%) and stratosphere (9.5%). The vertical structure of the atmosphere, showing the features that are most relevant to the problems covered in this chapter, is illustrated in Figure 1. The figure shows the stratosphere, troposphere and boundary layer (that layer which is closest to the Earth's surface). The difference between the layers is characterised by changes in temperature and pressure with height.

The variability of conditions found in the troposphere leads to "the weather" as the layman understands it. The depth of the troposphere varies according to the amount of heat released at the Earth surface. For example, near the equator the mean height of the troposphere is 18 km while near the poles it is only 8 km deep. The temperature of the troposphere is determined primarily by heat release from the surface and thus temperatures decrease with height. Since warm air is less dense than cooler air, the air heated at the Earth's surface rises, causing convectional mixing or turbulence. Frictional forces, resulting from contact between the Earth surface and the atmosphere, act to increase turbulent mixing further and as a result the lower layers of the troposphere are typically well mixed during the day. As a result of the evaporation of water from the Earth surface, more than 95% of the total cloud formation and precipitation can be found in the troposphere. Water vapour in the troposphere however has a short residence time of the order of a few days and weeks.

The tropopause marks the boundary between the moist turbulent troposphere and the dry, stable, ozone-rich stratosphere. The stratosphere typically extends 50 km above the tropospause. The stratosphere is relatively cloud-free and considerably less turbulent – hence long distance passenger jets fly at stratospheric altitudes. In the stratosphere the temperature starts to increase again.

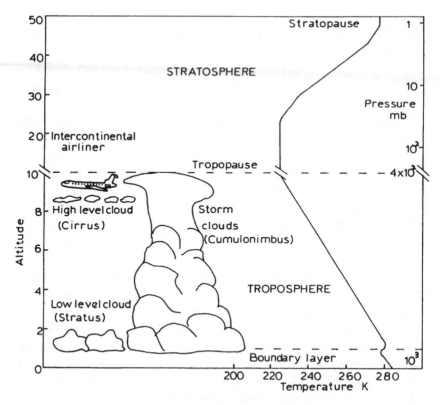

Figure 1 *The vertical structure of the atmosphere. The temperature profile would be typical for latitude 60° N in summer. Note the change of scale used for the upper half of the figure*

The stratosphere is primarily heated by the absorption of UV radiation from the sun by ozone (O_3), concentrations of which are highest between 15 and 50 km. This situation of a layer of warmer, less dense air over a layer of cooler, denser air is quite stable. Consequently, air is typically mixed across the tropopause very slowly unless special events such as tropospheric folding occur.

In the troposphere most pollutants have a fairly limited lifetime before they are washed out by rain, removed by reaction or deposited to the ground. However, if pollutants are injected directly into the stratosphere they can remain there for long periods because of slow downward mixing, resulting in noticeable effects over the whole globe. Thus, major volcanic eruptions injecting fine dust into the stratosphere can lead to a reduction in the amount of solar energy reaching the ground for more than a year after the event. Other global problems relating to events in the stratosphere such as the possibility of damage to the ozone layer are discussed later.

2.1.1.2 Atmospheric Circulation. Identifying the patterns of pollutant transportation within the atmosphere is fundamental to understand global and local environmental problems. The main driving forces for the circulation of the atmosphere are the energy received from incident solar radiation and the Earth's rotation. This results in two main types of global winds: *meridional circulation* patterns, which transport air north and south, and *jet streams*, which transport air primarily east to west around the globe.

Due to the angle between the sun and the Earth surface, the amount of solar energy falling on a given area varies with latitude so that the poles are cold and the equatorial regions warm. Warm air rises at the equator and flows North and South towards the poles. At about 30° North and South the air becomes cooler than the surrounding air and starts to sink. The resulting pressure gradient sets up a weak return flow near the surface as air moves from areas of high pressure in the so-called "horse latitudes" back towards the lower pressure areas near the surface at the equator. A similar situation occurs at the Poles where cool air sinks from aloft towards the surface resulting in high pressure at the surface and air moves along the surface towards the polar front located in the mid-latitudes (50–60°). This results in a second rotational cell in each hemisphere. The jet streams (bands of fast-moving latitudinal air flows aloft) typically occur at the junction between these cells.

The rotation of the Earth affects the circulation patterns in a fundamental way due to an effect called the Coriolis force. For example, in the North Hemisphere, air moving South towards the equator gives the impression of being influenced by a force in the westerly direction. The net result is the tendency for air to circulate in large-scale eddies around the "low" and "high" pressure regions on synoptic weather charts.

The proportions of incident radiation reflected back to space, or absorbed by the land or sea and reradiated at a longer wavelength, varies from place to place. This also affects the temperature distribution and circulation patterns. This part of the energy balance is crucial to the determination of the global climate and is considered in more detail in Section 2.1.2. The processes of evaporation of water, cloud formation and precipitation also affect the energy balance and circulation patterns.

Frictional forces, from the presence of the ground, only affect the lowest levels of the atmosphere and thus have little effect on the overall patterns of atmospheric circulation. At most altitudes therefore, air movements approximate to those of a non-viscous fluid. The theoretical wind speed can be calculated from the pressure gradient and the rotational velocity of the Earth – the so-called geostrophic wind speed. The pressure gradient is reflected on a weather chart by the closeness of the isobars, lines of constant pressure. If the isobars are close together the wind speed will be high.

2.1.1.3 The Boundary Layer. Near to the ground the situation is more complicated due to the effects of frictional and buoyancy forces. Mechanical forces generate turbulence as air flows over uneven ground features such as hills, buildings or trees. The ground may also warm or cool the air next to it resulting in up-currents and down-currents. In the language of fluid mechanics, the turbulent transport of momentum and energy corresponds to velocity and temperature gradients in the vertical direction. Consider the variation of wind speed with height over the lowest few hundred metres of the atmosphere. This variation is greatest over rough surfaces (*e.g.* a city) where the effect could be a reduction of 40% of the wind speed aloft, that is, the geostrophic wind. Over smooth surfaces (*e.g.* sea, ice sheets) the effect is less and the reduction may be only 20%. The changing effect of friction with height also causes a variation of wind direction with increasing height from the Earth's surface *i.e.* "wind shear". A plume from a tall chimney may therefore appear to be travelling in a different direction to the ground level wind direction.

Within the troposphere a boundary layer can be defined within which surface effects are important. This is of the order of 1 km in depth (Figure 1) but varies significantly with meteorological conditions. Vertical mixing of pollutants within the boundary layer is largely determined by the atmospheric stability, which relates to the intensity of the buoyancy and frictional effects previously mentioned. This is the subject of a later section. As a generalisation, mixing within the boundary layer is relatively rapid whereas mixing through the remainder of the troposphere is slower. This gives rise to the idea of a mixing depth within which pollutants are retained and may be transported long distances. So, for example, models of pollutant transport from the UK to the rest of Europe involve a distance scale of about 1000 km and often assume vertical mixing depth of perhaps 1 km with the pollutants uniformly distributed within this layer. Table 1 indicates the time and distance scales involved in the dispersion of pollutants emitted from the ground. No account is taken in this table of the rates of removal of any pollutant by reaction, deposition to the ground, *etc*.

2.1.2 Greenhouse Gases and the Global Climate

2.1.2.1 The Global Energy Balance. The amount of energy that reaches the Earth from the sun, and the absorption and loss of radiation from the Earth and

Table 1 *Time and distance scales for atmospheric dispersion of emissions*

Time of travel	Typical distances	Area affected
Hours	Tens of km	Throughout the boundary layer
Days	Thousands of km	Pollutant escaping from boundary layer into free troposphere
Weeks	Round the earth	The whole troposphere in one hemisphere. Transport to other hemisphere beginning
Months	Round the earth	Whole global troposphere. Some penetration into lower stratosphere

its atmosphere, determines our climate. The atmosphere plays a very important role in determining the amount and characteristics of radiation received at the Earth surface. If the Earth had no atmosphere, the mean surface temperature would be 255 K, well below the freezing point of water. Thus, the atmosphere serves to retain heat near the surface and the Earth is thereby made habitable. This accounting for incoming and outgoing energy is called the global radiation balance and could potentially be upset by any significant change to the Earth's atmosphere.

Most of the radiant energy from the sun lies in or near the visible region of the spectrum (*i.e.* at short wavelength ca $0.6\,\mu$m) with some in the UV region. The stratosphere absorbs UV radiation primarily due to the ozone present and this results in warming above the tropopause as shown in Figure 1. The lower atmosphere is transparent to visible light so it gains relatively little energy from incoming radiation. Some of the transmitted radiant energy penetrates to the ground and is absorbed. Some is reflected unchanged from clouds or from the ground (especially by snow or ice). The fraction of reflected light is termed the albedo and is over 0.5 for clouds but below 0.1 for the oceans. The global average albedo is about 0.3. Figure 2 shows the amounts of radiation for different components of the overall energy balance.

The radiation emitted from the ground lies in the infra-red region of the spectrum (long wavelength, ca 10–$15\,\mu$m) and several atmospheric constituents absorb radiation at these wavelengths. Carbon dioxide, water vapour and ozone

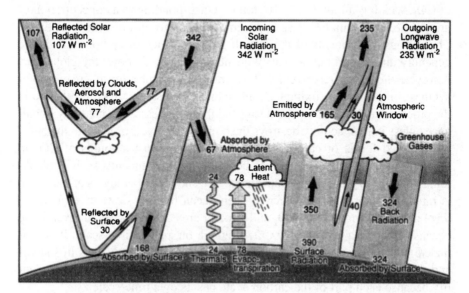

Figure 2 *The Earth's radiation and energy balance for a net incoming solar radiation of $342\ Wm^{-2}$*
(Reproduced with permission from the Intergovernmental Panel on Climate Change[2])

are the most important of these. Methane, nitrous oxide and chlorofluoro-carbons (CFCs) are also significant. Some of the absorbed energy will still be re-radiated back to space but a part will be returned to the ground or retained in the atmosphere. The net effect is that more energy is retained near the surface of the Earth and the mean temperature is therefore higher (global average 288 K). This is described as the "greenhouse effect" by analogy with the properties of glass. Glass is largely transparent to solar radiation while absorbing completely radiation in the infra-red at wavelengths greater than 3 μm. In fact the most important function of a greenhouse is to prevent the circulation of air, inhibiting the normal cooling processes, but the term "greenhouse effect" has none the less been retained.

The final factor that results in surface-to-atmosphere transfer of energy is direct warming of the air nearest to the ground together with evaporation/condensation processes.

2.1.2.2   The Carbon Dioxide Cycle.   Carbon dioxide is of major concern as a greenhouse gas because there is little doubt that man's activities are leading to a gradual increase in the atmospheric CO_2 level. This suggests that human activity may eventually modify the global climate. Fossil fuel burning is the main contributor to the global annual emissions, which have increased by a factor of about 10 since 1900 to an enormous mean of 6.3 $\times$ 10^9 tonnes per annum in 1990–1999.[1] Deforestation adds about another 1.6 $\times$ 10^9 tonnes per annum.[2] This must be considered in relation to the total atmospheric content of CO_2, which is about 750 $\times$ 10^9 tonnes corresponding to a concentration of around 367 ppmv in 1999[3] as opposed to 280 ppmv in pre-industrial times. The various components of the overall global balance of carbon dioxide are generally understood but not easily quantified. Figure 3 shows the global carbon cycle and carbon reservoirs. CO_2 is removed from the atmosphere by photo-synthesis in plants thus fixing CO_2 into a biomass reservoir. CO_2 is released in the processes of respiration and decay and these processes are naturally in balance unless human activity destroys the biomass reservoir or leads to the burning of fixed forms of carbon. The oceans contain vast amounts of CO_2 in inorganic form as well as in association with living organisms such as plankton. Exchange of gas between the atmosphere and the upper layers of the ocean is rapid and subsequent transfer to deep ocean regions slow. In some areas there may be net release of CO_2 and in other areas net removal but overall the oceans represent a net sink for CO_2 although on a slow time-scale. It is estimated that the time taken for the atmosphere to adjust to changes in sources and sinks of CO_2 is between 50 and 200 years although this is difficult to quantify because each part of the carbon cycle has its own time-scale.

2.1.2.3   Global Warming.   The rate of concentration change of CO_2 and other greenhouse gases is shown in Table 2. What is important however is not

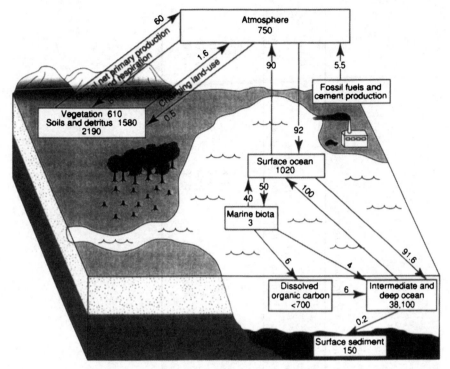

Figure 3 *The carbon dioxide cycle, showing the reservoirs (in GtC) and fluxes (GtC per year) relevant to the anthropogenic perturbation averages over the period 1980 to 1989* (Reproduced with permission from the Intergovernmental Panel on Climate Change[2])

just the rate of increase but the effect each species could have on global warming. Molecule-for-molecule changes in CH_4, N_2O and the CFCs have more effect than changes in CO_2 although their overall concentrations are lower. The Global Warming Potential (GWP) is a quantified measure of the relative effects of each species on radiative forcing of the atmosphere, including both direct and indirect effects. The index is defined as a cumulative radiative forcing between the present time and some specified time in the future caused by a unit mass of gas emitted at present relative to CO_2.[2]

The total global warming effect of each gas is then determined by multiplying by the amount of gas emitted. A typical uncertainty in the figures is about 35%. Table 2 demonstrates the GWPs for a number of greenhouse gases. Although CO_2 is the most important contributor, the other gases taken together contribute about half the overall radiative forcing. Some of the species represent CFCs and their replacements following the Montreal Protocol. Although these species have high GWPs their concentrations are small and their total impact less than 3%. These species are covered further in Section 2.1.3.

Table 2 *Concentration changes, lifetimes and global warming potentials of Greenhouse gases*[2,3]

Species	Pre-industrial concentration	Concentration in 1998	Rate of concentration change (% per year)	Atmospheric lifetime (years)	GWP (time horizon 20 years)	GWP (time horizon 100 years)
CO_2	280 ppmv	368 ppmv	0.4	50–200	1	1
CH_4	700 ppbv	1745 ppbv	0.6	12	56	23
N_2O	275 ppbv	314 ppbv	0.25	120	278	296
CFC-11	zero	268 pptv	0	50	6300	4600
CF_4	zero	80 pptv	12	50,000	3900	5700
HCFC-22 (a CFC substitute)	zero	132 pptv	5	12	4800	1700

The situation with ozone is quite complex and depends on its vertical distribution. In the stratosphere it absorbs UV and so a reduction in lower stratospheric ozone has a negative effect on global warming because it allows less radiation to reach the tropospheric system. In the troposphere there appears to be a gradual increase in the level of ozone due to emissions of NO_x and hydrocarbons and this will have a positive effect on global warming. The net effect of changing ozone levels is predicted to be positive although small and may vary from region to region.

Aerosols (including particles, small droplets and soot) may also effect global warming by either scattering or absorbing radiation or through their effects on clouds.

Although the effects of aerosols show a complex dependency on their size and distribution, there have been significant advances in quantifying their contribution to global warming, and current models predict that they have a negative (*i.e.* a cooling) overall effect. Aerosols are typically very short-lived species and hence their impact on radiative forcing will change quickly with variations in emissions. An example of this is the cooling effect caused by the 1991 volcanic eruption of Mt Pinatubo.

2.1.2.4 Climate Change. There now seems to be some consensus that mean surface temperatures have been increasing since the late 19th century at a rate over and above natural variability. However, the climatological consequences of global warming are still not well understood. Modelling the effect of increased greenhouse gas levels on the global climate is an enormously complex problem requiring high performance computers. Three-dimensional ocean atmosphere global circulation models (OAGCMs) describe the vertical, latitudinal and longitudinal variations in conditions and attempt to compare the present situation with various scenarios for future emissions based on predicted population and economic growth, energy availability, *etc.*[3]

The feedback processes described above are increasingly represented more accurately in the models as is the coupling between atmosphere and ocean. Predictions show that changes in surface and atmospheric temperatures, cloud cover, evaporation and precipitation, *etc.* are all affected by the changed radiation balance but the effects are not equally distributed over the globe.

OAGCMs predict that global temperatures will rise by 1.5–4.5°C as a result of doubling the CO_2 concentrations in the atmosphere. Despite the improvements in modelling capability, this range is very similar to that predicted in 1990. The uncertainties arise from our limited understanding of (and hence ability to model) the complex feedback processes, which operate in the atmosphere-ocean system, particularly those involving clouds and aerosols. It is interesting to note however that there is more variability between two different OAGCMs which utilise the same emissions scenario than from output using the same model run twice, once including the effects of sulfate aerosols (thought to reduce warming) and once excluding their effect.

Our ability to accurately predict global climate change is also limited by the accuracy of our assumptions regarding future economic and social changes which have a direct impact on fossil fuel use and hence emissions. For this reason the latest Inter-Governmental Panel on Climate Change (IPCC) report focuses on four different emissions scenarios (Special Report on Emissions Scenarios or SRES). These are based on different assumptions about human behaviour. As shown in plate 1, different assumptions can have a significant impact on regional trends in warming.

The top figure shows the scenario which relates to a heterogeneous world (SRES A2 in the IPCC report).[3] The emissions estimates are based on the assumption that populations continue to increase and economic development is regionally orientated. In comparison, the lower figure is based on an emissions scenario in which emphasis is placed on developing local solutions to economic, social and environmental sustainability (SRES B2 in the IPCC report).[3] Although populations continue to grow they do so at a slower rate than in scenario A2. This figure shows less warming, particularly in the Southern Hemisphere, which is consistent with reduced emissions. Both figures are based on the output from a number of different OAGCMs.

Model output is even less consistent as to the effect of increased CO_2 concentrations on global precipitation. Models typically suggest that the tropics will become wetter and the sub-tropical regions become drier. This might be critical for some central African regions, which already suffer severe drought conditions.

Changes in sea level arising from the melting of ice-caps, glaciers and thermal expansion of the oceans are predicted at 0.09–0.88 m by 2100. A significant thinning of the Arctic ice seems to have occurred already and is currently being studied. Changes in ocean circulation patterns are a further possible consequence of global warming. For example, changes in the thermohaline circulation (an ocean current found in both the Northern and Southern Hemispheres which plays an important role in transporting heat between the tropics and the poles) is expected. A weakening of this current could result in significant cooling in north-west Europe and north-eastern North America. However, again model results are inconsistent and this represents one of the important feedback mechanisms that require further study.

2.1.2.5 International Response.   International discussions have been taking place for some years with a view to limiting the emissions of greenhouse gases. The second World Climate Conference met in Geneva in 1990. It had as its technical basis a report[5] from the UNIPCC, an international body of

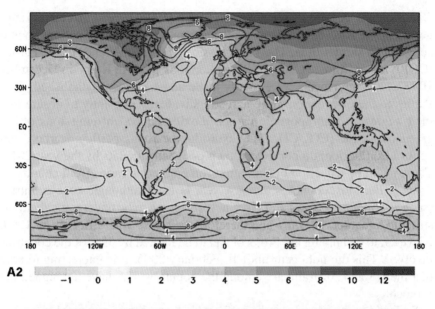

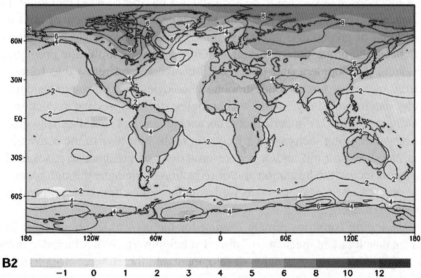

Plate 1 *The annual mean change of the temperature (colour shading) and its range (isolines)*
(Unit: °C) for the Special Report on Emissions Scenarios (SRES) scenario A2 (upper
panel) and the SRES scenario B2 (lower panel) (see text for explanation). Both SRES
scenarios show the period 2071 to 2100 relative to the period 1961 to 1990 and were
performed by Ocean Atmosphere Global Climate Models
(Reproduced with permission from the Intergovernmental Panel on Climate Change)

300 scientists. A follow up meeting in Berlin (1995) agreed that targets suggested by the Rio summit were inadequate and industrialised nations should act within a shorter time. The Third session of the Conference of the Parties (COP)[4] to the Climate Change Convention took place in Kyoto in December 1997 where agreements were finally made and a Protocol established.[6] The Parties to the Convention, have agreed individually or jointly to ensure that their aggregate anthropogenic carbon dioxide equivalent emissions of the greenhouse gases (CO_2, CH_4, NO_2, hydrofluorocarbons, perfluorocarbons and sulfur hexafluoride) do not exceed their assigned amounts by the year 2010. The amount of reduction or limitations on emissions varies from country to country. For most countries the base year is 1990, although some countries (such as Hungary, Poland and Ukraine) are agreed to be undergoing the transition to a market economy and therefore their base year differs from 1990. However, to be binding the Protocol had to be ratified by 55% of the countries involved. This did not occur until 16 February 2005. It is interesting to note that the US and Australia are among the countries who have not signed the Protocol.

Reductions are expected to be achieved in a number of ways including: the enhancement of energy efficiency; the protection and enhancement of sinks and reservoirs of greenhouse gases *e.g.* through sustainable forest management practices; the promotion of sustainable forms of agriculture; the promotion, research, development and increased use of new and renewable forms of energy, of carbon dioxide sequestration technologies and of advanced and innovative environmentally sound technologies; the progressive reduction or phasing out of market imperfections, tax exemptions and subsidies in all greenhouse gas emitting sectors that are counter to the objective of the convention; and measures to limit and/or reduce emissions of greenhouse gases in the transport sector; the limitation and/or reduction of methane through recovery and use in waste management, as well as in the production, transport and distribution of energy.

Developing countries remain exempt for the present, although there is a "clean development mechanism" aimed at helping parties not included in the Protocol to achieve sustainable development and therefore contribute to the ultimate objective of the Convention. There are also clauses in the Protocol aimed at encouraging the transfer of environmentally sound technologies from developed to developing nations through both technological and financial aid. Without such a transfer of technology it is likely that emissions from developing nations will rise steeply and may account for 60% of emissions over the next two decades. The transfer of emission reduction units from one party to another is allowed by the Protocol as long as the reduction processes used are additional to those which would already occur, and that all parties are already complying with the requirements of the Protocol.

The wide spread use of natural gas as a substitute for coal has reduced the mass of CO_2 emitted per unit of heat released: gas 0.43, oil 0.62, coal 0.75 k tonne (Mw yr)$^{-1}$. Increased reliance on renewable energy sources, such as solar power, wind power, hydroelectric power sources and nuclear energy have the potential to reduce CO_2 emissions from the power industry in the UK still further. However, such sources of energy remain expensive and have other environmental impacts and thus remain controversial.

However, many developing nations such as India and China have large coal reserves. The other alternatives are the use of renewable energy sources such as wind, solar, wave and tidal power or the further development of nuclear energy. This seems unlikely in the short term for both economic and environmental reasons. The use of biomass is also a possibility on a local scale since the replanting of biomass fuels makes it a sustainable energy source. National emissions of CO_2 for several countries are listed below. Figures for China and former USSR are calculated estimates.

UK (1994) 551 Tg[7] Germany (1994) 874 Tg[7] France (1994) 308 Tg[7]

USA (1995) 4786 Tg[8] Former USSR (1995) 3804 Tg[9] China (1995) 2389 Tg[9]

2.1.3 Depletion of Stratospheric Ozone

2.1.3.1 The Ozone Layer. Although ozone occurs in the troposphere and plays an important role in air pollution chemistry, about 90% of the total ozone content of the atmosphere occurs in the stratosphere at altitudes between 15 and 50 km. The ozone layer acts as a filter for UV radiation from the sun, removing most of the radiation below 300 nm. This serves to protect humans from the adverse effects of UV, which become significant below 320 nm since decreasing wavelength corresponds to higher energy photons, which can cause sunburn and types of skin cancer. Any depletion of stratospheric ozone would therefore lead to a larger amount of UV radiation incident on the Earth's surface and an increased risk of cancer.

Concern was first expressed about this risk in the early 1970s in connection with emissions of nitrogen oxides from supersonic aircraft such as Concorde, which fly in the lower stratosphere. Nitrogen oxides are potential catalysts for the destruction of ozone. This particular effect is now thought to be relatively minor and attention switched in the 1980s to halogen compounds, especially CFCs or freons. Freons are a group of chlorofluorocarbons, which have been used as aerosol propellants, refrigerants and as gases for the production of foamed plastics. Their attraction lies in the fact that they are non-toxic, non-flammable and chemically inert. Global production of the two commonest gases, CFC 11 ($CFCl_3$) and CFC 12 (CF_2Cl_2), rose rapidly from below 50,000 tonnes per annum in 1950 to 725,000 tonnes per annum by 1976 decreasing slightly to 650,000 tonnes in 1985. About 90% was released directly to the atmosphere while the remainder, representing the refrigerant use, will be released when the equipment is eventually discarded.

The actual concentration of CFCs in the atmosphere is extremely small, (less than 1 ppb, see Table 2), but rose dramatically the last century. For example, concentrations of CFC 11 rose from 0 to 268 ppt and CFC 12 to 533 ppt in 1998. Evidence suggests that concentrations of CFC 11 may now have peaked as the trends in concentration in the 1990s showed a net decrease of 1.4 ppt per year. However, continued release of CFC 12 has resulted in a net increase in the 1990s of 4.4 ppt per year, which correlates well with known emissions. This rise in CFCs in the 1980s has clearly affected stratospheric ozone levels via the processes described below and has been the subject of control measures in the 1990s which will be discussed in a later section. Concern over the global concentrations of these gases is also raised due to the high global warming potentials of many of these gases. However, the exact role these compounds play in global warming is dependent on historic patterns of emissions, residence time and the amount of chlorine and bromine in each molecule.[10]

Since they are chemically inert, CFCs are resistant to attack by molecules, radicals or the UV radiation present in the troposphere and are not subject to significant dry deposition or rain-out. The higher energy UV radiation in the stratosphere can however lead to photodissociation forming chlorine atoms which can in turn lead to the destruction of ozone. Despite the slow exchange of air between the troposphere and the stratosphere this effect is now known to be significant.

2.1.3.2 Ozone Depletion. The chemistry of ozone depletion is complex but a basic outline of the important processes is as follows. Ozone is formed from the dissociation of molecular oxygen by short wave length UV radiation in the upper stratosphere:-

$$O_2 \xrightarrow{\quad UV \quad} O^{\bullet} + O^{\bullet} \tag{1}$$

$$O^{\bullet} + O_2 + M \rightarrow O_3 + M \ (M = \text{inert third body}) \tag{2}$$

However, ozone itself is rapidly photodissociated

$$O_3 \xrightarrow{\quad UV \quad} O_2 + O^{\bullet} \tag{3}$$

and the so-called "odd oxygen" species and O_3 may interconvert many times before they destroy one another by

$$O^{\bullet} + O_3 \rightarrow O_2 + O_2 \tag{4}$$

In fact, measurements of the ozone profile in the atmosphere suggest that ozone destruction must be considerably faster than could be achieved by reaction (4) alone and that other reactions must be involved. These other mechanisms can be represented by

$$X + O_3 \rightarrow XO + O_2 \qquad (5)$$

$$XO + O^{\bullet} \rightarrow X + O_2 \qquad (6)$$

$$\text{Net effect} \qquad O^{\bullet} + O_3 \rightarrow O_2 + O_2$$

X may represent a range of molecules including Cl, Br, NO, OH and H and is not consumed in the overall ozone destruction process. If $X = NO$, the reactions form and destroy NO_2; if $X = Cl$, the reactions form and destroy ClO, but because this is a catalytic cycle small concentrations of X can have a significant effect on ozone levels. Other sets of reactions involving NO and Cl simply achieve the interconversion of O_3 and O and therefore have no effect on the net ozone levels. The reactive NO_x and Cl species can be removed by the formation of the relatively stable "reservoir" molecules HNO_3 and HCl or the somewhat short-lived chlorine nitrate $ClONO_2$. About half the stratospheric content of NO_x is stored as HNO_3 and about 70% of the chlorine as HCl. Although these may be reactivated by conversion back to NO_x and Cl, they may eventually be transferred back to the troposphere and removed to the ground by rain-out.

Case Study 1: The Antarctic Ozone "Hole"

In 1985 Farman *et al.*[11] published the results of ground-based measurements in Antarctica showing very significant depletions, of the order of 50%, in the total column ozone content of the atmosphere. The Antarctic ozone "holes" of 1992 and 1993 were the most severe on record, with ozone being locally depleted by more than 99% between about 14–19 km in October, 1992 and 1993. Subsequent aerial surveys and analysis of satellite data confirmed this phenomenon and led to a complete re-appraisal of the chemistry and meteorology involved in ozone depletion in the Antarctic.

During the dark, cold Antarctic winter upper stratospheric air moving from low to high latitudes subsides and as it does so develops a strong westerly circulation pattern. This produces a vortex, which effectively isolates the air in the lower stratosphere over the Antarctic continent from the air at lower latitudes. Within the vortex the temperature falls progressively until below about $-80°C$ polar stratospheric clouds (PSCs) may form. These are composed of very small particles ($1\,\mu m$) of nitric acid trihydrate ($HNO_3\,3H_2O$). A further drop in temperature of about $5°C$ may result in water ice crystals being formed. These are rather larger ($10\,\mu m$). It is the heterogeneous reactions involving these cloud crystals, which dramatically alters the chemistry of the stratosphere. Basically these reactions convert chlorine from its inactive, reservoir forms (HCl, $ClONO_2$) into forms, which are active ozone depletors (Cl, ClO). HCl is readily incorporated into ice crystals and can undergo reaction with $ClONO_2$

$$HCl + ClONO_2 \rightarrow Cl_2 + HNO_3 \tag{7}$$
$$\text{ice} \qquad \text{gas} \qquad \text{gas} \quad \text{ice}$$

and

$$H_2O + ClONO_2 \rightarrow HOCl + HNO_3 \tag{8}$$

The nitric acid is left in the ice phase. The chlorine remains in the gas phase until the polar spring when the sun reappears and photodissociates it to chlorine atoms.

$$Cl_2 + h\nu \rightarrow Cl^{\bullet} + Cl^{\bullet} \tag{9}$$

The Cl atoms rapidly react with ozone generating ClO.

$$Cl^{\bullet} + O_3 \rightarrow ClO^{\bullet} + O_2 \tag{10}$$

In the winter the stratosphere is thus chemically "preconditioned" by heterogeneous reactions so that in the spring very rapid ozone depletion occurs. In addition to the ozone destruction cycle represented by equations 5 and 6 with X = Cl it is now recognised that chlorine monoxide dimers are also important. This was realised because the oxygen atom concentrations in the lower stratosphere are too low to account for the observed ozone destruction rates.

$$ClO + ClO + M \rightarrow Cl_2O_2 + M \tag{11}$$

$$Cl_2O_2 + h\nu \rightarrow Cl^{\bullet} + ClOO \tag{12}$$

$$ClOO^{\bullet} + M \rightarrow Cl^{\bullet} + O_2 + M \tag{13}$$

These reactions by-pass the ClO + O reaction as a route for reconversion of ClO back to Cl. ClO plays an important role in determining ozone concentrations within the Antarctic stratosphere is illustrated in Figure 4. Reactions of bromine atoms in addition to those of chlorine atoms are now thought to account for about 20% of the ozone depletion. Bromine emissions occur in the form of methyl bromide which has natural and man-made sources such as soil fumigation, biomass burning, and the exhaust of automobiles using leaded gasoline, plus another family of halocarbons the Halons – Halon 1301 is CF_3Br, Halon 1211 is CF_2BrCl (Table 3). Methyl bromide is currently being targeted for phase out and is covered under amendments to the Montreal Protocol.[15] Current models of ozone depletion also consider the effects of CO_2. Increased CO_2 levels would lead to a lowering of temperatures in the stratosphere which may serve to slow down the destruction reactions: $O + O_3$ and $NO + O_3$. The role of sulfate aerosols also has to be considered in the lower stratosphere.[16]

But is the "hole" in the Antarctic ozone layer significant for the rest of the globe? Several facts suggest that it is. As the Antarctic spring progresses the vortex breaks up and the ozone depleted air can then be transported to lower latitudes. Such an event has been observed in Australia. Large increases of surface UV are observed in Antarctica and the southern part of South America during the period of the seasonal ozone "hole." In the Northern Hemisphere airborne studies of the Arctic winter were carried out in 1988/1989. Although the temperatures are not as low as in the Antarctic and the occurrence of stratospheric clouds not as common, nevertheless they are formed and the existence of a similar chemistry with high ClO concentrations has been demonstrated. The extent of ozone depletion is less marked – perhaps 15–20% in the range 20–25 km altitude representing a reduction of some 3% in total column ozone, with the worst years again 1992/1993 as in the Antarctic. The link with chlorine and bromine levels has been proven although is more difficult to quantify in the Arctic due to a larger uncertainty in the dynamics of the Arctic vortex. Data from the network of ground level monitors for column ozone show a clearly decreasing trend in the Northern Hemisphere since 1970 although the considerable variability in the data makes the precise percentage decrease sensitive to the start date assumed. For Europe and North America the decrease is 2.5 to 3.5% per decade with an indication that the trend has accelerated in the last decade, in parallel with the worsening conditions in the Antarctic. Data from Environment Canada shown in Plate 2 illustrates the ozone depletion in the Northern Hemisphere[17]. It is interesting to note the scale of the hole compared to the characteristics of the inter-annual variability in stratospheric ozone concentration.

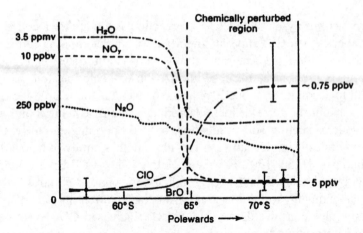

Figure 4 *Profiles of ClO and other species in the Antarctic stratosphere, 18 km altitude, near the boundary of the chemically perturbed region. The decreases in water vapour and NO$_x$ are due to condensation of water and nitric acid in polar stratospheric clouds followed by gravitational settling*
(Reproduced with permission from UK Review Group on Stratospheric Ozone[12])

Table 3 *Atmospheric lifetimes and ozone depletion potentials for halogenated compounds*

Compound name	Chemical formula	Atmospheric lifetime per year[2,13]	Ozone Depletion Potential[14]
CFC 11	CFCl$_3$	50	1.0
CFC 12	CF$_2$Cl$_2$	102	1.0
CFC 113	C$_2$F$_3$Cl$_3$	85	1.1
Carbon tetrachloride	CCl$_4$	42	1.08
Methyl chloroform	CH$_3$CCl$_3$	4.9	0.12
HCFC 22	CH$_3$CCl$_3$	12.1	0.05
HFC 134a	C$_2$H$_2$F$_4$	14.6	0
Halon 1301	CFBr	65	12.5
Halon 1211	CF$_2$BrCl	20	3.0
Methyl bromide	CH$_3$Br	1.2	0.6

2.1.3.3 Effects of International Control Measures. The UN Convention on the Protection of the Ozone Layer (the "Vienna Convention") was agreed in 1985 and subsequently measures to reduce the emissions of various halocarbons were incorporated into the "Montreal Protocol" in September 1987. Further meetings in London in 1990 and Copenhagen in 1992 further tightened the restrictions under the Protocol, which has as its final objective the elimination of ozone depleting substances. More than 160 countries are now Parties to

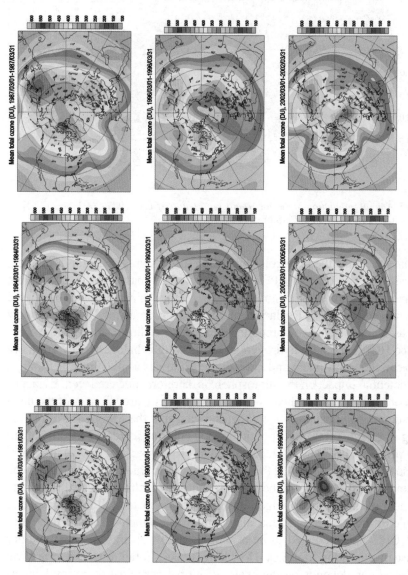

Plate 2 *Ozone concentration in the stratosphere in March over the Northern Hemisphere from 1981–2005*[17] (Reproduced with permission from Environment Canada)

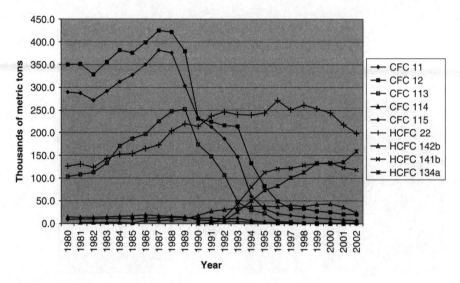

Figure 5 *Annual production of fluorocarbons reported to AFEAS 1980–2002*
            (Data from www.AFEAS.org[18])

the Convention and the Protocol[14] with some special agreements for developing countries. The use of both CFCs and Halons were phased out in 2000 along with the use of carbon tetrachloride (CCl_4). Methyl chloroform will be phased out by the end of 2005. Replacement chemicals such as the HCFCs are now being used and the change in reported emissions[18] is illustrated in Figure 5.

Since HCFCs contain hydrogen atoms as well as halogens they are more reactive in the troposphere and have shorter lifetimes as illustrated in Table 3. Their potential impact on the stratosphere is therefore much reduced. Evidence is now emerging which indicates that the atmospheric growth rates of the main ozone depleting substances are slow, showing that the Montreal Protocol is having an effect. Total tropospheric organic chlorine increased by only about 60 ppt per year (1.6%) in 1992, compared to 110 ppt per year (2.9%) in 1989.[19] Even so the expectation is that the total stratospheric chlorine loading will peak early in the 21st century at about 4 ppb and only decline slowly through the remainder of the century.[13] Figure 6 shows the predicted fall in stratospheric chlorine in the 21st century but also demonstrates the sources of this chlorine.[18] CFCs are predicted to be a major source well into the next century because of their long lifetimes. How long it takes to fall to the pre-war level of below 1 ppb depends on the extent of global participation in implementing restrictions and the extent of future use of alternative chlorine-containing compounds such as the HCFCs (Table 3). Global ozone losses and the Antarctic ozone "hole" are predicted to recover in about the year 2045 as long as the Montreal Protocol and amendments are closely adhered to.

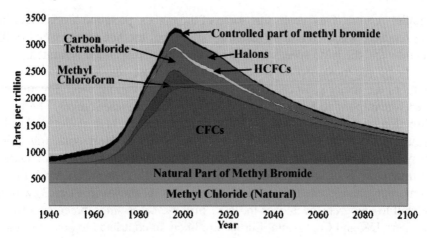

Figure 6 *Predicted chlorine loading for next century by source*
(Graph from http://www.afeas.org/atmospheric_chlorine.html)

2.2 ATMOSPHERIC TRANSPORT AND DISPERSION OF POLLUTANTS

2.2.1 Wind Speed and Direction

The previous section dealt with global pollution problems. This chapter focuses on the local pollution problems that directly affect the air quality around us. Local air quality is significantly influenced by wind speed and atmospheric stability, which determine the rate of mixing of pollutants within an air mass. In general low wind speeds result in high pollutant concentrations and *vice versa*. If the wind blows across the top of a chimney emitting smoke at a constant rate, the volume of air into which the smoke is emitted is directly proportional to the wind speed. The concentration of smoke in the air is thus inversely proportional to the wind speed. A similar description can be applied to a source distributed uniformly over a wide area (*e.g.* domestic emissions from a city). The concentration of pollutants in the hypothetical box of air into which the pollutants are mixed is proportional to the emissions rate and inversely proportional to the wind speed. In practice this picture is grossly oversimplified and the concentration of pollutants measured in urban areas rarely decreases with wind speed as rapidly as predicted.

For reasons that will be discussed later, the wind speed at ground level tends to drop overnight and rise again during the morning, especially during cloud-free conditions. Of course, emissions also tend to drop overnight – fewer fires, boilers and furnaces are alight, fewer cars are on the roads. For reasons discussed in the next section, the depth of the boundary layer is also much lower during the night and early morning so the pollutants are mixed through a much smaller volume. This means that some of the highest pollution levels occur in the morning when emissions increase rapidly but the boundary layer is still shallow and the wind speeds have yet to pick up so dispersion is limited.

The people most affected by air pollution are those who are situated downwind of the major sources. A knowledge of the prevailing wind direction is therefore important in predicting the likely impact of these sources. The wind direction at a given point is not sufficient to identify high pollution levels resulting from the long-range transport of pollutants. Horizontal transport of pollutants over the scale of hundreds to thousands of kilometres equates to time periods of the order of 1–3 days. During this time span wind directions may change, creating curved trajectories as shown in Plate 3, which related to smog episodes over the UK. In these cases smog precursors from southern UK and Continental Europe are likely to contribute to poor air quality over the UK. Air mass trajectories, which bring air masses to the UK from the Atlantic are generally consistent with much lower pollution concentrations.

2.2.2   Atmospheric Stability

In addition to wind direction, the extent of vertical mixing in the atmosphere also affects pollutant concentrations. This is related to the stability of the atmosphere, which is dependant on many factors such as the time of day, the synoptic weather conditions, the characteristics of the Earth's surface, *etc.*

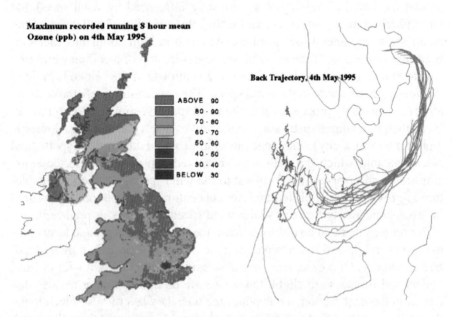

Plate 3 *Trajectory modelling study showing a smog episode across the UK*
(Reproduced with permission of NETCEN)

2.2.2.1 The Lapse Rate. The roughness of the ground produces a certain amount of turbulence in the lowest layer (boundary layer) of the atmosphere which promotes the mixing and dispersion of pollutants. This effect increases with the scale of the surface roughness and is greater for a city with large buildings than for open ground with few obstructions. However the dominant factor affecting atmospheric stability and turbulence is thermal buoyancy.

The pressure in the atmosphere decreases exponentially with height. Ascending air expands as the pressure decreases and as it expands it cools. A simple calculation based on the properties of gases leads to the conclusion that temperatures should decrease with height by $9.8°C \, km^{-1}$ or about 1°C for every 100 m of dry air. The figure is somewhat lower for moist air. The variation of temperature with height is called the lapse rate and the calculation for the ideal case leads to what is known as the adiabatic lapse rate (a.l.r.). In the real atmosphere the lapse rate can be greater than, smaller than or close to the adiabatic lapse rate. This fundamentally affects the extent of vertical mixing of air as the following two examples show:

Case (a) Temperature decreases with height more rapidly than the a.l.r., Figure 7a. Air that is slightly warmer than its surroundings starts to rise and to cool at the a.l.r. The temperature difference between the rising parcel of air and its surroundings increases with height, thus the air parcel always remains warmer than the surrounding air and continues in an upward movement due to thermal buoyancy continues. In this scenario the atmosphere is unstable. Upwards flowing currents in one location are balanced by downwards flowing currents elsewhere and rapid vertical mixing of air occurs, promoting rapid dispersion of pollutants. The strength of the up currents (a function of surface heating and the temperature difference between the air parcel and the surrounding atmosphere) will determine how high the air parcel travels and thus the depth of the mixing layer. Unstable situations occur with bright sunlight warming the ground to a temperature above that of the air. The air adjacent to the ground is subsequently warmed and rises due to its buoyancy. Such situations are common during daytime in summer, especially when the wind speed is low.

Case (b) Temperature decreases with height less rapidly than the a.l.r. or actually increases with height, Figure 7b. Air that is slightly warmer than its surroundings starts to rise and cool at the a.l.r. and the temperature difference between the rising parcel of air and its surroundings soon decreases to zero. Upward movement due to thermal buoyancy ceases. The atmosphere is stable since any vertical movement of air tends to be damped out. Lower polluted layers stay near the ground and pollutant concentrations will be high. Stable conditions typically occur at night when the ground cools at a faster rate than the air above.

High wind speeds tend to lead to neutral conditions with the lapse rate close to the adiabatic value (Figure 7c).

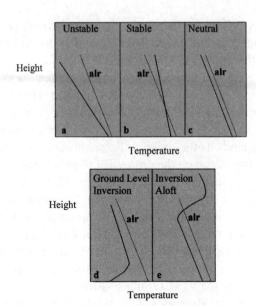

Figure 7 *Schematic illustration of the atmospheric lapse rate for various stability categories. Full lines: actual temperature profile; dashed lines: adiabatic lapse rate (a.l.r.)*

2.2.2.2   Temperature Inversions.   At night radiative cooling of the ground often leads to low level temperature inversions on clear nights. As shown in Figure 7d this means that the temperature of the atmosphere initially increases with height but after a certain point the temperatures start to decrease again. This situation is very stable as cooler dense air lies underneath warmer air. The effect is that ground level emissions become trapped in the stable inversion layer, which may not be more than 100–200 m deep. Emissions from tall industrial chimneys, however, may be above the inversion layer and be vertically dispersed by a different set of atmospheric conditions that occur independently aloft. The following day the inversion layer maybe gradually eroded by the warming effect of the sun until, by mid-morning, the temperature profile has returned to that of neutral conditions (Figure 7c). At this point trapped pollutants are effectively released to be dispersed to higher levels through a greater volume of air, however, any pollutants emitted above the stable layer may also be mixed to the surface.

Another factor contributing to high pollution levels during inversion conditions is the lowered wind speed. The surface layers become isolated from the wind flows aloft thus there is reduced likelihood of the downward transport of momentum. The air closest to the surface may therefore become stagnant, particularly if there is insufficient solar radiation (due to time of year) to break down the inversion during the day. In the low temperature conditions, dew frost or fog formation may occur. Fog adds to the problem by slowing down the break-up of the overnight inversion layer because the sun's energy is reflected by its upper surface and does not reach the ground. The ground therefore stays cool rather than warming. In extreme cases fog may persist for several days as happened in the London 1952 smog. Polluted fogs are more persistent than clean fogs because the chemicals dissolved in the water droplets prevent complete evaporation even when the relative humidity drops well below 100%.

The other main type of inversion occurs during anticyclonic conditions and is described as a subsidence inversion. Within an anticyclone air is diverging from a high-pressure region and at the centre subsides from a high level to lower levels of the atmosphere. As it subsides it warms with decreasing height and increasing pressure resulting in the development of an elevated inversion layer as illustrated in Figure 7e. Below the inversion layer the air may be neutral or even unstable so that good mixing occurs but only up to the inversion height. Subsidence inversions are often associated with warm, dry weather in the summer and very cold dry weather in the winter. They provide the ideal conditions for a long-range transport of pollution. Summer haze conditions in which the UK receives already polluted air from Europe before the addition of our own emissions, can lead to very high particulate matter concentrations. At the same time the levels of photochemical oxidants such as ozone are also high as shown in Plate 3.

In the discussion so far, no reference has been made to the geographical situation in which air pollution levels are being considered. Towns situated in valleys are particularly susceptible to pollution problems. Cool air will tend to flow downhill into the valley so aggravating the problem of low level inversions. Mixing between the air in the valley and the air above is reduced. Fogs will persist longer. Often in winter, a layer of polluted air over the town with cleaner air above can be clearly seen from the higher ground.

Case Study 2: Los Angeles

Los Angeles is one example of a city whose geographical location exaggerates pollution problems caused by high emissions. It is situated in a basin area surrounded by large hills, leading to a high incidence of temperature inversions, which limits the mixing of pollutants out of the city. Towns such as Los Angeles that are situated by the sea may be subject to sea breezes. The proximity of relatively warm ground to the cool water surface results in a circulation of air from sea to land. This can result in shallow mixing layers, the re-circulation of pollutants at a diurnal scale as well as the prevalence of sea mists, which may blow inland aggravating the general discomfort. Inversions are also common due to the presence of subsidence conditions over the Pacific. In fact, five of the worst polluted cities of the world are situated in the Pacific basin region.[20]

2.2.3 Dispersion from Chimneys

2.2.3.1 Ground Level Concentrations. The effluent gases leaving a chimney gradually entrain more and more air and the plume width both in the vertical and in the horizontal directions increases with downwind distance. In the cross wind direction there is a rapid fall-off in concentration away from the plume centre line which is defined as being along the average wind direction. The ground level concentration very close to the chimney will be zero because it takes some time for vertical mixing to drive the plume to the ground. Some distance downwind the dispersing plume reaches ground level and the concentration rises rapidly to a maximum value. Thereafter the concentration falls off with distance as the plume becomes more and more dilute (Figure 8). The distance at which maximum concentrations occur depends on stability conditions. Unstable conditions (significant vertical mixing) bring the plume down to ground rapidly, that is closer to the chimney, but the rate of dilution is large so the concentration falls rapidly from its maximum value with larger distances. Stable conditions result in the plume dispersing very slowly. It may remain visible for a considerable downwind distance. The point of maximum ground level concentration is a long way from the chimney. Above the chimney top, the vertical dispersion of the plume can be hindered by an inversion layer. The pollutants are then trapped between the inversion height and the ground. At sufficient distance downwind, the concentration may be virtually constant at all heights as the plume becomes well dispersed within the mixing layer.

2.2.3.2 Plume Rise. The basic models of plume dispersion suggest that the maximum ground level concentration depends on the square of the chimney

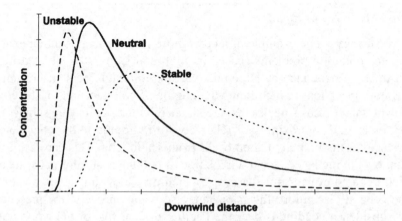

Figure 8 *The variation in ground-level concentration along the plume centre line downwind of a chimney for various stability classes*

height. A 40% increase in chimney height should roughly halve the ground level impact. However, the height of the plume depends on both the chimney height and the plume rise caused by the exit conditions (such as temperature) of the release. The major factor is the thermal buoyancy of the plume. The hot flue gases will rise relative to the cooler surrounding air and the hotter the release the greater the plume rise. A lesser factor is the vertical momentum of the gases due to their efflux velocity out of the chimney top. Plume rise can mean that the effective chimney height is double the physical chimney height under low wind speed conditions and so is a considerable asset in achieving effective dispersion. Conversely any control technology for pollutant reduction, which also reduces the flue gas temperature (such as a scrubber) results in poorer dispersion of the plume. If plume rise is taken into account then the effects of raising the stack height to lower ground level concentrations is lower than would otherwise be predicted.

Case Study 3: Pollution from Eggborough Power Station

One of the primary pollutants emitted from Eggborough Power Station, a 2000 Mw power station in Yorkshire[21] (Northern England), is sulfur dioxide. This is emitted from a 200 m tall chimney. Measurements of this pollutant made by the Central Electricity Generating Board suggest that the maximum ground level pollution from this chimney occurs in a radius about 8 km from the chimney. At this point the peak 3 min concentrations of sulfur dioxide due to the power station were approaching $1000 \, \mu g \, m^{-3}$. Such peaks were extremely rare and 95% of the 3 min averages were below $10 \, \mu g \, m^{-3}$. The highest daily averages were around $100 \, \mu g \, m^{-3}$ and the highest monthly averages around $10 \, \mu g \, m^{-3}$. The overall annual average contribution of the power station to the ambient SO_2 concentration was $2\text{–}3 \, \mu g \, m^{-3}$ in an area where the prevailing concentration from other sources was about $40 \, \mu g \, m^{-3}$.

2.2.3.3 Time Dependence of Average Concentrations. The above description of plumes is really a simplification since atmospheric turbulence is unpredictable making the dispersion process irregular. Visible plumes will often look ragged and from the point of view of an observer making spot measurements on the ground, one minute the concentration may be close to zero and the next very high. The longer averaging time taken, the less the variability of the results.

2.2.4 Mathematical Modelling of Dispersion

Air quality models are effective tools for calculating the concentration of pollutants at a given point in time and space from a known source, and can also be used to investigate more about the behaviour of different components of the system. These models are based on mathematical representations of the atmospheric processes that determine the rate at which clean air is mixed with polluted air and the pollutants are transported away from the source area. There are a number of techniques for modelling the dispersion and reaction of atmospheric pollutants. The models vary in sophistication, but all include some simplification of atmospheric dispersion processes. More complex models attempt to represent all the physical processes that are relevant in the atmosphere. These are called prognostic models. Simpler tools rely on statistical parameterisations of an individual or a group of processes. These are empirical or statistical models.

Space does not allow us a full study here but the main methods will be introduced. Choosing a particular type of model depends on many factors such as the temporal and spatial scale of the problem *e.g.* urban roadside, plume dispersion, regional scale smog modelling, global circulation modelling, *etc.* and the resolution of the data available to run the model and validate the output. The need for models, which include chemical reactions and deposition should also be considered, as well as external factors such as the skill of the operator, the time available and the financial and computer resources available. Available methods include Gaussian formulations,[22,23–25] Eulerian grid modelling,[26–28] Lagrangian trajectory modelling[29] and turbulence models which try to account for apparently random fluctuations.[30–33]

The simplest approach is the Gaussian model. This model assumes that when pollutant concentrations are averaged over a period of time (usually an hour) the peak concentration will occur near the centre of the plume with concentrations dropping in the horizontal and vertical plane according to a bell-shaped curve, otherwise known as a normal or Gaussian distribution (Figures 9a and b). The curve is symmetrical about the mean and tends towards zero at the tails. A Gaussian Plume dispersion model is based on the assumption that an empirical relationship between atmospheric parameters such as wind speed and direction or stability (which determines dispersion) and the distortion of the Gaussian distribution can be formulated to describe the concentration of pollutants downwind of a source. For a ground level source in very stable conditions, high ground-level concentrations will occur near the source of the pollution whilst in unstable conditions dispersion is much more effective and more dilute concentrations will be measured over a wider area. For elevated sources stable conditions may lead to the plume grounding some distance away from the point of emission as discussed above (Figure 8). For stack sources it is unstable and neutral conditions that often lead to the highest pollutant concentrations close to the stack as shown in Figure 8.

Fast screening type calculations of the dispersion of a pollutant from a point source such as a large chimney are usually based on the Gaussian Plume

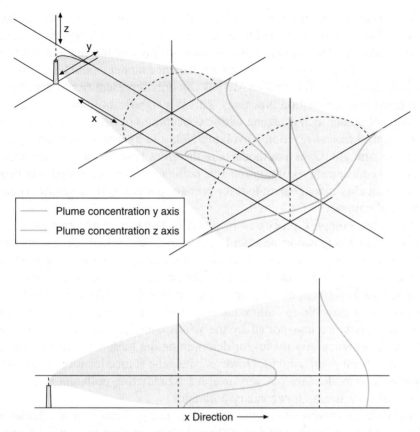

Figure 9 *Diagram to show the dispersion of pollutants from an industrial plume as predicted by a Gaussian Plume Model, a) pollution concentrations in the x, y and z directions, b) vertical slice through the plume*
(Figure drawn by K. Burchill)

Model.[22] This is because it is a simple, computationally efficient model. It treats dispersion as a statistical process, rather than attempting to represent the individual turbulent motions of the atmosphere. These models typically only require simple input data sources such as wind speed and direction, atmospheric stability class and temperature for example. Gaussian Plume Models can also be adapted to treat line sources (such as roads) and area sources (such as wind-blown dust from a stockpile or odours from a sewage works). Urban areas can be modelled as a sum of area sources representing domestic and commercial emissions plus larger point sources such as factories or power stations. Gaussian Plume Models are particularly useful for calculating long term average distributions around a source using statistically averaged meteorological data.

The disadvantage of Gaussian Plume Models is that they are steady state models. They assume that conditions remain similar over the averaging period. This means that the Gaussian model cannot represent the random fluctuations present in a real plume (for this a turbulence model would be required). In their standard form Gaussian Models can only represent first order chemical reaction or deposition processes. This is a significant limitation for secondary pollutants such as ozone whose concentrations depend on a series of non-linear chemical reactions involving NO_x and hydrocarbon species, and levels of sunlight. Gaussian models are therefore best suited to determining averaged concentrations of non-reactive pollutants. They are commonly used as an air quality management tool for primary pollutants from roadways and industrial chimneys.

The more complex processes involved in the long range transport and reaction producing smog and acid rain, which also involve a number of source types, require numerical simulation on large computers. Here either Lagrangian or Eulerian models maybe chosen. Lagrangian models treat the atmosphere as a series of air parcels moved around within a wind field. In this way it is possible to follow the trajectories of a single or multiple air parcels as they are transported by the winds over long distances. Such an approach is particularly useful for determining the long-range dispersion of pollutants from point sources. However, since the source location is not well known these models are not very useful for predicting pollutant dispersion from diffuse sources, or secondary pollutants.

The Eulerian approach treats the atmosphere as a grid made up of volumes or boxes, whose properties change with time. Usually, perfect mixing is assumed within each grid box and transport between each box is modelled. In this case the equations representing transport and chemical transformation have to be solved at each grid point, rather than at every point along the trajectory of the air parcel as for the Lagrangian models. Eulerian models are better at simulating dispersion from area sources such as urban pollution or secondary pollutants such as ozone. They give better coverage of the modelling domain but at the expense of computer simulation time. Often, large parallel computers are used for Eulerian simulations[26–28] *e.g.* smog models, global climate models, *etc.*

There are advantages and disadvantages of both methods. The choice of model type depends on the required outcome and accuracy needed. The accuracy of all types of model however, depends heavily on the input data. Detailed knowledge of emissions data for a range of sources is needed along with an accurate representation of meteorological data such as wind-speed and direction, stability class and precipitation. The resolution of the input data significantly affects the results, and care needs to be taken to ensure that the data are representative of the modelled region. Detailed modelling of urban pollution requires an emissions inventory on a grid basis, for example, 1 km^2, along with point source information. Emissions inventories for the most common pollutants are now available for the whole of Europe on a national basis,[7] on the 50×50 km grid used by the European Monitoring and Evaluation Programme (EMEP)[34] and on grid sizes down to 1×1 km for the UK.[35,36]

Model output must also be validated using appropriate data. This requires the temporal and spatial scales of the model to be matched to the data set. For example, the averaged area output from a Eulerian model cannot be accurately compared to a single point measurement from a local air pollution monitor.

2.3 EMISSIONS TO ATMOSPHERE AND AIR QUALITY

2.3.1 Natural Emissions

2.3.1.1 Introduction. The primary components of pure dry air are nitrogen, N_2 (78.1%), oxygen, O_2 (20.9%), argon, Ar (0.9%) and carbon dioxide, CO_2 (approximately 0.035%). Water vapour is present in amounts, which typically range from 0.5–3% at ground level, depending on temperature and relative humidity. Analysis of air samples reveals the presence of hundreds of other substances in trace amounts. Some of these substances can be described as pollutants. However, the definition of an "air pollutant" is problematic. If a substance is acutely harmful and does not occur naturally in the atmosphere, then it can easily be described as an "air pollutant". However, many substances that are described as "air pollutants" may be introduced into the atmosphere as a result of natural as well as anthropogenic activities. Sulfur dioxide and carbon dioxide are good examples of this. In these cases it is more helpful to define "pollution" as the presence of the pollutant above the natural background concentrations resulting in unacceptable adverse consequences to human health and/or the natural environment.

Within densely populated areas of land, the levels of most pollutants are dominated by the contributions for which human activities are responsible. However, on a global scale the natural emissions may be comparable to human emissions. This is illustrated in Table 4. Pollutants may be emitted directly into the atmosphere. These are known as primary pollutants. Other pollutants may be formed within the atmosphere as a result of chemical processes. These secondary pollutants, including gases like ozone and particulate compounds such as sulfates, are dealt with in Section 2.4. Here the focus is on primary pollutants.

2.3.1.2 Sulfur Species. The largest natural source of sulfur in the form of SO_2 and some H_2S is emitted from volcanoes (Table 4). Significant amounts of sulfur are also emitted from biological processes. In the absence of air, biological decay results in emissions of hydrogen sulfide (H_2S) and organic compounds such as dimethyl sulfide (DMS). Carbonyl sulfide (OCS) emissions also occur together with small amounts of carbon disulfide (CS_2) and

Table 4 *Natural emissions of S and N compounds*[38,39]

Source	Tg S per year	Source	Tg N per year
Volcanoes	9.3	Lightning	8
Biomass burning	2.2	Biomass burning	5
Marine biosphere	15.4	Soil – biogenic	7
Terrestrial biosphere	0.35	NH_3 oxidation	0.9
		From stratosphere	0.6
Natural total	27.25	Natural total	22
Anthropogenic	77*	Anthropogenic	22**

* More recent estimates place this value at 55 Tg S per year in 2000[37]
** More recent estimates place this value at 44 Tg N per year in 1997[39]

dimethyl disulfide (DMDS). The emissions related to phytoplankton from oceanic sources are primarily DMS. The grouping together of all sulfur compounds is appropriate since H_2S and organic sulfur compounds are converted to SO_2 in the atmosphere. Various estimates place the total natural emissions at approximately 20–30 Tg sulfur per year (1 Tg = 1 million tonnes). In 1992 estimates of global SO_2 from combustion emissions were approximately 77 Tg S per year,[38] thus natural sulfur emissions were of the order of one third of the anthropogenic emissions.

Although global concentrations peaked in 1989 following the OPEC crisis and introduction of newer technologies, the increasing prominence of Asian emission sources slowed the general rate of global decline. However, with the diffusion of abatement technologies, Asian emissions peaked in 1996 and global concentrations have since declined rapidly.[37] Although a range of estimates exists, a conservative value placed anthropogenic emissions at 55 Tg S per year in 2000.[37]

This assessment omits the largest single component of the atmospheric sulfur budget, which is represented by the sulfate content of sea salt aerosols. There is obviously a sharp vertical gradient of these aerosols and a large proportion return to the sea quite quickly.

2.3.1.3 Nitrogen Species. Biological processes in soil lead to the release of all of the common nitrogen oxides, nitric oxide (NO), nitrogen dioxide (NO_2) and nitrous oxide (N_2O). The amounts involved are very uncertain but for NO and NO_2 are of the order of 5–10 Tg N per year. Lightning and biomass burning are other major sources. Oxidation of (ammonia) NH_3 to NO occurs in the troposphere and some nitrogen in the form of HNO_3 (nitric acid) is transferred to the troposphere from the stratosphere. These sources total 20–30 Tg N per year.[39] In comparison the anthropogenic emissions of $NO + NO_2$ from combustion are about 20 Tg N per year. Lee *et al.*[39] estimate the total global NO_x emissions to be 44 Tg N per year with an uncertainty range of 23–81 Tg N per year.

The major source of nitrous oxide (N_2O) is the release from soil especially in situations where fertiliser has been added. Smaller releases occur from the oceans and from combustion processes. The sources are not as well understood as the main loss mechanism, which is decomposition in the stratosphere (6–10 Tg N_2O per year). Industrial emissions occur during the production of nitric acid and adipic acid (an intermediate in the production of nylon). Total UK emissions were estimated at 130–140 thousand tonnes in 2002.[40]

The other important nitrogen species is ammonia (NH_3). Its sources may be considered partly natural and partly anthropogenic since they are dominated by animal excreta. But there are also significant releases from biomass burning, crops and the oceans. Several estimates[41] of global emissions place the total at around 50 Tg N per year. The sources of ammonia are difficult to quantify since the Earth's surface can act as a source and a sink and the overall emission rate depends on soil and climatic conditions. In the UK the predominant source of ammonia is agriculture. Of the estimated 290 kt per year emitted in 2001, 224 kt per year are from animal wastes and a further 35 kt per year from non-livestock agriculture.[42]

Currently ammonia emissions in Europe are estimated at 2.7 M tonnes per year for the countries of the European Union and 6.4 M tonnes per year for all the European countries covered by the EMEP area (European Monitoring and Evaluation Programme).[43]

2.3.1.4 Hydrocarbons. The importance of methane as a greenhouse gas has been discussed earlier. The largest natural sources of methane are anaerobic fermentation of organic material in rice paddies and in northern wetlands and tundra, plus enteric fermentation in the digestive systems of ruminants (*e.g.* cows). Methane is also released from insects, from coal mining, gas extraction and biomasss burning. Total emissions are 300–550 Tg per year and appear to be increasing at a rate of 50 Tg per year. Within the UK, animals (29%), mining (8%), landfill gas (46%) and gas leakage (9%) account for most of the 3.9 Tg per year emissions (1994 estimate[44]).

Heavier hydrocarbons, such as isoprene, α and β-pinene and other terpenes, are released directly to the atmosphere from trees and can contribute to the formation of photochemical smog in high emission areas (see Section 2.4). Global emissions of these Biogenic Volatile Organic Compounds (BVOCs) are estimated to be 1150 Tg C per year of which 88% is from trees and shrubs and 10% from crops.[45]

2.3.2 Emissions of Primary Pollutants

2.3.2.1 Carbon Monoxide and Hydrocarbons. Efficient combustion can be achieved in most stationary combustion appliances providing they are properly adjusted, and CO and hydrocarbon concentrations do not give serious cause for concern. Faulty or improperly adjusted appliances can produce dangerous amounts of CO (several percent in the flue gas) usually due to some abnormal limitations of the air supply.

CO emissions from internal combustion engines are more of a problem. The combustion takes place under high pressure instead of atmospheric pressure and the peak temperatures are higher than in a normal boiler. However the time available for combustion is limited by the engines cycle to a few milliseconds instead of a second or more. In petrol engines lacking any control devices, incomplete burnout of the fuel leads to high carbon monoxide and significant hydrocarbon emissions, especially during idling and deceleration. Installation of catalytic converters results in reductions of 90% for CO and hydrocarbons although significant emissions may take place under cold start conditions before the catalyst "lights off". Diesel engines have much lower carbon monoxide and hydrocarbon emissions. Permitted emissions from European vehicles have been progressively reduced. The standards for 2000 and 2005 are indicated in Table 5.

The term Volatile Organic Compounds (VOCs) is used to describe organic material in the vapour phase excluding methane. There are many non-combustion sources of VOC emission of which the most important is the use of solvents, including those released from paints. Evaporative losses of gasoline during storage and distribution are also significant.

VOCs are important in atmospheric chemistry for the formation of photochemical smog and this is discussed in Section 2.4. Various control measures have therefore been implemented so as to enable the UK to meet its agreed obligation under a 1991 UNECE Protocol of reducing VOCs by 30% by 1999 relative to 1988 emissions. However, due to delays ratifying the agreement the Protocol did not come into affect until 1997 and by 2003 although the net European emissions were reduced by 30%, seven of the 21 countries which ratified the Protocol had not achieved their targets.

2.3.2.2   Nitrogen Oxides.   Although there are small emissions of NO_x from industrial processes such as nitric acid production the main emissions are from combustion. There are several chemical routes to the formation of the nitrogen

Table 5 *Present European Standards for passenger cars applicable to all models in 2000 (Euro 3) and 2005 (Euro 4) (EC directive 1999/96/EC)*

Vehicle type	HC g/km	PM mg/km	NO_x (g/km)
Petrol (2000)	0.20		0.15
Diesel (2000)		50	0.5
Petrol (2005)	0.10		0.08
Diesel (2005)		25	0.25

oxides NO and NO_2, together described as NO_x. One involves the combination of nitrogen and oxygen in the air at the peak flame temperatures to form nitric oxide, NO. This is termed "thermal NO_x". Another starts with the nitrogen originally present in the fuel (1–2% by weight for coals and heavy fuel oils) – "fuel NO_x". Depending on the conditions, some of the fuel-nitrogen will be converted to NO and some to nitrogen gas, N_2. Factors such as the burner design, the intensity of combustion, the overall shape and size of the furnace and the amount of excess air all influence NO formation and can be modified to achieve a certain measure of control. However this falls considerably short of eliminating the emissions. Typical flue gas concentrations for industrial coal-burning are about 550 ppm but with new designs of low-NO_x burners this is reduced by about 50%. Nitrogen dioxide, NO_2, forms only a small fraction of the waste gases from combustion (usually less than 10%) so the description "NO_x emissions" actually refers mainly to NO emissions. Once in the atmosphere, oxidation of NO to NO_2 occurs, as described later, and the relative proportions of the two oxides may then be comparable.

There is negligible nitrogen in gasoline or diesel fuels so the NO_x arises from the thermal route. Nitrogen oxide emissions are high due to the high temperatures and pressures and are at a maximum during acceleration and minimum during idling. Diesel engines have at least comparable NO_x emissions to a petrol engine and have higher permissible emissions for Euro 3 and Euro 4 vehicles as shown in Table 5. While Euro 5 standards are still under debate it is anticipated that future diesel vehicles will be expected to match the NO_x emission standards for petrol engines.

2.3.2.3 Sulfur Oxides. Sulfur dioxide, SO_2, arises from the sulfur present in most fuels amounting to 1–2% wt in coal, 2–3% in heavy fuel oils and decreasing amounts in lighter oil fractions. There are also non-combustion emissions from sulfuric acid plant and from the roasting of sulfide ores during non-ferrous smelting, although these tend to be very localised and are not very significant for the UK.

The limit for the sulfur content of diesel fuels has been progressively reduced and now stands at 0.05% in the USA and 0.035% within most of Europe, although lower in the UK. Typical sulfur content of both gasoline (petrol) and diesel in Europe is below 50 ppm (0.005%), thus very little SO_2 is emitted by road traffic.

During combustion, conversion of sulfur dioxide is virtually complete although about 10% may be retained by coal ash. For the highest sulfur fuel oils (3% S) the flue gas concentrations can reach 2000 ppm while for a typical power station coal of 1.6% S the concentration is about 1200 ppm. No control over SO_2 emissions can be achieved by modification of combustion conditions and reductions must be sought by pre-treatment of the fuel or desulfurisation of the flue gases after combustion.

The 1994 "Protocol to the Convention on Long-Range Transboundary Air Pollution on Further Reduction of Sulfur Emissions" reflected a change in focus on attempts to control pollution emissions. This Protocol placed increased emphasis on the concept of "critical loads" (see Section 2.5.4.3). The main objective of the Protocol was therefore not to reduce emissions to some numerical limit but to ensure that emissions were reduced to the extent that rates of sulfur deposition that would not cause long-term damage to ecosystems or other receptors. This was therefore one of the first examples of attempts to control pollution according to environmental capacity.

2.3.2.4 Particulate Matter.   Dust can arise by the disturbing action of outdoor industrial activity on the ground or on raw materials. Quarrying, open cast mining, tipping, digging, the action of heavy lorries or simply a strong wind acting on stock piles can lead to grit and dust blowing beyond the site and causing a nuisance to others in the area. These are described as "fugitive" emissions and are particularly difficult to quantify.

During combustion, formation of *soot* commonly accompanies carbon monoxide formation and is generally due to inadequate air supply. Soot particles produced from gas-phase fuel/air reactions are commonly sub-micron ($<1 \mu$m) and because they are of comparable size to the wavelength of light they are effective both at light absorption and light scattering. A relatively small mass concentration will therefore render the exhaust of flue gases opaque and give a dark stain on a filter paper. This is the basis of the reflectance technique used for measurement of "smoke".

In the combustion of coal, volatile material is released and burns as a gas, the remaining char burns more slowly and a final residue of ash is left. Smoke can be produced from incomplete combustion of the volatile matter. Anthracite, coke and the other manufactured smokeless fuels contain low levels of volatile material and so avoid the problem. Incomplete combustion of volatile matter is also the major problem with wood smoke and smoke from bonfires or biomass burning. This can lead to serious problems in developing countries where the use of biomass is still common for domestic cooking and heating in cities causing high emissions at low level.

Smoke formation is not a significant problem with spark ignition engines because the petrol and air are well mixed before entering the cylinders. Diesel engines suffer from the disadvantage of producing more smoke, especially under heavy load or acceleration conditions, because of the relatively poor mixing of air with the fuel injected as a spray into the cylinder. This produces regions that are too rich in fuel for complete combustion, leading to soot formation.

Five important terms relating to particulate matter in the atmosphere are:

Smoke – the particulate material assessed in terms of its *blackness* or *reflectance* when collected on a filter, as opposed to its *mass*. This is the historical method of measurement of particulate pollution in the UK. The size of particles collected into the sampler is below 10–15 μm.

Total suspended particulate (TSP) matter – Mass concentration determined by filter weighing usually using a "Hi-Vol" sampler which collects all particles up to about 20 μm depending on wind speed.

PM10 – this corresponds to particulate matter that is *inhalable* into the human respiratory system. It is measured using a size selective inlet that has 50% efficiency for particles of 10 μm aerodynamic diameter.

PM2.5 – particulate matter below 2.5 μm. This is closer to, but slightly finer than the definitions of *respirable* dust, which have been used for many years in industrial hygiene to identify dusts which will penetrate the lungs.

Ultra-fine particles – particulate matter less than 100 nm in diameter.

Nano particles – particulate matter less than 50 nm in diameter.

Particulate vehicle emissions, for example, diesel smoke, require careful description. The *total number* of particles emitted is dominated by ultrafine particles in the 10–50 nm range, however a high proportion of the *total mass* is contributed by larger particles – one 0.5 μm particle has the same mass as one thousand 50 nm particles.

Within urban environments the concentrations of ultrafine particulates is determined by the complex interactions between intermittent (primary and secondary) formation processes and turbulent transport mechanisms. The main source of primary ultrafine particulates is direct tail pipe emissions and the freshly nucleated particles that form from the condensation of gas phase exhaust emissions. The rate of formation depends on traffic flow and vehicle characteristics as well as ambient temperature and humidity.[48] Secondary particulates may also be formed within the urban atmosphere as a result of the condensation of processed gas phase pollutants, although this process is more likely to occur in less polluted areas.

Current evidence suggests that combustion derived ultrafines, which often contain trace metals or organic compounds may be potentially more toxic per unit mass than their larger counterparts within the respirable fraction.[49,50] This may be because ultrafine particles are small enough to pass directly from the lung into the blood stream generating a systemic response.[51] Recent work in the US has shown a significant relationship between PM2.5 concentrations and cardiopulmonary mortality, lung cancer mortality and all cause mortality rates and a weaker relationship with PM10 concentrations.[52,53] However, few studies have looked at the effects of ultrafines alone and PM2.5 concentrations are not always well correlated with ultrafine concentration.

2.3.2.5 Emissions Limits. Emissions from industrial processes are controlled via emissions limits set by international bodies such as the European Union or by national organisations such as the UK Environment Agency or US

EPA. Processes subject to control receive an authorisation, which includes the permitted levels of emission. For example, large European municipal incinerators burning more than 3 tonne h^{-1} of waste must comply with the following emissions standards in $mg\,m^{-3}$ of dry gas as a daily average.[54]

Particulates	10	CO	50
SO_2	50	HCl	10
HF	1	VOC	10 (as total C)
Pb,Cr,Cu,Mn,Ni,As	0.5 (new plant)	Hg	0.05 (new plant)
Pb,Cr,Cu,Mn,Ni,As	1.0 (existing plant to 2007)	Hg	0.1 (existing plant to 2007)

Dioxins and furans ($ng\,m^{-3}$ toxic equivalent see Section 2.3.3.9)

2.3.2.6 Emissions Inventories.

The use of *spatially* disaggregated emissions inventories in modelling has been mentioned above. Knowledge of particular emission sources directs attention to the best targets for control and emissions reductions. Table 6 shows the UK national breakdown of emissions by *sector* for various pollutants while Table 7 illustrates *national* emissions for several European countries. These tables will be referred to in following discussions.

Table 6 *Estimated UK emissions of primary pollutants by UNECE source category for 2002 Thousand Tonne*[42]

	PM10	CO	NO_x^1	SO_2	NMVOC[2]
BY UN/ECE CATEGORY					
Combustion in energy prod					
Public power	10	71	379	680	9
Petroleum refining plants	2	4	31	66	1
Other combustion & trans.	0	26	64	5	1
Combustion in comm/res					35
Residential plant	28	208	74	39	
Comm/agricul combustion	4	17	28	12	
Combustion in industry					8
Iron & Steel combustion	3	253	9	10	
Other Ind. combustion	14	137	139	127	
PRODUCTION PROCESSES	38	154	2	33	175
Extr./Distrib. of fossil fuels	0	1	1	1	277
Solvent Use	0	0	0	0	389
Road Transport	39	1916	711	3	211
Other Trans/Machinery	8	423	139	25	59
Waste	1	18	4	1	20
Agriculture & Land Use Change	14	10	0	0	0
Nature	0	0	0	0	178
BY FUEL TYPE					
Solid	38	525	325	772	27
Petroleum	43	2345	904	149	271
Gas	10	68	308	11	23
Non-Fuel	70	300	45	71	1043
TOTAL	161	3238	1582	1002	1364

[1] NO_x is expressed as NO_2 equivalent
[2] Volatile organic compounds excluding methane

Table 7 *Expert estimated emissions inventory data for European Countries from the EMEP database in 1990 and 2002 (Gg per annum)*[47]

Country	NMVOC		NO_x (as NO_2)		SO_2	
	1990	*2002*	*1990*	*2002*	*1990*	*2002*
Germany	3591	1478	2845	1499	5326	611
France	2499	1542	1897	1352	1325	537
UK	2419	1186	2771	1582	3721	1002
Italy	2041	1467	1919	1317	1748	709
Spain	1591	1459	1206	1339	2097	1507
Poland	831	576	1280	805	3210	1564

* Note total anthropogenic emissions (excluding category S11 – other sources and sinks – in EMEP data base)
http://www.emep.int/

Table 8 *Air quality standards for European Community and UK*

Pollutant	Measure	Concentration ($\mu g\,m^{-3}$)	Date effective
PM10	24 hr mean – not to be exceeded more than 35 times a year	50	31st December 2004
	Annual mean	40	31st December 2004
SO_2	24 hr – not to be exceeded more than 3 times a year	125	31st December 2004
	1 hr – not to be exceeded more than 24 times a year	350	31st December 2004
	15 min mean – not to be exceeded more than 35 times a year	266	31st December 2005
NO_2	Hourly means – not to be exceeded more than 18 times a year	200	31st December 2005
	Annual mean	40	31st December 2005
CO	Running 8 h mean	11.6	December 2003
Lead	Annual mean	0.5	31st December 2004
	Annual mean	0.25	31st December 2008
Benzene	Running annual mean	16.25	December 2003
1,3-Butadiene	Running annual mean	2.25	December 2003

Units: 100 ppb corresponds to 266.1 $\mu g\,m^{-3}$ for SO_2, 191.2 $\mu g\,m^{-3}$ for NO_2, 116.4 $\mu g\,m^{-3}$ for CO, and 199.5 $\mu g\,m^{-3}$ for O_3 at a reference temperature of 20°C
Sources: The Air Quality (England) Regulations 2000, Environmental Protection, England, 2000 No. 928

2.3.3 Air Quality

2.3.3.1 Air Quality Standards. Air quality is assessed by reference to air quality standards. Those applicable in Europe and the UK are indicated in Table 8. The standards support the new responsibilities for air quality management of the Local Authorities following the 1995 Environment Act. They also form the focus of the government's National Air Quality Strategy.[55]

2.3.3.2 Air Quality Monitoring. A critical activity in air quality management is the monitoring of the pollutant concentrations. Continuous monitoring data for common pollutants such as CO, NO_x, O_3, SO_2 and particulate matter (PM10) are recorded at many sites around the world. Information from an increasing number of these sites is available on the World Wide Web.[56] Sample data from Leeds in the UK is illustrated in Figure 10 and shows diurnal cycles of the major pollutants.

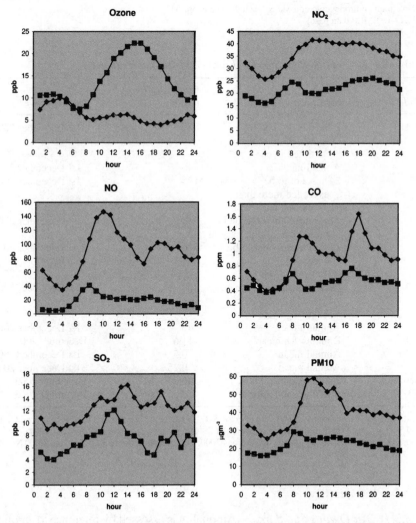

Figure 10 *Diurnal variation of pollutant concentrations for the Leeds AUN site[56]. Each point represents the monthly average for that hour of the day.* ◆ *January 1997,* ■ *July 1997*

Table 9 *Air quality data for megacities of the world. Annual average data in residential areas*[75]

City	Population 2005 (thousands)	PM_{10} ($\mu g\,m^{-3}$) 1999	SO_2 ($\mu g\,m^{-3}$) 1995–2001[a]	NO_2 ($\mu g\,m^{-3}$) 1995–2001[a]
Tokyo	35,327	43	18	68
Mexico City	19,013	69	74	130
New York	18,498	23	26	79
London	7,615	23	25	77
Beijing	10,849	106	90	122
Moscow	10,672	27	109	-
Deli	15,334	187	24	41

[a]Data for most recent year available

A global monitoring network, the GEMS/AIR programme, was initiated by WHO in 1973, and in 1975 became part of the Global Environment Monitoring System GEMS,[57] now jointly managed by WHO and UNEP. GEMS/AIR was implemented to strengthen urban air pollution monitoring and assessment capabilities, to improve validity and comparability of data among cities and to provide global assessments on levels and trends of urban air pollutants and their effects on human health. Approximately 170 monitoring stations in 80 cities in 47 countries participate in the programme measuring SO_2 and suspended particulate matter (55 cities in 33 countries) with other stations measuring NO_2, CO, O_3 and Pb. An indication of the air quality in major cities of the world from this programme is given in Table 9.

2.3.3.3 Air Quality Trends. The most significant trends in air quality on a national basis relate to a rather small number of factors such as

- changes in the patterns of fossil fuel usage,
- growth in vehicular transport,
- improved technology for vehicles giving lower emissions, and
- tighter control of industrial emissions, especially those from large scale combustion for power generation.

There is however a significant difference between industrialised and developing nations with respect to the major contributors to air quality problems. A UNEP report has shown that cities in developing countries seem to be following the same industrial trends as in other countries. Before industrialisation, pollution is mainly from domestic sources and light industry. Concentrations are generally low but increase as the population increases. As industrial development and energy usage increase air pollution levels begin to rise dramatically and emission controls have to be introduced.[20] Emissions and air quality

trends will be illustrated in the next few sections by reference to several of the
more common pollutants. Ozone, as a secondary pollutant is dealt with later.

2.3.3.4 Vehicular Emissions – CO and Hydrocarbons. A major feature of
the development of most countries in the last few decades has been growth
in vehicular traffic. The number of cars in the UK increased by 70% from
15.5 million in 1982 to 26.2 million in 2003, an average growth rate of 3.3%
per annum.[58] But in other countries the increase has been even more rapid –
China for example, has seen growth of 23.3% per annum expanding to 5.34
million in 1999 from 816,000 in 1990.[59]

The importance of vehicular emissions of carbon monoxide relative to other
sources is shown in Table 6. Carbon monoxide emissions are predominantly
from road vehicles while about 30% of VOCs also come from vehicles.
Historically, CO emissions increased as traffic density increased through the
1980s but are now on a downward trend arising from the progressive introduction of catalytic converters (Figure 11). The significance of vehicular emissions
is shown in the diurnal variation of the CO levels in major cities as illustrated in
Figure 10. The double-humped curve corresponding to the morning and
evening rush hours is clearly seen. Carbon monoxide levels beside busy roads
can under very adverse conditions reach approximately half the occupational
exposure limit of 50 ppm but general urban levels in much of Europe and the
US are usually below 10 ppm and annual averages only 1–2 ppm.[60]

2.3.3.5 Nitrogen oxides. Emissions of nitrogen oxides in the UK gradually increased until the early 1990s due to increased traffic density (Figure 11)
but gradually reduced during the 1990s as new vehicular emissions regulations began to take effect and the NO_x emissions from power stations also
decreased. The reduction is NO_x emissions has however flattened out in the
early part 21st century. Countries which have high coal consumption and/or
high vehicle populations tend to have high NO_x emissions. Table 7 and
Figure 12 illustrate this for several European countries.

Since nitrogen dioxide is the main health hazard, most attention has been
paid to monitoring urban levels of NO_2 as opposed to total NO_x. Annual average concentrations in urban areas of the UK range from 15 to 30 ppb. A significant wintertime episode of NO_2 pollution was experienced in London in
December 1991 when the levels exceeded 300 ppb for 8 h over one night and
exceeded 100 ppb continuously for 3 days. The current EU standard for
hourly averages is $200 \mu g\,m^{-3}$ (104.6 ppb). Throughout Europe there are a
number of cities with long term average NO_2 levels above the EU guideline
of $40 \mu g\,m^{-3}$ (21 ppb). NO shows a similar diurnal variation to CO because
of the traffic contribution but NO_2, which can be considered as a secondary
pollutant, does not show as marked a trend (Figure 10). This is discussed later.

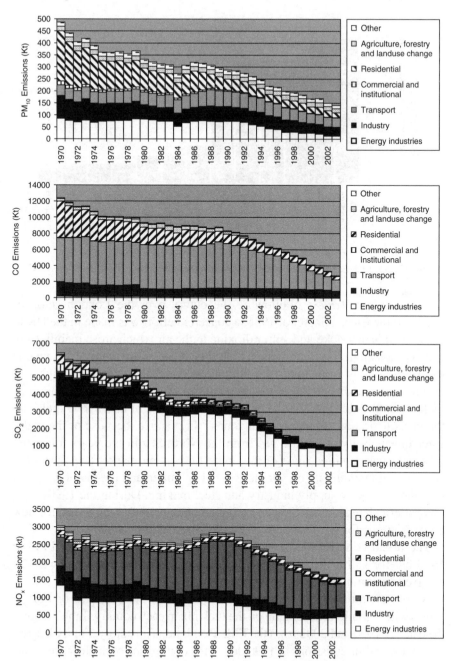

Figure 11 *UK annual emissions of pollutants by source type 1970–2003*
(Data from e-Digest of Environmental Statistics, published March 2005[46])

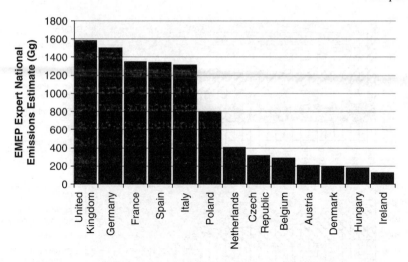

Figure 12 *2002 EMEP Expert National Emissions estimate for NO$_x$ for select countries*[47]

2.3.3.6 Sulfur Oxides. Following the outcry over the London smog of 1952 and the Clean Air Act of 1956, a concerted effort was made in the UK to reduce levels of smoke and sulfur dioxide in the air which arose mainly from coal combustion. The components of this effort were control of visible smoke emissions from industrial chimneys, control of chimney heights to ensure adequate dispersion of SO$_2$ and control of domestic smoke emissions by local authorities through the promotion of the use of alternative fuels and the introduction of "smoke control zones". In practice, because of the widespread availability of natural gas from the North Sea in the years following 1970, gas has captured most of the domestic heating market and shares the commercial market with oil. Domestic coal is limited and the major users are now the electricity supply industry and steel making. During the 1990s even this pattern changed with the generating companies switching to gas for new power stations for both economic and environmental reasons.

The urban concentrations of smoke declined dramatically from 1960 to 1980 but for sulfur dioxide the overall upward trend in emissions was not reversed until much later – 1970 was the peak year.

A comparison of sulfur dioxide emissions across the countries of Europe shows wide variations depending on the fuel mix in each country. Table 7 shows the data for 1990 and 2002 from the EMEP emissions inventory. France has relatively low sulfur emissions in 1990 due to the dominance of nuclear power for electricity generation. However, by 2002, improvements in abatement technology have evened out the differences between these countries.

2.3.3.7 Vehicular Particulates. Black smoke emissions were originally dominated by coal smoke and, as discussed in the section covering sulfur dioxide above, control efforts were targeted on this source and rapid improvements in air quality were achieved. However the more recent decline has not been as rapid because of the rise of smoke derived from vehicles, particularly diesels. Diesel engine soot is 3 times blacker than coal smoke and this is incorporated into the estimation of the "black smoke" emissions. In contrast the small amount of smoke from a petrol engine is much lighter than coal smoke because it consists largely of unburnt hydrocarbons rather than elemental carbon.

In terms of the *mass* of fine PM10 particles the picture is not so one sided as both petrol and diesel engines make a significant contribution. Although a given mass of diesel smoke is much blacker than petrol derived smoke and gives a much greater influence on the reflectance (and hence the "smoke" measurement) this is irrelevant if a direct mass measurement technique is used. Analysis of PM10 measurements suggest that on an average 40–50% of urban PM10 is contributed by vehicle exhausts in winter with higher proportions when the PM10 concentration itself is high and lower proportions in summer.[56] In addition to primary emissions indicated in Table 4, account must also be taken of the secondary particulate pollutants such as sulfates and nitrates and secondary organics. These are discussed in Section 2.5.1.

On a global scale, PM represents the biggest problem for the urban environment in terms on non-attainment of air quality standards. Many of the world's largest cities persistently exceed air quality standards, exposing large populations to a health risk.

The current lack of information regarding the quantitative relationship between adverse health effects and specific components of PM10 make it very difficult to determine the impact of reduced PM10 concentrations on human health.[49] Current WHO guidelines recommend the need for developing PM2.5 air quality standards. Indeed, in the US PM2.5 concentrations are now used as a basis for their air quality standards. However, health based guidelines for PM2.5 cannot simply be derived from scaling PM10 concentrations due to the varied correlation between the two size classes and different particle compositions.

2.3.3.8 Heavy Metals. Emissions of lead (Pb) from automobile exhausts used to arise from the use of lead tetra-alkyl anti-knock additives to improve the octane rating of the petrol. Increasing concern over the potentially harmful effects of lead on health, particularly the health of children resulted in a gradual decrease in the maximum permitted amount, which can be added to petrol. All new cars must now run on unleaded gasoline. The emissions of lead in 1995 were about 14% of what they were in 1975 and all gasoline sales in the UK are now of unleaded petrol. The decrease in the use of lead has led to a marked reduction in the atmospheric concentrations. This may be compared with the EPAQS UK air quality standard of $0.5 \mu g \, m^{-3}$.

The decrease in use of lead has led to a marked reduction in the atmospheric concentrations in many countries, with Europe following some years behind the USA. For UK urban sites, long term averages have reduced from a typical $500 \, ng \, m^{-3}$ to around $100 \, ng \, m^{-3}$ in 1998, mirroring the trend in the USA where concentrations were around $50 \, ng \, m^{-3}$ in 1995. This trend is mirrored in Europe, where lead was banned from petrol in 2000. However, UNEP reported that several of the world's megacities still exceed WHO limits for lead, with Cairo and Karachi showing the highest exceedences in 1992.[20]

2.3.3.9   Toxic Organic Micropollutants (TOMPS).   This phrase is used to describe several groups of organic pollutants which are present in the atmosphere in very small concentrations but which by inhalation or ingestion of contaminated food products could result in human health effects. TOMPS may be present in the vapour phase or adsorbed onto particles, the distribution between vapour and particulate phases depending on temperature. They include polynuclear aromatic compounds (described variously as PNA, PAH or PAC) – high molecular weight hydrocarbons often found in association with soot, some of which are carcinogenic. Benzo(a)pyrene is one example. Following the US EPA, analyses are commonly made on a group of 16, 2–6 ring compounds.

There has been a long-standing interest in the presence of such carcinogens in the general environment and in the levels emitted from combustion processes. Attention has been switched from coal smoke to diesel engine exhaust smoke as a source of these compounds but it remains to be established whether there is a significant health hazard at the very low concentrations to which the general public are exposed.

The other main group of TOMPS is the polychlorinated aromatic compounds. These include polychlorinated biphenyls (PCBs) which have entered the environment following wide scale use as insulating fluids in electrical transformers, plus polychlorinated dibenzo-dioxins and dibenzo-furans ("dioxins", PCDDs; "furans", PCDFs) which have combustion origins, especially from incinerators unless they are very well controlled, as is typically the case. There are numerous isomers of each of these classes of compounds depending on the number and locations of the chlorine atoms in the molecules.

Extensive monitoring of air, water and soil samples for TOMPS have been undertaken in recent years. The average air concentrations in London between 1991 and 1995, for example, were as follows:[44]

PAHs $50–150 \, ng \, m^{-3}$; PCBs $1.0–1.5 \, ng \, m^{-3}$; Dioxins $100–200 \, fg \, TEQ \, m^{-3}$

The US EPA carried out a thorough review of dioxins in the environment and their health effects, published in 1994.[61] Airborne concentrations data reported in the USA and Europe also indicated average air concentrations of the order of 100 fg TEQ m^{-3}.

Note: 1 ng = 10^{-9} g, 1 fg = 10^{-15} g. TEQ = Toxic Equivalent – in summing the masses of different dioxins each is weighted by its relative toxicity with 2,3,7,8 tetrachlorodibenzo-p-dioxin taken as the reference.

2.4 GAS PHASE REACTIONS AND PHOTOCHEMICAL OZONE

2.4.1 Gas Phase Chemistry in the Troposphere

2.4.1.1 Atmospheric Photochemistry and Oxidation. Although the emissions patterns and dispersion discussed in the previous sections give a part of the picture of where high levels of pollutants can be found, this must be combined with an understanding of other atmospheric processes to see how secondary pollutants such as ozone and acid rain are formed. These processes including dispersion, chemical reaction and deposition are illustrated in Figure 13. The chemical processes involved are complex and changes both in the gas phase and in the aqueous droplet phase are important. Not only must the transformation of the primary pollutants be considered but also the formation of secondary pollutants such as ozone which can have adverse effects on the environment and on human health.

Of basic importance to an understanding of the gas phase chemistry is the effect of the sun. Photons of ultra-violet light provide a means of initiating chemical reactions which would other-wise not take place. In addition to stable molecules, photochemical reactions involve free radicals such as hydroxyl

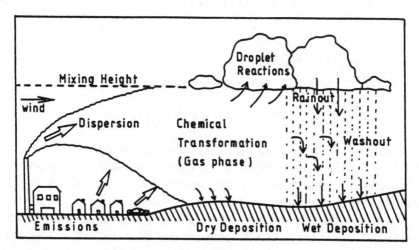

Figure 13 *Processes which may be involved between the emission of an air pollutant and its ultimate deposition to the ground*

OH, hydroperoxy HO_2 and methyl CH_3. Free radicals are extremely reactive and have very short lifetimes. Their concentrations in the atmosphere are small but nonetheless significant. For example, OH concentrations in polluted atmospheres may be in the range 10^6–10^7 radicals per cm^3, *i.e.* one radical for every 10^{13} nitrogen molecules.

One of the most important overall processes to describe is oxidation, that is, the combination of atmospheric oxygen with the primary pollutants. For the three commonest inorganic pollutants the overall results are

carbon monoxide   CO $\rightarrow$ CO_2 carbon dioxide

nitrogen oxides   NO, NO_2 $\rightarrow$ HNO_3 nitric acid

sulfur dioxide SO_2 $\rightarrow$ H_2SO_4 sulfuric acid.

Later the involvement of the two acids in particles and droplets will be considered, but as far as the gas phase chemistry is concerned these species mark the end point of the process. For organic hydrocarbon species there may be a number of intermediate stable molecules formed but the overall process is rather like combustion with the end product being carbon monoxide; for example:

methane (CH_4) $\rightarrow$ formaldehyde (HCHO) $\rightarrow$ carbon monoxide

However, the time taken to complete this process can be very long. In all these cases the main species initiating the sequence of reactions is the hydroxyl radical.

$$CO + OH^{\bullet} \rightarrow CO_2 + H^{\bullet} \tag{14}$$

$$NO_2 + OH^{\bullet} \rightarrow HNO_3 \tag{15}$$

$$SO_2 + OH^{\bullet} \rightarrow HSO_3^{\bullet} \tag{16}$$

$$CH_4 + OH^{\bullet} \rightarrow CH_3^{\bullet} + H_2O \tag{17}$$

In the case of nitric acid formation, there are no remaining free radicals to continue the chain of reactions. In the other cases the hydroxyl radical is eventually regenerated but only after several further steps, which interlink the chemistries of the various species (Table 10). Carbon monoxide oxidation is a slow process and the lifetime of CO in the atmosphere is several years. The oxidation rate of SO_2 can be around 2% h^{-1} in urban air or a factor of 10 lower in clean air resulting in an overall lifetime of a few days. Radicals other than OH such as HO_2, CH_3O_2 or other hydrocarbon peroxy radicals will also attack SO_2 but at a slower rate and their contribution to the overall oxidation is thought to be relatively small.

2.4.1.2 Ozone. The question of where the hydoxyl radicals come from in the first place must now be addressed. One source present even in non-polluted atmospheres at a back-ground level of 20–40 ppb is ozone. Ozone is a secondary pollutant formed through photochemical reactions and can have a harmful effect on human health causing respiratory problems, and on crop yields. It is the primary constituent of photochemical smog. Ultra-violet light of wave length below 310 nm can dissociate ozone producing electronically excited oxygen atoms (O^*) which rapidly split molecules of water vapour:

$$\text{UV light}$$

$$O_3 \rightarrow O_2 + O^* \qquad (18)$$

$$H_2O + O^* \rightarrow 2OH^{\bullet} \qquad (19)$$

Aldheydes (RCHO, including formaldehyde HCHO) can also be photolysed producing hydrogen atoms, which eventually result in OH radicals via reactions already shown in Table 10.

Of basic importance to the understanding of polluted urban atmospheres is the photolysis of NO_2 and the subsequent formation of ozone above the background levels. It was noted earlier that only a small proportion of NO_x emissions are in the form of NO_2, the rest being NO. NO emitted into the atmosphere can be slowly oxidised to NO_2 by the reaction with molecular oxygen.

$$O_2 + 2NO \rightarrow 2NO_2 \qquad (20)$$

Table 10 *Chemical reactions for the atmospheric oxidation of CO, SO_2 and CH_4*

Carbon Monoxide		
$CO + OH^{\bullet}$	$\longrightarrow$	$CO_2 + H^{\bullet}$
$H^{\bullet} + O_2 + M$	$\longrightarrow$	$HO_2^{\bullet} + M$
Sulfur dioxide		
$SO_2 + OH^{\bullet}$	$\longrightarrow$	$HSO_3^{\bullet}$
$HSO_3^{\bullet} + O_2$	$\longrightarrow$	$SO_3 + HO_2^{\bullet}$
$SO_3 + H_2O$	$\longrightarrow$	H_2SO_4
Methane		
$CH_4 + OH^{\bullet}$	$\longrightarrow$	$CH_3^{\bullet} + H_2O$
$CH_3^{\bullet} + O_2 + M$	$\longrightarrow$	$CH_3O_2^{\bullet} + M$
$CH_3O_2^{\bullet} + NO$	$\longrightarrow$	$CH_3O^{\bullet} + NO_2$
$CH_3O_2^{\bullet} + O_2$	$\longrightarrow$	$HCHO + HO_2^{\bullet}$
HO₂ to OH conversion		
$HO_2^{\bullet} + NO$	$\longrightarrow$	$OH^{\bullet} + NO_2$

Photolysis of NO_2 by UV light below 420 nm produces ground state (unexcited) oxygen atoms and subsequently ozone:

$$\text{UV light}$$
$$NO_2 \rightarrow NO + O^{\bullet} \tag{21}$$

$$O_2 + O^{\bullet} \rightarrow O_3 \tag{22}$$

However the process is reversed by the reaction:

$$O_3 + NO \rightarrow O_2 + NO_2 \tag{23}$$

so the net result is an ozone level in equilibrium with NO and NO_2 and dependent on the intensity of the solar radiation. The observed levels of ozone are higher than would be predicted on the basis of this limited scheme. High ozone levels imply a high NO_2/NO ratio or significant $NO \rightarrow NO_2$ conversion which cannot be achieved by the molecular reaction (20). The types of reactions responsible for increasing the rate of NO to NO_2 conversion have already been shown in Table 10; they are the transfer of an O-atom from VOCs such as HO_2, CH_3O_2 and other peroxy radicals. Simply put, the photochemical reaction of the VOC mixture catalyses the conversion of NO to NO_2 resulting in the build-up of ozone.

Hydrocarbon molecules differ in their photochemical ozone creation potential (POCP)[62–64] largely related to how quickly they react with the OH radical. Methane has a very low potential but other species including substituted benzenes, such as toluene, the xylene isomers and light unsaturated hydrocarbons such as ethene, propene and but-2-ene have high POCPs. All POCPs are conventionally calculated relative to ethene, which has a value 1. The aromatics are present in high concentration in gasoline and the unsaturated compounds are typical products of engine combustion. Table 11 gives some example of the rate of reaction with OH for selected hydrocarbon species along with associated POCP weighted emissions. These are calculated by multiplying the POCP by the annual emission and therefore indicate which VOCs are best targeted for control strategies.

Driven by the chemistry described above, the concentrations of O_3 and NO_2 are dependent on sunlight and emissions and therefore will vary diurnally as well as with the time of year. The diurnal variations of NO, NO_2 and O_3 typically detected are illustrated in Figure 10. The morning rush hour peak in the NO and hydrocarbon emissions is followed by the gradual conversion

Table 11 *Species lifetimes, k_{OH} and POCP weighted emissions for selected hydro-carbon species. POCP values are calculated relative to a POCP of 1 for ethene based on 1990 data*[62]

Species	$k_{OH} \times 10^{12}$	Lifetime (h)	POCP weighted emission (tonnes ethene per year)
methyl benzene (toluene)	5.96	58.3	85,423
ethene	8.52	40.7	78,925
1,3-dimethyl benzene	23.6	14.7	66,086
butane	2.54	136.7	58,649
propene	26.3	13.2	36,946
benzene	1.23	282.3	9,418
ethane	0.268	1295	3,017

to NO_2 and subsequent rise of O_3 which decays as the sun goes down in late afternoon.

The production of ozone is much greater in the summer than for winter due to higher photolysis rates. For NO and NO_2 the winter hourly averages are higher. In an air mass moving downwind from a city, the ozone peak may be worse in the surrounding countryside than in the city centre because of less destruction of O_3 by NO. During the night most of the reactions that have been described die down but there is an additional route for conversion of NO_2 to nitric acid via the nitrate radical NO_3 which is formed from reaction of NO_2 with O_3. NO_3 is photolytically unstable in daylight. Dry deposition is also an important loss process for O_3 and can account for up to 30% of its loss in surface layers of the atmosphere.

Space does not permit a detailed discussion of the hydrocarbon chemistry in the atmosphere which is extremely complex. In addition to reactions with OH radicals and molecular oxygen similar to those for methane shown in Table 10, hydrocarbon species are attacked by the oxygen atoms released in reaction (21) and by ozone. One result of the interaction of the hydrocarbon and NO_x chemistry is the formation of a group of lachrymatory substances including peroxyacetyl nitrate (PAN) and peroxybenzoyl nitrate (PBzN). The threshold for eye irritation for these compounds are 700 and 5 ppb respectively. They are formed through the reaction of NO_2 with oxidation products of aldehydes.

$$RCHO \xrightarrow{OH^\bullet} RCO^\bullet \xrightarrow{O_2} RCO - O - O^\bullet$$
$$\xrightarrow{NO_2} RCO - O - O - NO_2$$

$R = CH_3$ *gives peroxy acetyl nitrate*
$R = C_6H_5$ *gives peroxy benzoyl nitrate*

Case Study 4: Trends in Ozone Levels

In the UK elevated levels generally occur in summer, anticyclonic conditions such as those illustrated in Plate 3 when photolysis rates are high and wind-speeds low. Concentrations tend to be higher in southern England where polluted air masses from continental Europe have a significant effect. The highest recorded concentration in the UK was 258 ppb at Harwell in 1976 as compared with the highest of 450 ppb in the Los Angeles basin in 1979. Los Angeles provides an ideal environment for ozone formation due to high traffic emissions and its meteorological conditions. Currently ozone trends in LA are falling due to emissions legislation although the number of days a year on which the State limit of 90 ppb is exceeded was still of the order of 100 in 2004. In Europe there is evidence that ozone is slightly increasing throughout the troposphere. Over 1000 monitoring stations are now reporting ozone levels in Europe. In the summer of 1996 the EEC threshold for warning the public of $1 h > 360 \mu g\, m^{-3}$ (~180 ppb) was exceeded at three stations in Europe, two in Athens and one in Firenze in Italy. The threshold for informing the public of $1 h > 180 \mu g\, m^{-3}$ (~90 ppb) was exceeded in all EU states except for Ireland. In terms of episodes of elevated ozone levels, the UK and Scandinavia are the least affected out of all the European countries with Southern Europe being most affected. Urban areas show less frequent exceedences because of high NO conditions.

The formation of photochemical smog is governed by emissions of NO_x and volatile organic carbons (VOCs). Because the chemistry is complex and nonlinear, abatement strategies for ozone and smog are difficult to devise. Within Europe some regions will respond better to reductions in NO_x emissions and some better to reduction in VOCs. In general whether a region is NO_x or VOC limited depends on the NO_x: VOC ratio, with regions of high NO_x often responding better to reductions in VOCs. The UNECE Protocol required a 30% reduction in VOCs from 1988 to 1999 and an EC directive on ozone requires national monitoring networks to assess conditions according to the reference levels which state that ozone concentrations should remain below an 8 h mean of $110 \mu g\, m^{-3}$ (~55 ppb) and a 24 h mean of $65 \mu g\, m^{-3}$ (~32.5 ppb).

2.5 PARTICLES AND ACID DEPOSITION

2.5.1 Particle Formation and Properties

2.5.1.1 Particle Formation. The nitric and sulfuric acids formed in the gas reactions described earlier generally undergo further changes. They are both water-soluble and will be rapidly absorbed into water droplets if these are present. They may react with solid particles forming sulfates and nitrates. For example, limestone particles (calcium carbonate $CaCO_3$) can be converted to calcium sulfate, salt particles (NaCl) of marine origin, can be converted to sodium sulfate, Na_2SO_4 or sodium nitrate, $NaNO_3$, with the displacement of hydrogen chloride gas, HCl. However, the most common reactions are those involving ammonia.

$$NH_3 + HCl \rightleftharpoons NH_4Cl \qquad (24)$$

$$NH_3 + HNO_3 \rightleftharpoons NH_4NO_3 \qquad (25)$$

$$NH_3 + H_2SO_4 \rightarrow NH_4HSO_4 \qquad (26)$$

$$NH_3 + NH_4HSO_4 \rightarrow (NH_4)_2SO_4 \qquad (27)$$

Reactions 24 and 25 are actually reversible and the position of equilibrium depends on the concentrations and the temperature. Significant dissociation of NH_4Cl and NH_4NO_3 occurs during warm summer weather. One end product of the oxidation of SO_2 in the atmosphere is ammonium sulfate, a substance better known as a fertiliser.

Particles formed by gaseous reactions or condensation are initially very small ($<0.1\,\mu m$) but grow rapidly either by surface accumulation of material from the gas phase or by particle-particle coagulation. Once in the size range 0.1 to $2.0\,\mu m$, they become relatively stable in size and can remain airborne for periods of days. Most smoke emissions are in this category along with the sulfates and nitrates. At the other end of the size spectrum ($2-50\,\mu m$) is coarse dust either emitted from industrial processes or raised by the wind from the ground. The action of road vehicles is another mechanism for raising such dust. Sea salt left from the evaporation of spray is also fairly coarse.

2.5.1.2 Particle Composition. Figure 14 presents a typical breakdown of the components of the total suspended particulate matter for an urban area based partly on results from a survey in Brisbane, Australia.[65] The fine particles ($<2.5\,\mu m$) are dominated by the ammonium sulfate and nitrates plus carbonaceous material. About one-third of this is elemental carbon and the other two-thirds organic carbon (*i.e.* high molecular weight hydrocarbons), which together comprise "smoke". The coarser particles are dominated by wind-blown dusts (clays, silica, limestone, *etc.*) and include the sea salt component but have smaller amounts of ammonium sulfate and carbon. These results are similar to those reported in the UK[68] although the sea salt component is higher.

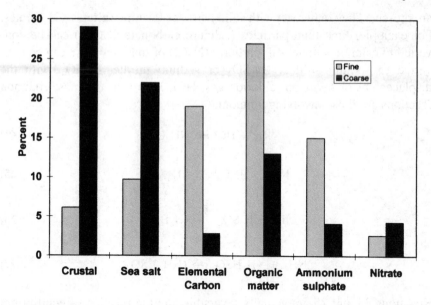

Figure 14 *Composition of coarse (>2.5 μm) and fine particles in Brisbane, Australia*[65]

Particulate sulfate can have associated free acidity in the form of sulfuric acid, H_2SO_4, or ammonium bisulfate, NH_4HSO_4. The degree of neutralization of the acids will depend on the supply of ammonia and in the UK sulfate particles are found to be well neutralised on average.[55] Further information on particle size and composition in the UK can be found in the Air Quality Expert Group report on Particulate Matter in the United Kingdom (2005).[49]

2.5.1.3 Deliquescent Behaviour. The particles of water soluble compounds such as sulfates, nitrates and chlorides will exist in the atmosphere either as solid particles or droplets depending on the relative humidity. For pure compounds the transition from solid to liquid is a sharp one. For example, it takes place with salt, NaCl at 75% relative humidity and with ammonium sulfate at 86%. Particles of mixed composition may continuously grow in size by water absorption from 60–70% up to 100% relative humidity. Insoluble carbonaceous particles are, of course, not subject to this phenomenon. Particles are important in cloud and fog formation since they act as condensation nuclei on which the much larger droplets may begin to be formed from water vapour.

2.5.1.4 Optical Properties. Fine particles in the 0.1–2 μm size range are effective at scattering light and some, especially soot, are also effective at absorbing light. These effects contribute to a reduction in visibility through the atmosphere. Quantitatively, the visibility is the maximum distance at which a large dark object can be seen against the horizon sky (sometimes called the visual range). In clean air this distance can be over 50 km but in air polluted by particles this can be severely reduced. A mass concentration (TSP) of 200–300 $\mu g \, m^{-3}$ will reduce the visibility to below 5 km.[66] The phenomenon of a mid-summer haze on a hot day reflects high particulate pollutant levels. It is a totally different phenomenon to an early morning mist, which is caused by water droplets at a relative humidity near 100% in Europe. Such highly polluted summer days are usually associated with anticyclonic conditions (*cf.* Plate 3). Wintertime visibility has been improved and the frequency of fogs reduced in urban areas by the measures taken to control smoke and SO_2 emissions. Any additional measures to reduce pollutant emissions, especially of NO_x and SO_2, will bring corresponding benefit in terms of improved visibility throughout the year. This trend has been confirmed by measurements taken throughout the USA where visibility in the East worsened between 1970 and 1980 but slightly improved by 1990 in line with SO_2 emissions.[60]

2.5.2 Droplets and Aqueous Phase Chemistry

Liquid water occurs in the atmosphere as clouds, mists and fog within which the concentration of water can be up to $1 \, g \, m^{-3}$. Smaller amounts of water are also present in association with deliquescent particles as discussed above. At relative humidities below 95%, this secondary amount will normally be less than $1 \, mg \, m^{-3}$ and will not be considered further.

Water droplets can accumulate pollutants by adsorption of either gases or particles and within the droplets chemical reactions can proceed changing the nature of the adsorbed species. Solution of SO_2 into water results in a mixture of the species SO_3^{2-} (sulfite), HSO_3^- (bisulfite) and H_2SO_3 (undissociated sulfurous acid) depending on the pH, which is defined as $-\log_{10}(H^+$ ion concentration). At typical cloudwater pH values the bisulfate ion is the dominant ion formed from:

$$SO_2 + H_2O \rightleftharpoons H^+ + HSO_3^- \qquad (28)$$

There are several mechanisms for the oxidation of bisulfate to sulfuric acid in the aqueous phase. In a cloud situation the most important oxidants appear to be ozone and hydrogen peroxide, which is formed from the reaction of two HO_2 radicals. The oxidants must first be absorbed into water from the gas phase and then result in the reactions:

$$O_3 + HSO_3^- \rightarrow H^+ + SO_4^{2-} + O_2 \qquad (29)$$

$$H_2O_2 + HSO_3^- \rightarrow H^+ + SO_4^{2-} + H_2O \qquad (30)$$

The sulfuric acid formed will be completely dissociated to ions H^+ and SO_4^{2-}. The ozone reaction is inhibited by low pH whereas the H_2O_2 reaction is not. In acidic droplets the oxidation by H_2O_2 is therefore dominant whereas at high pH the O_3 becomes more significant. In a cloud or fog situation where there has been a significant input of particulate pollution, the oxidation of SO_2 by atmospheric oxygen catalysed by metal ions (iron Fe^{3+} or manganese Mn^{2+}) or by soot can also be important. Although not indicated by the simple reactions written in eqs 29–30 free radical reactions are as important in solution chemistry as they are in the gas phase and many different species are involved. The results of a recent European Project in this area have been published.[67]

It is difficult to distinguish experimentally between the photochemical reaction mechanism leading to H_2SO_4 with subsequent absorption of the acid into water and the aqueous phase oxidation of SO_2. However, it appears that both routes are important. In the case of nitrogen oxides the routes for nitrate formation in clouds are

- dissolution of gaseous nitric acid vapour formed by reactions previously described in Section 2.4;
- dissolution of nitrate containing particles into droplets; and
- absorption of nitrogen oxides or nitrous acid HONO into droplets followed by oxidation of nitrite ions NO_2^- by oxidants such as H_2O_2.

Because of the low solubility of NO_2 and especially NO in water it is likely that the dominant processes are the first two.

Despite partial neutralization of dissolved acids by ammonia, the water in polluted fogs and clouds can be much more acidic than in collected rainfall. pH values down to 2 have been measured in urban fogs – more acid than vinegar. Similarly low values have been measured in the plumes of large power stations.

2.5.3 Deposition Mechanisms

2.5.3.1 Dry Deposition of Gases. As illustrated in Figure 13, the life cycle of an air pollutant normally involves emission, dispersion and transport, chemical transformation and finally deposition to the ground. Understanding the rates and mechanisms of deposition is important to the assessment of the environmental impact of many pollutants. Experimentally the concentration of the pollutant ($\mu g \ m^{-3}$) can be measured and the total rate of deposition ($\mu g \ m^{-2} \ s^{-1}$). The higher the ground level concentration, the more rapid the deposition but the ratio of these two quantities gives a useful measure of the efficiency of the deposition process. It is called the *deposition velocity*.

$$\text{Deposition velocity} = \frac{\text{deposition rate}}{\text{air concentration}} \ (\text{units: m s}^{-1})$$

Simply, the process of surface adsorption and downward mixing of the SO_2 by turbulent diffusion can be envisaged as the calculated deposition velocity. Different surfaces (water, soil, ice, *etc.*) will have correspondingly different deposition velocities. The description "dry deposition" is used in all cases even if the removal is to a wet surface.

Deposition to vegetation is rather more complex than deposition to a plane surface. The pollutant's progress may be retarded either by the slowness of diffusion through the air within the canopy of vegetation (*i.e.* between the leaves, *etc.*) or by the rate of transfer from air to leaf surface (the cuticle) or to the interior of the leaf via stomata. Moisture makes a considerable difference in that transfer to a wet surface is generally faster than to a dry surface. Based on experimental data, dry deposition velocities for SO_2 used in modeling large scale deposition are typically assumed to be in the range 2–$5\,mm\,s^{-1}$. Similar considerations apply to other pollutants, which are subject to a significant rate of adsorption at the ground or onto vegetation (NO, NO_2, HNO_3, *etc.*). NO_2 has a lower value of $1\,mm\,s^{-1}$ while HNO_3 deposits very rapidly and values up to $40\,mm\,s^{-1}$ are assumed.[43]

2.5.3.2 Wet Deposition. The term "wet deposition" is used to describe pollutants brought to ground either by rainfall or by snow. This mechanism can be further sub-divided depending on the point at which the pollutant was absorbed into the water droplets. In-cloud absorption followed by precipitation is termed "rain-out"; below cloud absorption, *i.e.* pollutants are collected as the raindrops fall, is termed "wash-out". The rate of removal of a pollutant by wash-out will increase in proportion to the rainfall rate. Overall a scavenging coefficient can be defined which is the fractional loss of the pollutant from the gas phase per second. For SO_2 the scavenging ratio is of the order of $10^{-6}\,s^{-1}$ in drizzle ($\leqslant 1\,mm\,h^{-1}$) and an order of magnitude greater in heavy rain. Half of the gas below the clouds can be removed in several hours of heavy rain. The rates for the nitrogen oxides are lower due to their reduced solubility in water. In remote areas the majority of the wet deposition of sulfur appears to be due to rain-out. Wash-out becomes relatively more significant near the sources of pollution where the gas concentrations are high.

Another mechanism of deposition is when fog or cloud droplets are removed directly to the ground or to vegetation. This is termed "occult deposition". It becomes significant at elevated locations such as mountains or hill tops. There is the potential for more severe damage to foliage than with acid rain since, as was previously mentioned, polluted fog or cloud droplets can contain much higher concentrations of acidic pollutants than raindrops.

For sulfur, dry and wet deposition are respectively 40 and 60% of the total, for nitrogen the split is 27% dry and 73% wet. Modelling work suggests that about 45% of the wet S deposition and 25% of the dry S deposition comes from other European countries. The UK is however, a net exporter of air pollutants, while other countries of Europe are net importers, for example, Norway and Sweden. The UK is the largest single contributor to sulfur deposition in southern Norway although emissions from many other countries are transported there. Deposition of reduced nitrogen (NH$_3$) is also important as discussed below.

Case Study 5: Acid Deposition in the UK

Dry deposition and wet deposition are both important in the total deposition of sulfur and oxidised nitrogen compounds to the UK. Dry deposition is most significant where ground level concentrations are high, in other words close to the sources. In general dry deposition and wet deposition are of similar importance for nitrogen compounds within the UK and wet deposition tends to dominate for sulfur compounds as shown in Table 12. The table also shows changing trends in the relative importance of sulfur and nitrogen compounds with respect to total deposition reflecting changes in emissions patterns. In 1986 sulfur compounds clearly dominated total acid deposition, whereas in 2001 sulfur and nitrogen compounds are of equal importance due to larger reductions in sulfur emissions.

Plate 4a shows the total deposition for the UK between 1999–2001[68] assuming moorland cover in the left figure and woodland cover in the right figure. Total deposition is highest over woodland and forest areas due to increased turbulence associated with trees in the vegetation layer. The figures show that deposition is generally greatest over England especially the Pennines, Lake District and Snowdonia. However, due to the different buffering capacity of the natural environment, the areas of high deposition are not always those areas where the most damage is done. For this reason the concept of 'critical loads' has been introduced in the evaluation of acid deposition (see Section 2.5.4.3).

Table 12 *UK sulfur and nitrogen budgets from the network measurements for 1986 and 2001 (kt S, N per year)[71]*

Year	Emis-sions S	Wet deposi-tion S	Dry deposi-tion S	Total deposi-tion S	Emis-sions N	Wet deposi-tion N	Dry deposi-tion N	Total deposi-tion N
1986	1939	252	415	667	797	106		106
2001	563	133	76	209	511	105	101	206

2.5.3.3 Deposition of Particles. The mechanism of deposition of particles depends on the particle size. Large particles with diameter greater than $10\,\mu m$ fall slowly by gravitational settlement. The larger the particles, the more rapid they fall. The sedimentation velocities for particles of density $2\,g\,cm^{-3}$ are as follows:-

Diameter μm	Velocity m s^{-1}	Diameter μm	Velocity m s^{-1}
5	1.5×10^{-3}	50	1.4×10^{-1}
10	6.1×10^{-3}	100	4.6×10^{-1}
20	2.4×10^{-2}	150	8.0×10^{-1}

Particles larger than $150\,\mu m$ diameter, falling at over $1\,m\,s^{-1}$ remain airborne for such a short time that they do not need to concern us as air pollutants. Particles less than $5\,\mu m$ have sedimentation velocities, which are so low that their movement is determined by the natural turbulence of the air, just as for gases.

Intermediate particles, between 1 and $10\,\mu m$ diameter, can be removed by impaction onto leaves and other obstacles. Particles in the 0.1 to $1\,\mu m$ range, which include most of the nitrates and sulfates, are only removed very slowly by dry deposition. The deposition velocities are of the order of $1\,mm\,s^{-1}$, much lower than for SO_2. The most likely route for their removal is rain-out following water vapour condensation and droplet growth in clouds. Wash-out is

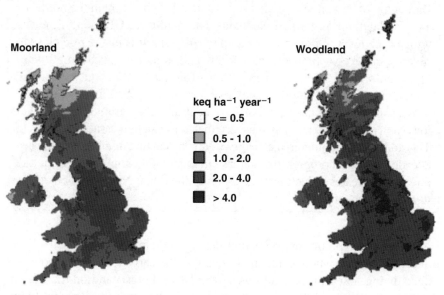

Plate 4a *Total acid deposition (non-marine sulfate plus oxidised and reduced nitrogen) for 1999–2001*[68]
(Map source: Centre for Ecology and Hydrology)

.cient for these fine particles although it becomes more significant ρarticles such as coarse dust.

Acid Rain

2.5.4.1 Rainwater Composition and Effects.   Even in the absence of air pollutants, rain-water is slightly acidic (pH 5.6) due to atmospheric carbon dioxide. "Acid rain" therefore refers to rain with a pH below about 5.

The phrase "acid rain" has come to be used very loosely, to mean almost anything to do with acidification whether or not rain is actually involved. Three particular effects have received most attention. Historically the first of these was the increased acidification of lakes and streams in Scandinavia leading to loss of fish and other aquatic organisms. This was attributed to the acidity of rain polluted by sulfur and nitrogen oxides and the UK was blamed as being the chief culprit. Later, the acidification of fresh waters in our own country was demonstrated.[70]

Case Study 6: Acid Rain in the UK

Annual average acidities over much of the UK correspond to pH values between 4.2 and 4.8, whilst the highest recorded acidities are below pH 3. Table 13 shows the annually averaged composition at three UK sites.[69] Barcombe Mills is in the southeast of England, Thorganby in the northeast and Eskdalemuir in southern Scotland. Thorganby is a UK monitoring site with a high level of precipitation acidity (pH 5.1). It is in a region close to several major coal fired power stations and is unusual in having a very high non-seasalt chloride arising from near-source HCl wet deposition. Excluding this site, sodium and chlorine in rain are predominantly provided by seasalt. The acid has been partially neutralized by ammonia and other ions such as Ca^{2+}, which may have originated as calcium carbonate, $CaCO_3$. The sulfate contribution to the acidity is larger than that of nitrate but the last decade has seen a progressive increase in the relative importance of nitrate due to the greater reductions in SO_2 emissions than NO_x emissions. This trend is expected to continue.

The second type of environmental damage is damage to forests.[72,73] The effects became very noticeable in Germany in the early 1980s with the worst effects being noted in the south-west (the Black Forest) and on the eastern border with the Czech Republic, a country which shares the same problem. Since then other countries have reported similar phenomena. Although the reasons for the damage have been the subject of much debate, it seems likely

Table 13 *Precipitation-weighted mean concentrations (*μeq l^{-1}) and mean annual rainfall for 3 UK sites, 2002*[69]

Site	Thorganby (NE England)	Barcombe Mills (SE England)	Eskdalemuir (S Scotland)
Rainfall mm	922	609	1780
pH	5.0	5.12	5.57
H	21	30	17
NH_4	20	44	15
Na	203	35	63
Mg	46	12	13
Ca	23	18	4
SO_4	49	46	21
NO_3	23	33	14
Cl	235	49	73

Note samples containing phosphate are excluded
* Micro-equivalents per litre (μeq l^{-1}) are defined in chapter 1

to be a combination of factors. Predisposing factors include drought and high altitude and in some cases disease plays a role. Some forests in the former Eastern European countries suffer from the effects of very high SO_2 levels but this is not the case elsewhere. Possible mechanisms include

- the effect of ozone initiating an attack on cell walls with subsequent further deterioration being due to acid rain or acid mists and fogs leaching nutrients and resulting in the breakdown of chlorophyll. Reduced root growth and nutrient uptake follows;
- acidification of the ground with consequent effects on the soil chemistry including elevation of mobile aluminium levels which can damage the roots; and
- excess deposition of nitrogen (as nitrate and ammonium) which can have a variety of effects. In the ground NH_4^+ can release H^+ during the process of being oxidised to NO_3^- by bacteria. The H^+ can then be leached out. From the point of view of acidification phenomena, ammonia should therefore not be regarded as an ally even though, prior to its transformation to nitrate, it reduces the rain-water acidity.

Given the complexity of the biological and chemical processes mentioned it is clear that control of the effects of "acid rain" on biological systems must focus on all the relevant pollutants – SO_2, NO_x, NH_3 and hydrocarbons (as precursors of ozone).

The third problem associated with acid rain is the attack on stonework and the decay of famous cathedrals and other buildings constructed of limestone.[74] Both wet and dry deposition of sulfur dioxide are involved. Under moist conditions SO_2 or sulfuric acid will convert calcium carbonate to gypsum, $CaSO_4.2H_2O$. Since the sulfate is more soluble than the carbonate, the reacted stone can be removed by dissolution. The solid gypsum also occupies a larger volume than the original carbonate and this leads to spalling of material from the surface. A combination of these factors leads to a rate of loss of stone, which depends on the deposition rate of SO_2 to the surface.

Deposition to a moist surface is more rapid than to a dry surface so the fraction of time the surface is wetted as well as the SO_2 concentration is important. It is generally assumed that in urban areas the dry deposition of SO_2 gas is the major factor rather than the acidity of rain itself.

2.5.4.2 Patterns of Deposition and Critical Loads Assessment. In order to highlight areas where acidity is causing damage, and therefore to enable more effective control measures, a *critical loads* approach has been introduced. A critical load for a particular receptor-pollutant combination is defined as the highest deposition load that the receptor can withstand without long-term damage occurring. A gridded critical loads map can be prepared based on the geology and other factors affecting the response of fresh waters or terrestrial ecosystems to pollutant input. Future patterns of exceedance of critical loads can be predicted on the basis of expected reductions in pollutant emissions. The needs for more stringent control measures can then be assessed. A similar critical loads approach is being used in other European countries and underlies the current European approach to emissions control.

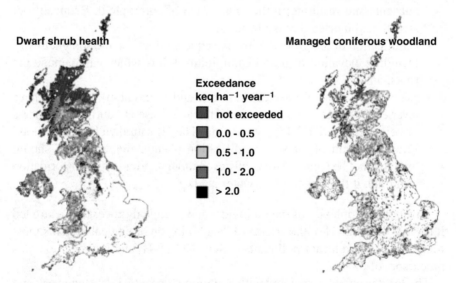

Plate 4b *Exceedance of acidity critical loads by total acid deposition (non-marine sulfate plus oxidised and reduced nitrogen) for 1999–2001*[68]
(Map source: Centre for Ecology and Hydrology)

Case Study 7: Critical Loads for Acidity in the UK

Plate 4b shows where in the UK the critical loads for acidity have been exceeded for dwarf shrub heath and managed coniferous woodland (two of the most sensitive ecosystems) respectively.[68] For dwarf shrub heath the areas of maximum exceedance are generally in highland areas such as the Pennines, the Scottish highlands and Wales reflecting areas of high rainfall and the low buffering capacity of the underlying soils. Temporal trends show a gradual decline in critical load exceedances for these areas reflecting the decline in sulfur and nitrogen emissions discussed earlier (see Table 13). Following emissions reductions of SO_2 within the EU the total area of Europe with exceedances of critical loads for sulfur have been reduced by more than 50% since 1980. However, these improvements have been offset by continued high rates of NO_x and ammonia deposition. The situation is similar in the US where a decreasing trend in sulfate ion concentrations has been seen in recent years along with a slight increase in nitrate ions.

REFERENCES

1. G. Marland, T.A. Boden and R.J. Andres, Global, regional, and national CO_2 emissions, in *Trends: A Compendium of Data on Global Change*. Carbon Dioxide Information Analysis Center, Oak Ridge National Laboratory, U.S. Department of Energy, Oak Ridge, TN, U.S.A. 2003.
2. J.T. Houghton, L.G. Meira Filho, B.A. Callander, N. Harris, A. Kattenberg and K. Maskell (eds), *Climate Change 1995: The Science of Climate Change*, Cambridge University Press, Cambridge, 1996.
3. J.T. Houghton, Y. Ding, D.J. Griggs, M. Noguer, P.J. van der Linden and D. Xiaosu (eds), *Climate Change 2001: The Scientific Basis Contribution of Working Group I to the Third Assessment Report of the Intergovernmental Panel on Climate Change (IPCC)*, Cambridge University Press, UK, 2001, 944.
4. http://www.ipcc.ch/, The Intergovernmental Panel on Climate Change (IPCC) home page, December 1997.
5. F.T. Houghton, G.J. Jenkins and J.J. Ephraums (eds), *Climate Change, The IPCC Scientific Assessment*, Cambridge University Press, 1990.
6. http://www.bna.com/prodhome/ens/text4.htm, The Kyoto Protocol, December 1997.
7. http://www.aeat.co.uk/netcen/corinair/94/corin94.html European topic centre on air emissions, December 1997.

8. http://www.epa.gov/globalwarming/inventory/, US EPA global warming emissions inventories, December 1997.

9. http://cesimo.ing.ula.ve/GAIA/countries/co2_total.html.

10. http://www.ipcc.ch/, IPCC Special Report Safeguarding the ozone layer and the global climate system: Issues related to hydrofluorocarbons and perfluorocarbons. Summary for policymakers, April 2005.

11. J.C. Farman, B.J. Gardiner and J.D. Shanklin, *Nature*, 1985, **315**, 207.

12. UK Review Group on Stratospheric Ozone, Stratospheric Ozone 1988, HMSO, London, 1988.

13. UK Review Group on Stratospheric Ozone, Stratospheric Ozone 1990, HMSO, London, 1990.

14. http://www.unep.ch/ozone/home.htm, The Ozone Secretariat WWW Home Page: UNEP, December 1997.

15. http://www.epa.gov/ozone/mbr/mbrqa.html, The US EPA Methyl Bromide Phase Out Web Site, December 1997.

16. Scientific Assessment of Ozone Depletion: 1994, World Meteorological Organisation, Global Ozone research and Monitoring Project, Report No 37, WMO.

17. http://es-ee.tor.ec.gc.ca/cgi-bin/selectMap, Environment Canada Experimental Studies Group Ozone Maps, June 2005.

18. http://www. AFEAS.org AFEAS, Alternative Fluorocarbons Environmental Acceptability Study, June 2005.

19. http://www.al.noaa.gov/WWWHD/pubdocs/WMOUNEP94.html, Executive summary of the WMO/UNEP Scientific Assessment of Ozone Depletion: 1994 document, December 1997.

20. D. Mage, G. Ozolins, P. Peterson, A. Webster, R. Orthofer, V. Vandeweerd and M. Gwynne, *Atmos. Environ.*, 1996, **30**, 681. This is a follow up report to the WHO/UNEP report Urban Air Pollution in Megacities of the World Published in 1992 in which detailed data are included.

21. A.J. Clarke, in *Environmental effects of utilising more coal*, F.A. Robinson (ed), Royal Society of Chemistry, London, 1980.

22. J.R. Martin (ed), *Recommended guide for the prediction of the dispersion of airborne effluents*, Am. Soc. Mech. Eng., New York, 3rd edn, 1979, p. 87.

23. John H. Seinfeld and Spyros N. Pandis, *Atmospheric chemistry and physics, from air pollution to climate change*, Wiley, New York, 1998, 916–919, 1193–1285.

24. D.G. Atkinson, D.T. Bailey, J.S. Irwin and J.S. Touma, Improvements to the EPA Industrial Source Complex dispersion model, *J. App. Meteorol.*, 1997, **36**, 1088–1095.

25. D.J. Carruthers, H.A. Edmunds, K.L. Ellis, C.A. McHugh, B.M. Davies and D.J. Thomson, The atmospheric dispersion modeling system

(ADMS) – comparisons with data from the kincaid experiment, *Int. J. Environ. pollut.*, 1995, **5**, 382–400.

26. Z. Zlatev, J. Christensen and A. Eliassen, Studying High Qzone Concentrations by using the Danish Eulerian Model, *Atmos. Environ.*, 1993, **27**, 845–865.

27. C. Pilinis and J.H. Seinfeld, Development and Evaluation of an Eulerian Photochemical Gas Aerosol Model, *Atmos. Environ.*, 1988, **22**, 1985–2001.

28. J.S. Chang, R.A. Brost, I.S.A. Isaksen, S. Madronich, P. Middleton, W.R. Stockwell and C.J. Walcek, A 3-Dimensional Eulerian Acid Deposition Model – Physical Concepts and Formulation, *J. Geophys. Res – Atmos.*, 1987, **92**, 14681–14700.

29. W.J. Collins, D.S. Stevenson, C.E. Johnson and R.G. Derwent, Tropospheric ozone in a global-scale three-dimensional Lagrangian model and its response to NO_x emission controls, *J. Atmos. Chem.*, 1997, **26**, 223–274.

30. R.I. Sykes and R.S. Gabruk, A second-order closure model for the effect of averaging time on turbulent plume dispersion, *J. Appl. Meteorol.*, 1997, **36**, 1038–1045.

31. J.C. Weil, L.A. Corio and R.P Brower, A PDF Dispersion Model for Buoyant Plumes in the Convective Boundary Layer, *J. Appl. Meteorol.*, 1997, **36**, 982–1003.

32. D.M. Lewis and P.C. Chatwin, A three-parameter PDF for the concentration of an atmospheric pollutant, *J. Appl. Meteorol.*, 1997, **36**, 1064–1075.

33. F. Pasquill and F.B. Smith, *Atmospheric diffusion*, 3rd edn, Ellis Horwood, Chichester, 1983.

34. G. Lövblad, L. Tarrasón, K. Tørseth and S. Dutchak, (eds), EMEP Assessment Part I: European Perspective, ISBN 82–7144–032–2, Oslo, 2004.

35. H.S. Eggleston and G. McInnes, Methods for the compilation of UK air pollutant emission inventories, Report LR 634(AP), Warren Spring Laboratory, Stevenage, 1987.

36. http://www.naei.org.uk/, National Atmospheric Emissions Inventory for the UK, April 2005.

37. D.I. Stern, Global sulfur emissions from 1850 to 2000, *Chemosphere,* **58,** 163–175, 2005.

38. T.S. Bates, B.K. Lamb, A. Guenther, J. Dignon and R.E. Stoiber, Sulfur emissions to the atmosphere from natural sources, *J. Atmos. Chem.*, **14**, 315–337, 1992.

39. D.S. Lee, I. Kohler, E. Grobler, F. Rohrer, R. Sausen, O. Gallard, L. Klenner, J.G.J. Oliver and F.J. Dentener, Estimates of global NO_x emissions and their uncertainties, *Atmos. Environ.*, **31**, 1735–1749, 1997.

40. http://www.naei.org.uk, National Atmospheric Emissions Inventory for the UK, April 2005.
41. A.F. Bouwman, D.S. Lee, W.A.H. Asman, F.J. Dentener, K.W. van der Hoek and J.G.J. Olievier, A global high-resolution emission inventory for ammonia- Global Biogeochemical Cycles **11**, 561–578, 1997.
42. C.J. Dore, J.D. Watterson, J.W.L. Goodwin, T.P. Murrells, N.R. Passant, M.M. Hobson, S.L. Baggott, G. Thistlethwaite, P.J. Coleman, K.R. King, M. Adams and P.R. Cumine, UK Emissions of Air Pollutants 1970 to 2002 Atmospheric emissions report, AEA Technology, August 2004.
43. Acid Deposition in the United Kingdom 1992–1994. Fourth Report of the Review Group on Acid Rain, Department of the Environment, Transport and the Regions, 1997.
44. Digest of Environmental Statistics No. **18**, 1996. HMSO, London.
45. A. Guenther, C.N. Hewitt, D. Erickson, R. Fall, C. Geron, T. Graedel, P. Harley, L. Klinger, M. Lerdau, W.A. Mckay, T. Pierce, B. Scholes, R. Steinbrecher, R. Tallamraju, J. Taylor and P. Zimmerman, A global model of natural volatile organic compound emissions, *J. Geophys. Res.*, **100**, No D5, 8873–8892, 1995.
46. Source publication: e-Digest of Environmental Statistics, Published March 2005, Department for Environment, Food and Rural Affairs, http://www.defra.gov.uk/environment/statistics/index.htm.
47. V. Vestreng, M. Adams and J. Goodwin, 2004, Inventory Review 2004. Emission data reported to CLRTAP and the NEC Directive, EMEP/EEA Joint Review Report, EMEP/MSC-W Note 1, July 2004.
48. D.B. Kittelson, Engines and Nanoparticles: A Review, *J. Aerosol Sci.*, 1998, **29**(5/6), 575–588.
49. Air Quality Expert Group report on Particulate Pollution in the UK, 2005.
50. K. Donaldson, X.Y. Li and W. MacNee, Ultrafine (nanometre) Particle Mediated Lung Injury, *J. Aerosol Sci.*, 1998, **29**, 533–560.
51. A. Seaton, W. MacNee, K. Donaldson and D. Godden, Particulate Air Pollution and Acute Health Effects, *The Lancet*, 1995, **345**, 176–178.
52. C.A. Pope, R.T. Burnett, M.J. Thun, E.E. Calle, D. Krewski, K. Ito and G.D. Thurston, *J. Am. Med. Assoc.*, 2002, **287**, 1132–1141.
53. D.W. Dockery, C.A. Pope, X. Xu, J.D. Spengler, J.H. Ware, M.E. Fay, B.G. Ferris and F.E. Speizer, An Association Between Air Pollution and Mortality in Six U.S. Cities, The New England Journal of Medicine, 1993, **329**(24), 1753–1759.
54. P.T. Williams, *Waste treatment and disposal*, 2nd edn, Wiley, London, 2005.

55. The Air Quality Strategy for England, Scotland, Wales and Northern Ireland, Department of the Environment, The Stationery Office Ltd, 2001.
56. UK Air quality data is available at http://www.airquality.co.uk, July 2006.
57. http://www.gsf.de/UNEP/gemsair.html, December 1997.
58. http://www.transport2000.org.uk/factsandfigures/Facts.asp, July 2006.
59. http://www.china-embassy.org/eng/zt/zgrq/t36662.htm, July 2006.
60. http://www.epa.gov/oar/aqtrnd96/toc.html National air quality and Emissions Trends Report 1996, December 1997.
61. Estimating Exposure to Dioxin-like Compounds (3 Vols) and Health Assessment Document for 2,3,7,8-Tetrachlorodibenzo-*p*-dioxin (TCDD) and related compounds (3 Vols), US Environmental Protection Agency, Office of Research and Development, Washington, 1994.
62. UK Photo-chemical Oxidants Review Group. Ozone in the United Kingdom, Department of the Environment, 1993.
63. R.G. Derwent and M.E. Jenkin, Hydrocarbon involvement in photo-chemical ozone formation in Europe. AERE – R13736, UK Atomic Energy Authority, Harwell.
64. UK Photo-chemical Oxidants Review Group, Fourth Report. Ozone in the United Kingdom, Department of the Environment Transport and the Regions, 1997.
65. T.C. Chan, R.W. Simpson, G.H. McTainsh and P.D.Vowles, *Atmos. Environ.*, 1997, **31**, 3773.
66. A.P. Waggoner, R.E. Weiss, N.C. Ahlquist, D.S. Covert, S. Will and R.J. Charlson, Optical characteristics of atmospheric aerosols Atmospheric Environment, **15**, 1891–1910, 1981.
67. P. Warneck (ed), Heterogeneous and Liquid-Phase Processes, Springer, Berlin, 1996.
68. http://www.defra.gov.uk/environment/statistics/airqual/aqacidd.htm, July 2006.
69. G. Hayman, K.J. Vincent, H. Lawrence, M. Smith, M. Davies, S. Hasler, M. Sutton, Y.S. Tang, U. Dragosits, L. Love, D. Fowler, L. Sansom and M. Kendall, Management and Operation of the UK Acid Deposition Monitoring Network: Data Summary for 2002, AEA Technology, 2004.
70. UK Acid Waters Review Group, Acidity in United Kingdom Fresh Waters, HMSO, London, 1986, and Second Report, 1988.
71. Acid Deposition Processes, D. Fowler, DEFRA report, AS 04/06 01/05/2004.
72. UK Terrestrial Effects Review Group, The Effects of Acid Deposition on the Terrestrial Environment in the United Kingdom, HMSO, London, 1988.

73. Air Pollution and Tree Health in the United Kingdom. Second Report of the UK Terrestrial Effects Review Group, Department of the Environment, 1993.

74. UK Building Effects Review Group, The Effects of Acid Deposition on Buildings and Building Materials, HMSO, London, 1989.

75. World Development Indicators (WDI), 2005 International Bank for Reconstruction and Development, The World Bank http://www.worldbank.org/data/wdi2005/wditext/Section3.htm, July 2006.

CHAPTER 3

The World's Waters: A Chemical Contaminant Perspective

JAMES W. READMAN

Plymouth Marine Laboratory, Prospect Place, The Hoe, Plymouth,
PL1 3DH, United Kingdom

3.1 INTRODUCTION

At an early stage, we are taught that water is vital to living organisms. In fact, it makes up about 80% of cell contents. The critical functions are often summarised[1] as:

- water is a metabolic reactant in reactions, such as photosynthesis and hydrolysis;
- most reactions take place in aqueous solution;
- transport of substances, such as oxygen and glucose requires water as a solvent; and
- the latent heat of vaporisation of water is a critical feature in temperature control.

These rudimentary features merely scratch the surface. Indeed, water has proven a major feature in evolutionary processes.

We are also taught about the purification mechanism that ensures adequate supplies of fresh water are regenerated to sustain the ecology. The hydrological cycle involves evaporation of the sea with subsequent precipitation of distilled water onto terrestrial ecosystems, draining through catchments and transitional waters back to the sea. This remarkable design affords, theoretically, global sustainability – brilliant!

Why then, should we be concerned about water quality? The problem arises through human intervention. Anthropogenic emissions cause contamination of, in particular, supplies of fresh water, which percolate back through the system to potentially affect all ecosystems on the planet. The purpose of this chapter is to investigate which contaminants are important and it concentrates on water contamination by chemicals rather than investigating biogeochemical cycling of pollutants and their ecological effects; topics covered by other chapters within this volume. The task of assessing the relative importance of contamination of the world's waters is vast, but this chapter, not intending to provide a fully definitive evaluation, will address key areas of concern, their causes and attempts to redress threats.

Before the chapter commences, it is worthwhile to consider the relative distribution of water within the global system. Approximately 96% of water on earth is saline, 3% is frozen in the ice caps and glaciers and the remaining 1% is primarily groundwater. Rivers and lakes account for only 0.0001% and 0.01%, respectively. Of the renewable freshwater on a global basis, approximately 8% is withdrawn and used by humans each year (further details are given by Farmer and Graham[2]). Concerns over contamination are generally associated with the precious fresh water resources, transitional (estuarine) waters and coastal locations, which are subject to land-based sources of pollution.

Contamination can occur from point sources (for example, domestic wastewater or industrial outfalls) or diffuse sources (such as agrochemical run-off or deposition of atmospheric contaminants). Point sources are comparatively easy to regulate, whereas diffuse sources are usually complex and are more difficult to control. This chapter will address both the types.

3.2 STRATEGIES TO ASSESS AND REGULATE POLLUTION

The basic strategy to assess and regulate pollution is exemplified in Figure 1. The contaminant input with time (flux) is estimated together with the volume of the receiving water, and hence dilution. After accounting for removal processes (such as degradation, sedimentation, *etc.* – see Figure 2), an ambient level can be determined and can be checked through analytical chemistry (monitoring). The critical question is: does the contaminant have a biological/ ecological effect (*i.e.* is it a pollutant)? By investigating environmental/ biological perturbations and undertaking appropriate risk assessments, an assimilative capacity can be calculated, and levels that do not affect the environment can usually be determined. Regulations can then be drafted to ensure that contamination remains below the "safe" level. Monitoring is then once again invoked to ensure that the management is effective.

Most legislation is based on these types of criteria, and variations in regulations generally relate to the way in which biological threats are measured or perceived. This is further discussed in the next section.

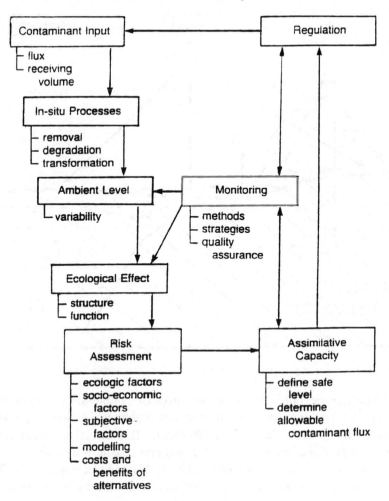

Figure 1 *A strategy to assess and regulate pollution*
(Adapted from an original by J. Albaiges)

3.3 REGULATIONS AND LEGISLATION

Contamination is generally *trans*-boundary and frequently International Laws need to be invoked to prevent pollution. Globally, these are often under the auspices of the United Nations through its specialised agencies, programmes and funds (*e.g.* The United Nations Environment Programme). In 1999, the United Nations Environment Programme reported that 200 scientists in 50 countries had identified water shortage as one of the two most worrying problems for the new millennium (the other was global warming).

More in depth, discussions relating to this topic are available within this series of publications[3] and by J. Kinniburgh within this book (Chapter 7).

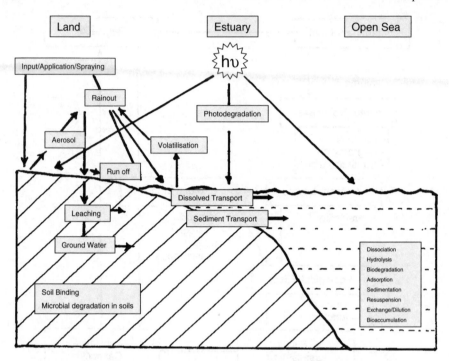

Figure 2 *Aquatic transport and degradation pathways of contaminants*

Within Europe, policy is set by the European Commission,[4] and currently legislation is undergoing a radical change with the introduction of the Water Framework Directive (WFD) (2000/60/EC).[5] This directive fundamentally changes policy and requires that all inland and coastal waters reach at least "good status" in terms of ecological, chemical and physical quality by 2015. It addresses both point and diffuse sources of pollution by establishing river basin management structures that demand stringent environmental objectives. The WFD is unlike previous legislation because it integrates a balance between environmental, economic and social considerations and, importantly, considers the condition of the organisms that live in the water and their sustainability. The timetable for implementation has strict and clear deadlines that are summarised in Table 1.[6] The priority substances (that include chemicals, plant protection products, biocides, metals and other groups such as polycyclic aromatic hydrocarbons (PAHs) and polybrominated diphenyl ethers (PBDEs)) that have been specifically identified for control are listed in Table 2.[7]

While we are clearly at the start of implementing the WFD, results are being achieved. For example, maps are being generated by the appropriate authorities (in England and Wales this is the Environment Agency) to identify vulnerable areas. Figure 3 shows rivers, groundwaters, estuaries and coastal waters in England and Wales that are unlikely to achieve the directive targets because

Table 1 *Timetable for implemention of the EU water framework directive*[6]

Year	Issue	Reference
2000	Directive entered into force	Art. 25
2003	Transposition in national legislation	Art. 23
	Identification of river basin districts and authorities	Art. 3
2004	Characterisation of river basin: pressures, impacts and economic analysis	Art. 5
2006	Establishment of monitoring network	Art. 8
	Start public consultation (at the latest)	Art. 14
2008	Present draft river basin management plan	Art. 13
2009	Finalise river basin management plan including progamme of measures	Art. 13 & 11
2010	Introduce pricing policies	Art. 9
2012	Make operational programmes of measures	Art. 11
2015	Meet environmental objectives	Art. 4
2021	First management cycle ends	Art. 4 & 13
2027	Second management cycle ends, final deadline for meeting objectives	Art. 4 & 13

Table 2 *EU list of priority substances in the field of water policy*[7]

Name of priority substance	CAS number	EU number
Alachlor	15972-60-8	240-110-8
Anthracene	120-12-7	204-371-1
Atrazine	1912-24-9	217-617-8
Benzene	71-43-2	200-753-7
Brominated diphenylethers	n.a.	n.a.
Cadmium and its compounds	7440-43-9	231-152-8
C_{10-13}-chloroalkanes	85535-84-8	287-476-5
Chlorfenvinphos	470-90-6	207-432-0
Chlorpyrifos	2921-88-2	220-864-4
1,2-Dichloroethane	107-06-2	203-458-1
Dichloromethane	75-09-2	200-838-9
Di(2-ethylhexyl)phthalate (DEHP)	117-81-7	204-211-0
Diuron	330-54-1	206-354-4
Endosulfan	115-29-7	204-079-4
Alpha-endosulfan	959-98-8	n.a.
Fluoranthene	206-44-0	205-912-4
Hexachlorobenzene	118-74-1	204-273-9
Hexachlorobutadiene	87-68-3	201-765-5
Hexachlorocyclohexane	608-73-1	210-158-9
Gamma-isomer, Lindane	58-89-9	200-401-2
Isoproturon	34123-59-6	251-835-4
Lead and its compounds	7439-92-1	231-100-4
Mercury and its compounds	7439-97-6	231-106-7
Naphthalene	91-20-3	202-049-5
Nickel and its compounds	7440-02-0	231-111-4
Nonyl-phenols	25154-52-3	246-672-0
4-(para)-nonylphenol	104-40-5	203-199-4
Octylphenols	1806-26-4	217-302-5

(Continued)

Table 2 *(Continued)*

Name of priority substance	CAS number	EU number
Para-tert-octylphenol	140-66-9	n.a.
Pentachlorobenzene	608-93-5	210-172-5
Pentachlorophenol	87-86-5	201-778-6
Polyaromatic hydrocarbons	n.a.	n.a.
Benzo(a)pyrene	50-32-8	200-028-5
Benzo(b)fluoroanthene	205-99-2	205-911-9
Benzo(g,h,i)perylene	191-24-2	205-883-8
Benzo(k)fluoroanthene	207-08-9	205-916-6
Indeno(1,2,3-cd)pyrene	193-39-5	205-893-2
Simazine	122-34-9	204-535-2
Tributyltin compounds	688-73-3	211-704-4
Tributyltin-cation	36643-28-4	n.a.
Trichlorobenzenes	12002-48-1	234-413-4
1,2,4-Trichlorobenzene	120-82-1	204-428-0
Trichloromethane (Chloroform)	67-66-3	200-663-8
Trifluralin	1582-09-8	216-428-8

of impacts from agricultural pesticides, sheep dips and antifouling agents (see also Section 3.7).

Although the WFD strives towards enhancing the quality and sustainability of our waters, there are some hurdles, which will need to be overcome. A major topic for present debate relates to how to judge status with respect to "control" locations (*i.e.* areas unaffected by people). These are extremely difficult to find.

3.4   CHALLENGES AND INDUSTRIAL IMPLICATIONS

Quality of life relies heavily on benefits afforded by chemicals. For example, pharmaceuticals and personal care products afford health and hygiene, petrochemicals provide fuels and manufactured products and agrochemicals aid food production. Life without them is unthinkable. Many chemicals, however, contaminate the environment and can threaten its sustainability.

In 1930, the global production of chemicals was 1 million tonnes. Today, it is approximately 400 million tonnes with about 100,000 different substances registered in the EU alone. Ten thousand of these are marketed in amounts greater than 10 tonnes and 20,000 between 1 and 10 tonnes.[8] Economically, the chemical industry is extremely important and, in 1998, the world chemical production was estimated at 1244 billion Euros (with 31% for the EU chemical industry and 28% for the USA). The industry employs 1.7 million Europeans directly, and about 3 million indirectly.[8]

It is important, however, to guard the environment and human health against contamination and detrimental effects for both the present and future generations. A balance, however, needs to be drawn to afford a competitive chemical industry and quality of life. To achieve these objectives the precautionary

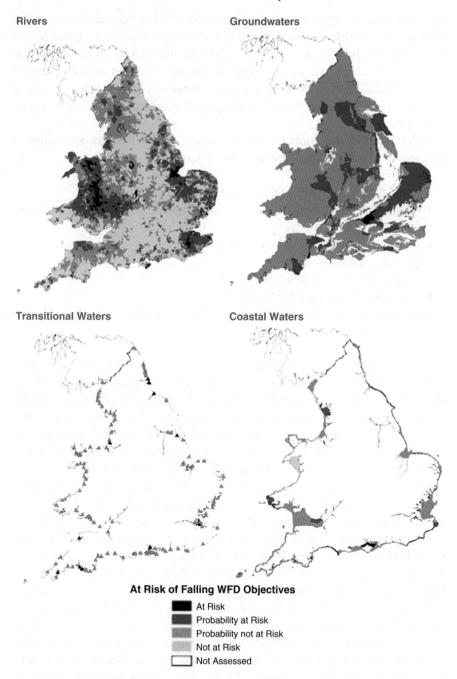

Figure 3 *Maps showing water bodies unlikely to achieve Water Framework Directive Objectives as a result of inputs from agricultural pesticides, sheep dips and antifoulants* (© Environment Agency, CEH, NERC, UK Hydrographic Office. Reproduced with permission)

principle is frequently invoked, which states that "whenever reliable scientific evidence is available that a substance has an adverse impact on human health and the environment, but there is still scientific uncertainty about the precise nature or the magnitude of the potential damage, decision-making must be based on precaution in order to prevent damage".[9] Also, another important objective is to substitute the use of dangerous substances by less dangerous alternatives.[8]

Within the EU, the future chemicals strategy is aimed towards sustainable development and has the following objectives:[8]

- protection of human health and the environment;
- maintenance and enhancement of the competitiveness of the EU chemical industry;
- prevent fragmentation of the internal market;
- increased transparency;
- integration with international efforts;
- promotion of non-animal testing; and
- conformity with EU international obligations under the World Trade Organisation.

To this end, the Commission has developed a single system to assess both current and future new substances.[8] The system is called REACH for the **R**egistration, **E**valuation and **A**uthorisation of **CH**emicals. The requirements (including testing) of the REACH system depend on the proven or suspected hazardous properties, uses, exposure and volumes of chemicals produced or imported. All chemicals (over 1 tonne) will be registered in an EU database. In addition, special attention will be given to long-term and chronic effects. Owing to the fact that so many chemicals are produced, imported and used, the task is substantial and the Commission has awarded priority to substances that lead to a high exposure or cause for concern by their known or suspected dangerous properties – physical, chemical, toxicological or ecotoxicological. It is envisaged that all such substances should be tested within 5 years and the burden for testing has been shifted away from the authorities. Industry has now been made responsible for safety; this is a radical change.

A problem that is particularly difficult to assess relates to the fact that only very rarely are contaminants or pollutants found individually within the environment. Normally, a highly complex cocktail of contaminants is present. This raises the question as to how mixtures of compounds are dealt with. In most instances, the effects are assumed to be additive, so that the concentration of each contaminant is measured and the potential toxicities are simply added. In medicine, however, often drugs are administered together and afford remarkable synergistic effects. Possibly, the same effect might occur in the environment to the detriment of the ecology. Basically, we do not know. This can be exacerbated by the presence of metabolites and degradation products that can also produce effects. Further research is clearly required. This topic is further discussed in the later section that addresses "Causality of Effects".

3.5 SELECTION OF CONTAMINANTS TO INVESTIGATE

Contaminants of concern are generally selected because:

- they are toxic at low concentrations;
- they bioaccumulate in organisms;
- they do not readily degrade and are persistent in the environment;
- they are mutagenic or carcinogenic; or
- they are frequently found in monitoring programmes.

Some contaminants (*e.g.* persistent organic pollutants (POPs); see Section 3.7) are found throughout the environment, *i.e.* they are ubiquitous, but more typically an assessment programme for a particular geographical region should be tailored to address those contaminants of relevance. Specific examples will be given later in this chapter. Basically, a survey should be conducted to investigate potential sources of contaminants by investigating chemical usage and discharges. These are likely to include domestic, urban and industrial point source emissions. Components of the outfalls should be investigated together with discharge volumes and dilution criteria. Agricultural inputs and other diffuse sources of contamination should also be considered. Together with the information on persistence and toxicity, priority ratings can then be estimated for the contaminants. These can then be considered with respect to development of criteria and legislation.

3.6 THE ROLE OF ANALYTICAL CHEMISTRY

The critical feature behind pollution assessment is protection of our ecosystems. This implies biological measures to gauge the health of the biota. Current legislation is moving further towards this objective by combining these with more traditional chemical measures (see Sections 3.8 and 3.10). However, biological assessments are inherently more variable and, for this reason, legislation in the past has generally been directed towards definitive chemical measurements; hence the influence of analytical chemistry. Daughton has reviewed the critical role of analytical chemistry with respect to the complexity of environmental mixtures and constraints associated with current analytical techniques.[10] If we investigate pollution concerns during the last few decades, they commenced primarily with heavy metals released from industrial sources. This coincided with our ability to measure them. Persistent organochlorines (such as DDT) followed, again associated with developments in gas chromatography (GC) and the introduction of the electron capture detector. Around the same time, much research into the composition of petroleum and petroleum products was undertaken using GC analyses. Concerns over pesticides and their persistence (and potential environmental effects) coincided with further developments in capillary GC, and a suite of selective detectors including mass spectrometry (MS). Ultra sensitive GC–MS and LC–MS now afford remarkably effective tools to measure organic contaminants spanning a very broad range of polarities to investigate environmental contamination. These analytical techniques warrant further consideration.

3.6.1   Gas Chromatography–Mass Spectrometry

Possibly, the most relevant and widely used instrument in measuring contamination of the environment by organic chemicals is GC–MS. A basic outline of the instrument is shown in Figure 4. Many contaminants that are known to have detrimental environmental effects are amenable to GC analysis because they are (or can be made) volatile below approximately 300°C. Typically, they are not particularly polar and are usually extracted from environmental matrices using solvents or, in the case of water samples, hydrophobic solid phase materials coated with, for example, long chain (C_8 or C_{18}) hydrocarbons. With these treatments, the contaminants are selectively removed into the organic phase. Solvents can then be concentrated by evaporation to provide an extract amenable to GC or GC–MS. Solid phase extraction (SPE) materials are eluted with solvents, once again, to provide a solvent extract for GC/GC–MS

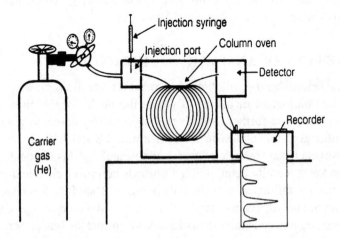

Figure 4 *Schematic diagram of a gas-chromatograph*
  (Adapted from Kenkel[11])

analyses. A small aliquot (typically 1 μL) of the solvent extract is injected into the GC, where it is volatilised and carried (by an inert gas) through a fused silica capillary column (approximately 30 m long with an internal diameter of 0.25 mm and coated with a methyl-silicone phase). Housed in an oven, the column is subjected to an increasing temperature through an appropriate temperature programme. As the components traverse through the column, their affinity for the vapour and methyl-silicone phases differ, and so they move through the column at different rates and are separated and elute sequentially from the column (with a characteristic retention time). In the case of GC–MS, the eluting compounds are ionised, often by impacting them with electrons from a filament, which breaks the molecules into fragments. Using charged lenses, the fragments are accelerated through a magnetic/ electro-magnetic field, which separates them according to their mass and charge. This, effectively, allows the fragments to be weighed and generates a "mass spectrum". This is akin to a fingerprint, affording identification of each compound. The detector is a photo-multiplier tube, which is very sensitive and is capable of responding to nanograms of compound. Mass spectra are usually generated every second, yielding very high-resolution detection of eluting materials.

Examples of applications including oils, pesticides, antifouling agents, *etc.* are given in later sections. In the case of more polar compounds, such as endocrine disrupting sterols, usually the polar portion of the molecules (such as the hydroxyl functional group) are removed and replaced by a non-polar moiety (*e.g.* a tri-methyl-silyl group), and this process is termed as derivatization. In the case of some highly polar materials, this is not feasible, and another technique (high performance (pressure) liquid chromatography (HPLC)–MS) is more appropriate.

3.6.2 High Performance (Pressure) Liquid Chromatography–Mass Spectrometry

Some contaminants have one or more of the following characteristics:

- high polarity;
- high molecular weight;
- thermal instability; and
- a tendency to ionise in solution.

This renders the molecules unsuitable for analysis by GC, and HPLC is the technique of choice. Most of us are familiar with separating pigments using a column packed with silica. HPLC is merely a more refined system that employs very small particles to increase resolution. By using smaller particles, the liquid eluent does not pass through the column easily and thus has to be pumped by inert and very high-pressure pumps; hence HPLC! Many different column types are available to suit the structures of the compounds in question. Frequently, a pre-concentration onto solid phase media or freeze-drying to remove water is appropriate to enhance detection limits. Usually, aqueous/polar solvents are injected into the system. Although the MS operates in a similar fashion to that identified in the previous section, a complication is that the aqueous eluent needs to be removed, rendering the interface with the MS complex, and usually involves high-volume pumping to isolate the determinants from the delivery eluent. Applications of this system have been described in numerous publications, many of which have been reviewed.[12]

3.7 CONTAMINANTS OF CONCERN

With respect to current concerns, possibly the most relevant water-borne agents that produce human health effects are microbes, especially faecal pathogens. These are not specifically dealt with in this chapter, which addresses chemical contaminants. In addition, eutrophication and acidification of our water resources are pertinent, but have been dealt with by numerous reviews and manuscripts. This leaves the chemical agents that threaten our society and its environment. I have selected those contaminants that rank among those that have been, and are currently affording most concern. These are dealt with using personal observations tempered with a list of reviews that afford the reader to follow-up areas of particular interest.

3.7.1 Heavy Metals

Decades ago, these contaminants were derived from "heavy" producing and manufacturing industries that offered a substantial threat to our environment. The elements of concern included: cadmium, lead, mercury, nickel, copper, zinc, chromium and aluminium. The decline in these industries and manufacture of related products, combined with importation, have to a large extent, shifted burdens of contamination to other geographical regions, especially those within less developed countries. Threats continue to exist where manufacture involving metals continues, and also where urban runoff contaminates. The UK Environment Agency has recorded substantial declines in heavy metal emissions, and although this might imply cleanup through legislative changes, in reality, it is the decline of heavy industry and metal production that has resulted in a cleaner environment. Today, fears relating to heavy metal contamination are lessened but are highly relevant to, for example, disused mine workings. Also, as it was noted in Table 2, Cd, Pb, Hg and Ni (together with their compounds), are still listed as priority substances of concern.

Removal of lead from petrol (gasoline) was essential to reduce pollution, but threats from organotins and methylmercury continue to be important.

Reductions in metal contamination (especially from lead, mercury and cadmium) are being addressed through using best available technologies in industrial sectors, reducing atmospheric emissions, and through the phase-out of lead in petrol (complete in most countries) and mercury in batteries.

3.7.2 Crude Oils and Petroleum Products

Contrary to media misconceptions, oil is not a single entity but comprises probably hundreds of thousands or millions of different compounds relating to the breakdown of components present in the biotic precursors. The variations in breakdown and degradation conditions relating to temperature, pressure and the presence of catalysts, renders all crude oils unique with respect to composition. This can be especially useful in petro-chemical industrial applications and in oil spill identification using, for example, sterane and terpane analyses.

Substantial literature has dealt with the prime compositional features of crude oil and its products. Much of this information has been gleaned through GC–MS.

So, what are the concerns? We have all seen the images of devastation that follow a sizable environmental oil spill, especially into aquatic environments. These need to be placed into perspective. Figure 5 summarises emissions of oil into the marine environment[13] and it is clear that effluents and urban/road runoff contribute by far the majority (over 60%). Major accidents, that appear so devastating because of their comparatively localised influence, represent a relatively minor and declining contribution to contamination, primarily owing to improved shipping regulations.

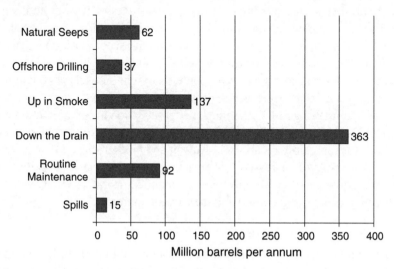

Figure 5 *Worldwide sources of oil to the oceans each year*
(Adapted from "Ocean Planet: Oil Pollution"[13])

Case Study 1: Oil Pollution of Gulf, 1991

The largest ever spill occurred during the 1991 Gulf War (see Figure 6), when about 1 million tonnes (or 6 million barrels) were released from (or near) the Sea Island terminal in Kuwait. An additional 500 million barrels were emitted (or ignited) from burning oil wells during the remainder of 1991, releasing oil aerosols, soot and toxic combustion products (e.g. PAH). This has been compared to approximately 1/3 of US consumption in one day. However, it was feared at the time, that this would decimate the entire Gulf ecosystem, which is relatively shallow and vulnerable. At the time, the author was working for the United Nations who conducted a rapid Gulf-wide survey to investigate the extent of the pollution.[14] By sampling, in particular, sediments and sentinel (stationary) bivalves, it was discovered that severe oil pollution was restricted primarily to the Saudi Arabian coast-line within approximately 400 km from the spillages (Figure 7). During the initial survey, access to the entire Kuwait coastline was forbidden owing to mines and unexploded ordnance. Concentrations outside the contaminated area in the vicinity of Bahrain were, surprisingly, lower than those recorded in pre-war (1983–1986) surveys, probably as a result of decreased tanker traffic and associated deballasting during and after the conflict. Concerning carcinogenic PAH combustion products, results revealed relatively low concentrations, even at sites, which were heavily impacted by the spill.

To investigate the recovery of the areas, which were affected by the spilled oil, during 1992 and 1993 subsequent surveys were conducted in Kuwait and Saudi Arabia to investigate temporal changes in contamination and recovery.[15] Sub-tidal surface sediments from the coastlines of Kuwait and Saudi Arabia were sampled and analysed for petroleum hydrocarbons and PAH. At most locations, by 1992 degradation had resulted in a composition (as determined by GC) dominated by an unresolved complex mixture (UCM) with only the most resistant resolved compounds surviving. Levels of contamination at impacted sites were generally shown to decrease by approximately 50% between 1991 and 1992. A much lesser reduction in contamination was recorded for the period 1992–1993 and an increase in hydrocarbon concentrations was reported for stations in Kuwait and northern Saudi Arabia, possibly as a result of increased tanker activity and associated deballasting. Concentrations of PAH were shown to remain comparatively low, with oil rather than combustion comprising their major source.

These studies serve to illustrate the behaviour and fate of spilled crude oil, which has been further detailed by Preston and Chester.[16] Basically, the freshly spilled oil is less dense than water and floats. It emulsifies and forms a mousse.

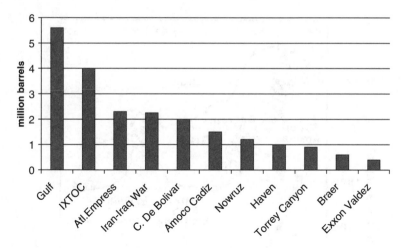

Figure 6 *Comparison of the magnitude of oil spills*

In the initial days, evaporation of volatile constituents and dissolution/dispersion of the smaller hydrocarbons occurs. It is at this stage where the oil coats and smothers coastlines and biota, such as the birds. The dissolved components (in particular the aromatics) poison the ecology but are oxidised rapidly. Detergents have been used to enhance dispersion of slicks. In the case of the Torrey Canyon oil spill, however, the dispersants proved to be more toxic to the coastal ecology than the oil. The described processes act to increase the density of the oil and within weeks/months, tar balls are produced together with tar-like residues on the coastlines. At this stage, on sandy coastlines the tar can become quite rapidly coated with sand improving the aesthetic quality, but the residues lurk and impede especially benthic fauna. The tar can also retain toxic properties.[17] In addition, such beaches lose their recreational amenity value and coat bathers with the tar for many years.

Oil spills appear quite rapidly forgotten by the general public, until the next occurs or if the event directly affects the person's amenities or livelihood.

Of perhaps more environmental relevance, albeit not as visually devastating, is the use of fuels and lubricating oils in motor vehicles. Associated with these are carcinogenic PAH (described in a later section). Road runoff, and the frequent disposal of oils through drains, can substantially elevate concentrations of toxic contaminants in receiving waters.

3.7.3 Fuel Oxygenates

During the last decade, the use of fuel oxygenates has increased dramatically, principally to replace the highly toxic alkyl-lead "anti-knock" compounds. Oxygenates include methyl *tert*-butyl ether (MTBE), and concerns have been

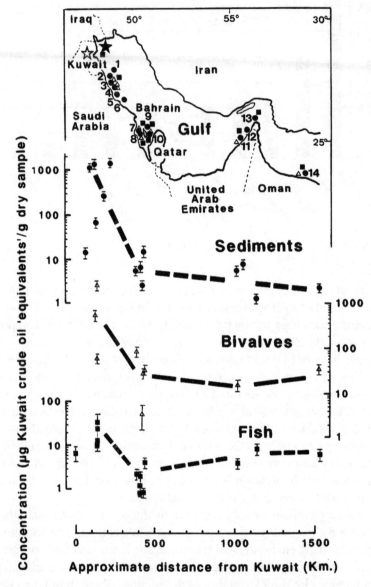

Figure 7 *Concentrations of petroleum hydrocarbons in the Gulf region following the 1991 Gulf War*
(Taken from Readman *et al.*[14])

raised over its environmental occurrence, fate and potential effects. Studies and assessments have been, and are currently being carried out in the United States through national programmes.[18,19] Many publications relating to sources and transport,[20–22] toxicity,[23] transport processes and fates[24,25] have been published.

Environmental levels of fuel oxygenates have increased following their introduction and wide use. The major concern about MTBE is focused on the tainting of ground water and drinking water supplies. Apparently, it tastes dreadful at very low concentrations. In Europe, the use of oxygenates as octane enhancers has seen more limited growth when compared to that observed in the United States. Also, concerns relating to MTBE in drinking water supplies are tempered because water supplies are not exclusively dependent on underground resources (in most of the European continental countries).

There are few data concerning MTBE concentrations in environmental waters for Europe when compared to studies undertaken in the United States. However, reviews concerning its occurrence in some countries (Denmark, Finland, France, Germany, Netherlands, Switzerland, UK and Spain) have been published.[26-28] A large range of environmental concentrations is reported and confirms its presence in different compartments. However, information on the occurrence and fate of MTBE in environments, such as estuaries and coastal waters are comparatively sparse. Guitart *et al.*[29] have, however, recently reported elevated estuarine levels associated with motor vehicle and boating activities.

3.7.4 Persistent Organic Pollutants

POPs initially gained their notoriety through Rachel Carson's book "Silent Spring", published in 1962. This exposed the hazards associated with DDT. DDT, developed in 1939, was the most powerful pesticide the world had ever known. In her book, Carson describes how DDT entered the food chain and accumulated in fatty tissues irrevocably harming birds and animals and contaminating the world's food supply. This brought about a new public awareness of the vulnerability of nature through human intervention. Others, however, point out that at the time of banning, DDT had saved more lives than any other chemical. The US National Academy of Sciences estimated that DDT prevented 500 million human deaths from malaria, since it came into use during World War II. This factor may well have influenced the time taken for international legislation, which addresses many compounds with similar properties, even though most others do not afford the medical benefits of DDT.

POPs are organic (usually halogenated) compounds or mixtures that share four characteristics.

- High toxicity
- Persistence
- Potential for bioaccumulation
- Ability for long-range transport.

In response to concerns relating to the protection of human health and that of the environment, the United Nations Stockholm Convention on POPs was adopted in 2001 and, following appropriate notification, became binding international law for those participating governments on May 17, 2004.[30] The Convention seeks the elimination, or restriction of production and use, of all intentionally produced POPs (*i.e.* industrial chemicals and pesticides). It also seeks the continuing minimization and, where feasible, ultimate elimination of the releases of unintentionally produced POPs, such as dioxins and furans. The Convention requires that stockpiles must be managed and disposed of in a safe, efficient and environmentally sound manner and imposes certain trade restrictions. It also includes a mechanism for expanding its list of POPs to new problem chemicals, such as perfluorinated compounds and brominated flame-retardants that are addressed in later sections. A list summarising the POPs specified in the Stockholm Convention is given in Table 3.

3.7.5 Carcinogens

The European Commission Consolidated List of C/M/R Substances (classified as Category 1 or 2 Carcinogens, Mutagens or toxic to Reproduction) is available from the web, though the complete document numbers approximately one hundred pages.[31] Equivalent listings from alternative, *e.g.* US sources, are also available via the web.[32]

This topic is complex owing to the various stages that are involved in producing malignant carcinomas. It is further complicated by the long-time scale involved in humans contracting cancer (cigarette smokers who develop lung cancer typically do so some 30 years after they take up smoking).

Often, evidence of possible human carcinogenicity of a chemical is indirect, and leaves a strong element of uncertainty. Such evidence may be that a chemical causes cancer when fed to rodents in high doses, but this is no guarantee that it will cause cancer in humans at low doses. Many chemicals cause genetic

Table 3 *POPs specified by the Stockholm Convention[30]*

Compounds	CAS registry number
Aldrin	309-00-2
Camphechlor	8001-35-2
Chlordane	57-74-9
DDT	50-29-3
Dieldrin	60-57-1
Endrin	72-20-8
Heptachlor	76-44-8
Hexachlorobenzene	118-74-1
Mirex	2385-85-5
Polychlorinated biphenyls	
Polychlorinated dibenzo-p-dioxins	
Polychlorinated dibenzofurans	
Toxaphene	8001-35-2

mutations in bacteria, which is an indication that they *may* be carcinogenic, but many are not.

Taking the EU listing it can be summarised, from an environmental viewpoint, that petroleum and combustion derived products probably pose the most threats. It is, however, noteworthy that some of the compounds listed as POPs in the previous section can also induce cancer (especially the dioxins and furans). The first-reported observation linking increased incidence of human cancer with exposure to environmental contaminants was that of Pott in 1775, who observed a higher incidence of scrotal cancer in chimney sweeps. This was, much later, linked to PAHs. PAH and their heterocyclic analogues are, through the metabolic chemical process of epoxidation, extremely potent carcinogens. Although present in many petroleum products, the prime source of the most carcinogenic isomers (with four or more conjugated benzene rings) is combustion. These are agents among those provoking lung cancer in cigarette smokers, and are present in soot and combustion products from burning of any carbonaceous materials. Motor vehicles emit PAH from their engines, PAH accumulate in sump oils and are present in tyre abrasion dusts. Hence, road and urban runoff contain substantial concentrations of carcinogens.

Other less ubiquitous carcinogens include the nitrosamines, which are used in chemical manufacture and are also combustion products; aromatic amines that are analogues of aniline used in the manufacture of dyes, and many natural products *e.g.* mycotoxins.

3.7.6 Pesticides

Pesticides are an important and integral part of our agricultural system. Although "organic" farming practices are on the increase, annually in England and Wales approximately 20,000 tonnes of pesticidal active ingredients are used in farming.[33]

About 80% of pesticides are used for improving agricultural and horticultural productivity and the aesthetic quality of the produce. The majority are synthetic compounds and the most widely applied are directed at insect pests and weeds. The other 20% of the total pesticides used are applied in, for example, domestic gardens and urban environments, as textile treatments, in timber preservation, as antifouling agents in boat paints and in personal care products (see later sections). In most countries, pesticides need to be approved for use. In the UK, the appropriate Minister approves sale, supply, usage, storage and advertising of products. In England and Wales, the Pesticides Safety Directorate (PSD)[34] deals with approval of agricultural pesticides, and the Health and Safety Executive non-agricultural pesticides. The issue is further constrained in the European Union, where, in the case of the UK, the PSD represents UK interests in the registration process. Currently, about 300 chemical compounds are approved for use as pesticides in the UK and their usage is carefully monitored.[33,34] It is the task of the Environment Agency to assess the potential for environmental impact through appropriate monitoring.[33] Maximum concentration limits for approximately 70 pesticides have been set so far. These are termed Environmental Quality Standards (EQS) and are determined individually for compounds based on their potential for environmental damage. Where specific standards for the aquatic environment have not been set, a value of $0.1 \, \mu g \, L^{-1}$ is used. Figure 8, produced by the UK Environment Agency, shows sites that have failed to meet standards for a variety of pesticides. For rivers and lakes, in 2003 approximately 6% of

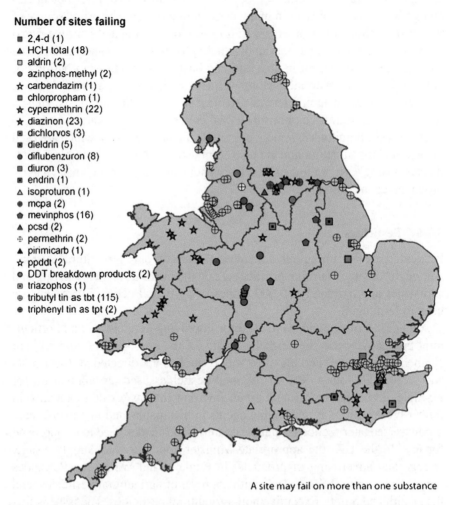

Number of sites failing

- 2,4-d (1)
- ▲ HCH total (18)
- ▫ aldrin (2)
- ● azinphos-methyl (2)
- ☆ carbendazim (1)
- ▣ chlorpropham (1)
- ★ cypermethrin (22)
- ★ diazinon (23)
- ▣ dichlorvos (3)
- ▪ dieldrin (5)
- ● diflubenzuron (8)
- ▣ diuron (3)
- ▣ endrin (1)
- △ isoproturon (1)
- ● mcpa (2)
- ⬠ mevinphos (16)
- ▲ pcsd (2)
- ⊕ permethrin (2)
- ▲ pirimicarb (1)
- ☆ ppddt (2)
- ● DDT breakdown products (2)
- ▣ triazophos (1)
- ⊕ tributyl tin as tbt (115)
- ⦿ triphenyl tin as tpt (2)

A site may fail on more than one substance

Figure 8 *Surface fresh water sites exceeding pesticide Environmental Quality Standards for England and Wales, 2003*
(© Environment Agency. Reproduced with permission)

sites failed on EQS at least once, whereas in coastal and estuarine water, 22% failed. The pesticides most frequently exceeding EQS were sheep dip compounds and the antifoulant tributyl tin (TBT) (which is discussed in the next section). Other frequently failing compounds included hexachlorocyclohexane (HCH), mevinphos and diflubenzuron.

Case Study 2: Pesticide Pollution in Mexico

In a previous section on POPs, problems associated historically with organochlorine pesticides were discussed. These compounds were phased out and replaced, initially by organophosphorous (OP) compounds, which are much less persistent, albeit very highly toxic. In some less developed countries, however, the efficiency of the chlorinated compounds has resulted in their continued usage. In more developed regions, OP pesticides continue widespread usage. In the case of Mexico, for example, conflicts have arisen in tropical coastal zones where horticulture surrounds lagoons, which host shrimp aquaculture. OP pesticides are applied to the crops, and if torrential rain follows application, residues are washed into the lagoons and, because shrimps are close relatives of the insect pests (both are arthropods), the aquaculture can be decimated. An example of such a study[35] is shown in Figure 9, and demonstrates that OP pesticides can be sufficiently persistent to contaminate lagoon ecosystems. Management options, such as holding reservoirs to afford degradation of such products, then need to be considered. Alternatively, as has been the case in the most developed nations, alternative products that are less persistent afford the most practicable solution. As has been indicated in Figure 9, however, should regular application be made, even though degradation is comparatively fast, the active ingredients can be constantly present in the environment and are then termed "Persistently present organic pollutants" (PPOPs).

3.7.7 Antifouling Agents

In a previous section, the incidence of exceedence of EQS levels for TBT was noted. TBT is an extremely toxic antifouling paint additive (although it is also used to a lesser extent in, for example, PVC plastics) used to inhibit biological growth on ships' hulls. At extremely low concentrations, (a few ng L^{-1}), TBT can induce severe deformities in molluscs. It is estimated that in Archachon Bay (France) alone, the use of TBT provoked a loss in revenue of 147 million US dollars through reduced oyster production.[36] The use of TBT has subsequently been severely restricted internationally.[36–38] Possibly the most publicised effect of TBT is the endocrine disruption in gastropod

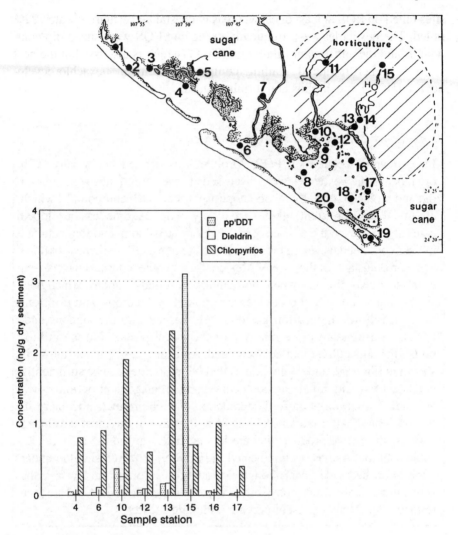

Figure 9 *Concentrations of organophosphorous and organochlorine pesticides in sediments*
of the Altata-Ensenada el Pabellon Lagoon system in Mexico
(Adapted from Readman *et al.*[35])

molluscs, where females develop a penis.[16] This is termed imposex and can occur at concentrations of less than $10 \, ng \, L^{-1}$ (see also the section on endocrine disruption).

The importance of antifouling agents in fuel consumption and performance of vessels is substantial and, in the case of shipping, substantially affects transport costs. The search for alternatives to TBT led primarily to the use of copper-based paints laced with additives to enhance their efficiency. These additional components are called "booster biocides". Products that are, or have been used are:

- 2,3,5,6-tetrachloro-4-(methyl sulfonyl) pyridine,
- 2-methylthio-4-tertiary-butylamino-6-cyclopropylamino-s-triazine (IRGAROL 1051),
- cuprous thiocyanate,
- 2,4,5,6-tetrachloro iso phthalo nitrile,
- 4,5-dichloro-2-n-octyl-4-isothiazolin-3-one (SeaNine 211),
- dichlorophenyl dimethylurea (Diuron),
- 2-(thiocyanomethyl thio)benzthiazole,
- zinc pyrithione,
- 4-chloro-meta-cresol,
- arsenic trioxide,
- *cis* 1-(3-chloroallyl)-3,5,7-triaza-1-azonia adamantane chloride,
- zineb,
- dichlofluanid,
- folpet,
- thiram,
- oxy tetracycline hydrochloride,
- ziram, and
- maneb.

The first publication on the contamination of coastal waters by booster biocides reported high concentrations of Irgarol 1051 along the Cote d'Azur[39] (Figure 10). A subsequent programme to appraise many of them was funded by the European Commission between 1999 and 2002 and was directed to assess:

- usage,
- analytical protocols,
- environmental distributions,
- toxicity, and
- prediction of effects.

This project was named "assessing antifouling agents in coastal environments" (ACE).[40] It concluded that the s-triazine Irgarol 1051 and diuron were widely distributed in European coastal waters. Global assessments have also been reviewed.[37] Indeed, toxic effects to epiphytes and phytoplankton have been shown to occur at below $50 \, \text{ng} \, \text{L}^{-1}$, below the levels reported in some coastal environments.[41,42] Subsequently, more recent legislation has followed to further restrict the usage of copper and booster biocides.[37,38] In the UK, Irgarol 1051 and diuron have now been banned on pleasure craft, and only zinc

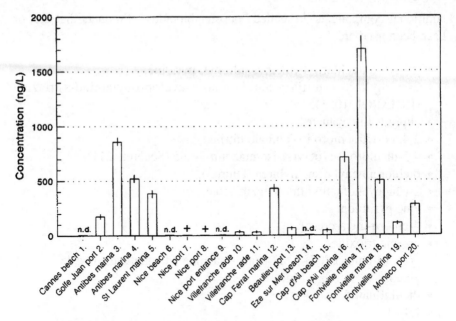

Figure 10 *Concentrations of Irgarol 1051 in coastal waters of the Cote d'Azur* (Adapted from Readman *et al.*[39])

pyrithione (primarily used in anti-dandruff shampoos), dichlofluanid and zineb are allowed as booster biocides in paints for small (<25 m) vessels.[40]

3.7.8 Flame-Retardants

Flame-retardants are important in saving lives through reducing the spread of accidental fires. However, recent evidence suggests that some may be harmful to human health.[43] Those particularly in question are the brominated flame-retardants. This diverse group include polybrominated biphenyls (PBBs), polybrominated diphenyl ethers (PBDEs), tetrabromobisphenol A (TBBPA) and hexabromocyclododecane (HBCD). Over the last 20 years, concern has been increasing with respect to environmental contamination and human exposure.[44] The environmental release and behaviour of the compounds has been reviewed,[43] and their occurrence in wildlife over recent years evaluated.[45]

The health and environmental risks have been assessed by various organisations including the United Nations through the World Health Organisation (WHO), the Organisation for Economic Cooperation and Development (OECD) and the European Commission. Most recently, concerns have accelerated due to marked increases in concentrations of PBDEs in human milk. While for many of the compounds, the prime source of human entry is probably diet, impacts on the worlds waters (and associated sediments and biota), are potentially important.[43]

The environmental behaviour and their potential to behave as POPs (see previous section) have widened concerns. Most data on environmental distributions are available for PBDEs and have demonstrated that they are widely distributed throughout the environment, not only in coastal and shallow seas, but also in deep water environments and oceanic food webs, with marked increases between predatory fish to marine mammals (even after normalising to lipid content). From temporal (dated sediment core studies), in temperate latitudes, PBDE concentrations began to rise earlier than in high latitudes. In many instances now, however, trends of increase have slowed and, in Europe, declines could be expected following cessation of use of the penta-mix formulations. This will not be the case in Arctic biota owing to the continued, and massive use of this formulation in the USA.[45]

3.7.9 Perfluorinated Compounds

Strategically located after the section on brominated flame-retardants, perfluorinated compounds such as perfluorooctane sulfonate (PFOS), are also considered to have properties, which render them POPs. While the concept of having stain and water resistant clothes, carpets and upholstery offers distinct benefits, PFOS, following environmental concerns, was voluntarily withdrawn from the market by the major manufacturer (3M, producers of Scotchgard®).[46]

The OECD undertook a hazard assessment[47] and concluded that, with regard to human health:

- PFOS is persistent, bioaccumulative and toxic (PBT) in mammals;
- PFOS has been detected in the blood serum of occupational and general populations;
- there is a statistically significant association between exposure to PFOS and bladder cancer; and
- there appears to be an increased risk of episodes of neoplasms of the male reproductive system, the overall category of cancers and benign growths and neoplasms of the gastrointestinal tract.

With regard to environmental effects, the OECD hazard assessment indicates that

- PFOS is persistent and bioaccumulative;
- PFOS is of high acute toxicity to honey bees and bioconcentrates in fish; and
- it has been detected in tissues of wild birds and fish, in surface water and in sediment, in wastewater treatment plant effluent, sewage sludge and in landfill leachate.

Extremely resistant to degradation in the environment, PFOS is a ubiquitous contaminant and publications have reported comparatively high levels in
polar bears in the Arctic,[46] dolphins in Florida, seals and otters in California,[48]
albatross in the mid-Pacific[49] and humans on a world-wide basis.[46] Unusually,
rather than bioaccumulating in fatty tissues, PFOS concentrates through
binding to proteins.[46]

With this general information, the UK government Minister (Alun Michael)
on 19 October 2004 announced the UK Government's steps towards a uni-
lateral ban on the use of PFOS and substances which break down to it.[50]

Other perfluorinated compounds continue to be used and are currently
attracting attention. Perfluorooctanoic acid (PFOA), a chemical used in the
production of Teflon, is currently undergoing investigation.[46]

3.7.10   Endocrine Disruptors

The revelations of environmental impacts associated with TBT (see the section
on antifouling agents) engendered awareness that exceptionally low concentrations of environmental contaminants can change hormone systems. Coupled
with the fact that sperm counts in man are generally on the decline,[51] speculation abounds as to whether or not chemicals in our environment can affect
endocrine systems in humans and wildlife. Basically, with regard to male
sperm counts, it is uncertain as to whether or not, for example, tighter under-
garments or chemical contaminants are to blame. We would be foolish to dis-
count options without thorough research. As with Rachel Carson's "Silent
Spring", Theo Colborn, together with co-authors Dianne Dumanoski and
John Peterson Myers, published a book called "Our Stolen Future" on this
topic that discusses threats towards our fertility, intelligence and survival.
This has fuelled debate. Their website[52] lists compounds that are widespread
pollutants with endocrine disrupting effects and provides the relevant references. It is worth a visit. The list includes

- persistent organohalogens,
- food oxidants,
- pesticides,
- phthalates,
- other compounds, and
- metals.

Legislative bodies, such as the European Commission, are taking the threats
seriously.[53] In reviewing the available literature, from a total of 564 compounds,
146 were selected for further evaluation. The most potent, together with
observed endocrine disrupting effects, are listed in Table 4.

Table 4 *Endocrine disrupting effects observed in category 1 substances (↑: increase; ↓: decrease)*
(Data from a European Commission DG ENV report.[53] Authors: Ch. Groshart & P.C. Okkerman)

Name	Effects
Chlordanes (2)[*]	Testicular toxicity
Kepone (chlordecone)	Sperm development ↓
Mirex	Testis descent ↓
Toxaphene	Thyroid tumours ↑
DDTs (3)[*]	Oestrus cycle ↑; ovulation ↓; eggshell thickness ↓; uterus weight ↑
Vinclozolin	Testis weight ↓; testosterone levels ↓; sexual potency ↓; sex organs malformation ↑
Maneb	Thyroid hormone synthesis ↓
Metam	Natrium neuroendocrine pituitary effects
Thiram	Thyroid hormone synthesis ↓
Zineb	Thyroid hormone synthesis ↓
Gamma-HCH lindane	Testis weight ↓; vaginal opening ↓; uterus weight ↓
Linuron	Sex organs weight ↓
Amitrol	Thyroid hormone synthesis ↓
Atrazine	Pseudopregnancies ↑; estrous cycle irregular; androgen receptors ↓
Acetochlor	Thyroid hormone levels ↓
Alachlor	Thyroid hormone levels ↓
Nitrofen	Thyroid effects
Hexachlorobenzene	Testicular effects; ovarian effects; testosterone levels
Tributyltin compounds (18)[*]	Imposex
Triphenyltin (2)[*]	Imposex
Tri-n-propyltin (TPrT)	Imposex
Tetrabutyltin (TTBT)	Imposex[**]
4-tert-Octylphenol	Vaginal opening ↑; uterus weight ↑
Phenol, nonyl-	Uterus weight ↑; testis weight ↓; vitellogenin level ↑
Butylbenzylphthalate (BBP)	Testes weight ↓; sperm production ↓; testosterone levels ↓
Di-(2-ethylhexyl) phthalate (DEHP)	Testes weight ↓; sex organs weight ↓; sperm production ↓; testosterone levels ↓; ovarian weight ↓
Di-n-butylphthalate (DBP)	Testicular atrophy; prostate atrophy
Bisphenol A	Skewed sex ratio; prostate size ↑; prolactin secretion ↑; persistent vaginal cornification; vaginal opening ↑
PCBs (9)[*]	Thyroid effects; uterus weight ↑; endometriosis; progesterone receptors ↑; uterus weight ↓; uterus weight ↑; T4 plasma levels ↓; estrous cycle length ↑
PBBs = Brominated biphenyls	Thyroid hormone levels ↓; sex hormone levels ↓
Dioxins/furans (3)[*]	Hepatic AHH induction; uterus weight ↓; sperm number ↓; thyroid effects; neoplasms[***]
3,4-Dichloroaniline	Androgen synthesis
4-Nitrotoluene	Uterus weight
Styrene	Prolactin secretion ↑; pituitary effects
Resorcinol T4/T3	metabolism ↓; thyroid effects

[*] In between brackets the number of individual substances of the group, is given.
[**] Tetrabutyltin is debutylated to TBT in both vertebrates and invertebrates. Therefore, same effects as TBT.
[***] Due to structural analogy all 2,3,7,8-substituted congeners have been categorised in category 1.

From an aquatic perspective, much work has focused on the induction of vitellogenin (an egg yolk precursor protein), which should only be found in female fish. It is, however, being induced in males, together with oocytes (eggs) in male gonads. Work through the Endocrine Disruption in the Marine Environment (EDMAR) programme describes the effects in detail[54] and attributes the observed effects primarily to natural human female hormones emitted from sewage outfalls.[55] Synthetic compounds [nonyl-phenol and *bis*(2-ethylhexyl) phthalate], did, however, contribute to the effects. The presence of plastics additives and the preponderance of these materials in modern life, fuels concerns.

3.7.11 Pharmaceuticals

Within our newspapers, we are frequently advised that medicinal drugs, such as anti-depressants are present in our drinking water supplies. In addition, antibiotics, pain relievers, beta-blockers and birth control drugs are becoming common contaminants – passed from humans into sewage treatment works and then, through outfalls into streams, rivers, estuaries and coastal waters. Because these compounds can cause highly directed responses within metabolic processes, these are an especially unusual group of contaminants.

There are approximately 3000 pharmaceuticals licensed for human use in the UK. A recent environmental assessment[56] for the 25 most used pharmaceuticals (see Table 5) in the National Health Service in England (during the year 2000), has estimated that predicted environmental concentrations (PECs) for the aquatic environment all exceed $1 \, ngL^{-1}$. Predicted no-effect concentrations (PNECs) based on aquatic toxicity data from the literature for eleven of the pharmaceuticals, and modelling for 12 of the remaining 14 were calculated. The PEC/PNEC ratios exceeded one for paracetamol, amoxycillin, oxytetracycline and mefenamic acid. Guidelines in Europe[57] published by the Committee for Propriety Medicinal Products of the European Medicines Evaluation Agency (EMEA), indicate that if the PEC is less than $0.01 \, \mu gL^{-1}$ or the PEC/PNEC ratio is less than one (and no bioaccumulation risk is identified), the risk assessment can cease and no management actions are required. The compound is deemed safe. If this is not the case, tiered actions commence. The risk assessment guidelines (for both the EU and the US Food and Drug Administration) have been reviewed.[58] The authors identify areas for improvement in certain cases, especially in the use of threshold values to require further investigations, chronic and mechanism specific toxicity screening and complications resulting from mixtures of compounds.

For further information on this topic, an article published in the *Lancet* entitled "Environmental Stewardship and drugs as pollutants"[59] is recommended.

Table 5 *The 25 most used pharmaceuticals by weight in England (2000)*
(Data from Jones *et al.*[56])

Compound name	Therapeutic use	Amount used per year (kg)
Paracetamol	Analgesic	390,955
Metformin hydrochloride	Anti-hyperglycaemic	205,795
Ibuprofen	Analgesic	162,209
Amoxycillin	Antibiotic	71,467
Sodium valproate	Anti-epileptic	47,480
Sulfasalazine	Anti-rheumatic	46,430
Mesalazine (systemic)	Treatment of ulcerative colitis	40,422
Carbamazepine	Anti-epileptic	40,348
Ferrous sulfate	Iron supplement	37,538
Ranitidine hydrochloride	Anti-ulcer drug	36,319
Cimetidine	H2 receptor antagonist	35,655
Naproxen	Anti-inflammatory	35,066
Atenolol	b-blocker	28,977
Oxytetracycline	Antibiotic	27,195
Erythromycin	Anti-inflammatory	26,484
Diclofenac sodium	Analgesic	26,121
Flucloxacillin sodium	Antibiotic	23,381
Phenoxymethylpenicillin	Antibiotic	22,228
Allopurinol	Anti-gout drug	22,096
Diltiazem hydrochloride	Calcium antagonist	21,791
Gliclazide	Anti-hyperglycaemic	18,783
Aspirin	Analgesic	18,106
Quinine sulfate	Muscle relaxant	16,731
Mebeverine hydrochloride	Anti-spasmodic	15,497
Mefenamic acid	Anti-inflammatory	14,523

3.7.12 Personal Care Products

Although often grouped with pharmaceuticals, personal care products are very different in that they are usually used and are introduced into the environment in far greater quantities. Generally, personal care products are the active ingredients or preservatives in products that are used to alter the odour, appearance, the feel (touch) or taste (*e.g.* toiletries, cosmetics or fragrances). With the exception of the preservatives, ideally they should not be biochemically active. As an example of usage, Daughton and Ternes[60] report that in Germany during 1993, 559,000 tonnes of personal care products (comprising bath additives; shampoos; skin, hair and oral hygiene products; soaps; sun screens and perfumes/aftershaves) were used. Comparatively, little data is available concerning the environmental behaviour of the ingredients.

In the substantive review by Daughton and Ternes,[60] a series of categories of personal care products are identified and discussed.

3.7.12.1 Fragrances (Musks). Used in virtually every cosmetic, detergent and toiletry, the synthetic musks are by far the dominant fragrances. These ingredients are ubiquitous in the environment, are persistent and bioaccumulate. The major musks used today are the polycyclic musks (substituted indanes and tetralins). Also broadly used are the nitro musks (nitrated aromatics). This latter group is under particular scrutiny owing to nitro musks persistence and possible environmental impact. Concentrations of various musks in human lipids have been likened to those of other ubiquitous bioaccumulative pollutants, such as PCBs.[61]

3.7.12.2 Preservatives. PARABENS (alkyl-p-hydroxybenzoates) are broadly used anti-microbial preservatives that are added to many cosmetics, toiletries and even foods. Although showing low acute toxicity, concerns have been raised with respect to the potential for environmental effects, and homologues (especially the butyl-parabens) have been shown to competitively bind to the oestrogen receptor affording one or two orders of magnitude greater potential for endocrine disruption than nonyl-phenol.[62]

3.7.12.3 Disinfectants/Antiseptics. A wide range of disinfectants and antiseptics are used domestically. Many are substituted phenolics. Active ingredients include biphenylol, 4-chlorocresol, chlorophene, bromophene, 4-chloroxylenol and tetrabromo-o-cresol.[63] Others include triclosan (2,4,4'-trichloro–2'–hydroxydiphenyl ether).

Although triclosan has been used for many years as a preservative/antiseptic in toothpaste, soaps and acne creams, it is registered in the US as a pesticide. The compound is also used as an agent to remove foot odour and as a slow release biocide in plastic kitchen utensils. Some concerns have been raised relating to bacterial resistance to triclosan.[64] In addition, it photodegrades to a dioxin.[65]

3.7.12.4 Sunscreen Agents. UV filters, such as methylbenzylidene camphor, 2–hydroxyl–4 methoxybenzophenone and 2–ethylhexyl–4–methoxycinnamate are important ingredients to protect human exposure to the sun's UV rays. They are, however, highly lipophilic and have been shown to accumulate in fish lipids[66] and have been detected in human breast milk.[67]

Other personal care products that have been implicated in environmental concerns include food supplements and herbal remedies (termed "nutraceuticals"). Many of these materials, although often natural products, are highly bioactive, potentially posing environmental threats.

3.7.13 Surfactants

Surfactants and detergents closely relate to personal care products and represent an essential contribution to our quality of life, affording cleanliness. Their usage is massive and therefore the potential for contamination of water is important.

These synthetic molecules are characterised by a "head" which is hydrophobic, and a "tail" which is lipophilic consisting of a C-based linear or branched chain. The head renders the detergent soluble in water and can be anionic, cationic or polar. It is the anionic detergents like salts of fatty acids (soaps) and alkyl-benzene sulfonates that are most common for domestic uses. The hydrophobic tail is responsible for attaching to oils or greases to solubilise them.

Whereas soaps form insoluble precipitates (scum) with Ca or Mg, synthetic detergents do not. The first popular detergents were the alkyl-benzene sulfonates and contained branched alkyl–groupings on the hydrophobic tail. These compounds, while being excellent surfactants, did not biodegrade and were toxic to fish. They also caused foaming in treatment facilities. The compounds were phased out in the early 1960s and were replaced by linear alkyl-benzene sulfonates, which degrade more rapidly and are less toxic.

An important group of surfactants are the alkyl-phenol ethoxylates, which have both domestic and industrial applications. The nonyl-phenol ethoxylates are by far the dominant active agents, although octyl-phenol ethoxylates are also widely used. Environmental concerns have, however, been raised concerning their breakdown products. Aerobic and anaerobic degradation produces shorter alkyl-phenol ethoxylates, alkyl-phenoxy carboxylates and alkyl-phenols. It is the alkylphenols, and particularly nonyl-phenol that has attracted attention owing to its oestrogenic potential[52] (Table 4). In addition, chlorine disinfection gives rise to a series of halogenated degradation products.

It is noteworthy that the surfactant active ingredients constitute only 10–30% of the products. The remainder, called detergent "builders", are added to increase the effectiveness of the product and include ingredients, such as complexing agents (*e.g.* polyphosphates to soften the water by removing Ca and Mg), bleaches, fabric softeners, enzymes, optical brighteners, corrosion inhibitors, foam stabilisers and inert suspended solids (*e.g.* carboxymethyl cellulose). Some of these also offer the potential for environmental contamination.

3.7.14 Sewage Outfalls

Concerns relating to sewage are primarily microbially driven. Where drinking water supplies become polluted with sewage-related pathogens such as the cholera and typhoid organisms, human health effects can be devastating. This is of particular concern in developing nations. Secondly, raw (untreated) sewage has an enormous oxygen demand thereby depleting watercourses and asphyxiating the biota. The associated nutrient inputs can also produce eutrophic conditions. This article, however, deals with chemical contaminants. In this context, virtually all of the potential pollutants identified in the previous sections become concentrated in sewage. While many degrade through treatment, certain contaminants that are highlighted in the chapter survive, and have the potential to impact our aquatic ecosystems.

Relative contributions of contaminants from the various sources will relate to the catchment of the particular treatment plant, *i.e.* whether or not it is urban or rural, which industries might contribute, population densities and even age distributions of the residents. Land use and a variety of socio-economic factors control inputs. Seasonal variabilities will also contribute to the changes.

3.7.15 Landfill Leachates

As with sewage, household wastes contain most of the products described within the previous sections and include most manufactured products used by man. Indeed, when any appliance ceases to function, it is disposed of. On a global scale, approximately 70% of municipal solid wastes are sent to landfill sites.[68,69] This rubbish includes an enormous array of toxic compounds in the form of paints, garden pesticides, pharmaceuticals, photographic chemicals, detergents, personal care products, unused oils, batteries, treated wood, broken electronic and electrical equipment … to name but a few. Slack *et al.*[69] have recently produced a thorough review of household hazardous wastes and the contaminants that leach out.

While hazardous industrial wastes are strictly controlled under *e.g.* the US Resource Conservation and Recovery Act (1976),[70] and the European Hazardous Waste Directive 91/689/EEC,[71] household wastes (including any hazardous materials) are disposed of together affording the potential for leaching. Leakage is, however, limited by, for example, the recent European Landfill Directive,[72] that enforces treatment of emissions, and a more recent European Council Decision[73] that sets out the criteria and procedures for waste acceptance at landfills.

The release of leachates, however, can occur,[74] especially in the case of older landfills. This offers the potential of contamination of watercourses by complex cocktails of pollutants. More than 200 organic compounds have been identified in landfill leachate,[74] and ironically, more than 1000 chemicals have been identified in groundwaters contaminated by landfills.[75,76] Slack *et al.*[69] provide a review of xenobiotic compounds found in leachates. These are summarised in Table 6 and, as pointed out earlier, incorporate most of the contaminants identified throughout this chapter. It is likely that further, and more stringent legislation will be introduced to address this issue in the future.

3.8 RAPID AND COMBINED ASSESSMENTS OF POLLUTION

In the section on strategies to assess and regulate pollution, the problem of identifying contamination together with subsequent biological effects was raised.

Table 6 *Organic compounds found in household waste and landfill leachates* (Data adapted from Slack *et al.*[69])

Compounds	Use
Halogenated hydrocarbons	
Bromodichloromethane	Chlorinated water, some as manufacture substrate
Chlorobenzene	Industrial solvent and substrate
1,4-Dichlorobenzene	Toilet-deodorisers and mothballs
1,3-Dichlorobenzene	Insecticide/fumigent; chlorophenol substrate
1,2-Dichlorobenzene	Pesticide, manufacture substrate, deodoriser, solvent
1,2,3-Trichlorobenzene	Insecticide, substrate, solvent
1,2,4-Trichlorobenzene	Insecticide, substrate, solvent
1,3,5-Trichlorobenzene	Chemical intermediate, explosives, pesticides
Hexachlorobenzene	Industrial by-product of solvent, pesticide and wood preservation
Hexachlorobutadiene	Manufacture of rubber/lubricants and industry
1,1-Dichloroethane	Paint solvent, degreasant, breakdown of 1,1,1-trichloroethane
1,2-Dichloroethane	Vinyl chloride manufacture: paint, adhesives, pesticides and leaning products: solvent to remove petrol lead
Tribromomethane	Degreasent and substrate – no longer used
1,1,1-Trichloroethane	Solvent especially paint and adhesive; cleaning products and aerosols
1,1,2-Trichloroethane	Solvent, unknown use: 1,1,2,2-tetrachloroethane breakdown product
1,1,2,2-Tetrachloroethane	Industrial solvent and substrate: was used in paint, pesticides and degreasant
trans-1,2-Dichloroethylene	Solvent and manufacture (pharmaceuticals, *etc.*)
cis-1,2-Dichloroethylene	Solvent (perfumes, *etc.*) and manufacture (pharma, *etc.*)
Trichloroethylene	Solvent, substrate, degreasant: solvent in tipp-ex, paint removers, adhesives and cleaners
Tetrachloroethylene	Dry-cleaning and degreasant
Dichloromethane	Solvent in paint stripper, aerosols, cleaners, photographics, pesticides
Trichloromethane	Solvent and substrate: forms from Cl in water
Carbon tetrachloride	All uses stopped? No longer a refrigerant, *etc.* Used for plastics?
Chloroethane	Plastics and vinyl production – house, drugs, *etc.*
Aromatic hydrocarbons	
Benzene	Multitude of uses – manufacturing of dyes, pesticides, drugs, lubricants and detergents
Toluene	Solvent in paint, paint thinners, nail varnish, *etc.*
Xylenes	Plastics manufacture: solvent in paints, nail varnish
Ethylbenzene	Pesticides, varnishes, adhesives and paints
Trimethylbenzenes	Solvent, substrate (paint, perfume, dye), fuel
n-Propylbenzene	Solvent and manufacture
t-Butylbenzene	Solvent and manufacture
Ethyltoluenes	Solvent and manufacture
Naphthalene	Moth repellent, toilet deodoriser, manufacture of dyes and resins

(Continued)

Table 6 *(Continued)*

Compounds	Use
2-Methylnaphthalene	Insecticides, chemical intermediate (dye/vit. K)
1-Methylnaphthalene	Insecticides, chemical intermediate (dye/vit. K)
Phenols	
Phenol	Slimicide, disinfectant, drugs and manufacture
Ethylphenols	Solvent, naturally occurring in some foods
Cresols	Wood preservatives, drugs, disinfectant and manufacture
Bisphenol a	Manufacture of epoxy resins, coating on food cans?
Dimethylphenols 2-Meth/ 4-methoxyphenol	Solvent Manufacture, antioxidants, drugs, plastics, dyes: flavouring
Chlorophenols	Pesticides, antiseptics, manufacture, Cl-treated water
2,4-Dichlorophenol	Manufacture herbicides, PCP: mothballs, disinfectant
3,5-Dichlorophenol	Manufacture herbicides, PCP: mothballs, disinfectant
Trichlorophenols	PCP and organochlorine pesticide metabolites
2,3,4,6-Tetrachlorophenol Pentachlorophenol	Pesticides, wood preservative Wood preservative no longer used in households
Polychlorinated biphenyls	Transformers and capacitors: b 1970s used in consumerables paint, adhesives, fluorescent lamps, oil
Alkylphenols	
Nonylphenol	Surfactants
Nonylphenol ethoxylate	Detergents, wetting/dispersing agents, emulsifier
Pesticides	
Aldrin/dieldrin	Banned insecticides
Ametryn	Herbicide
Ampa	Glyphosate
Atrazine	Herbicide – US licenced
Bentazon	Herbicide
Chloridazon	Pyridazinone herbicide
Chlorpropham	Carbinilate herbicide
DDT (DDD, DDE)	Banned insecticides
Dichlobenil	Herbicide
Dichlorvos	Insectide (indoor) and veterinary care
N,N-Diethyltoluamide	Insecticide (body)
Endosulfan (a/h)	Insecticide and wood preservative
Endrin	No longer used (insect/rodent/avicide)
Fenpropimorf	Morpholine fungicide
Glyphosate	Herbicide
Hexazinon	Non-agricultural herbicide
Hydroxyatrazin	Atrazine metabolite
Hydroxysimazin	Simazine metabolite
Isoproturon	Phenylurea herbicide
g-Hexachlorocyclohexane	Insecticide and lice treatment
Malathion	Insectide, flea and lice treatment
Mecoprop	Herbicide
Methyl parathion	Insecticide – agricultural

(Continued)

Table 6 *(Continued)*

Compounds	Use
MCPA	Herbicide
Propoxur	Acaricide/insecticide
Simazine	Herbicide
Tridimefon	Fungicide
Trifluralin	Herbicide
4-CPP	Herbicide
2,4-D	Herbicide
2,4,5-T	Herbicide (agent orange)
2,4-DP	Herbicide (alongside mecoprop)
Phthalates	
Monomethyl phthalate	Plastics
Dimethylphthalate	Plastics
Diethyl phthalate	All plastic consumables, insecticides, drugs, cosmetics
Methyl-ethyl phthalate	Plastics
Mono-(2-ethylhexyl) phthalate	Plastics
Di-(2-ethylhexyl) phthalate	All plastics including medical ware
Mono-butylphthalate	Plastics
Di-n-butylphthalate	PVC plastics and nitrocellulose lacquers (varnish)
Di-isobutylphthalate	Plastics
Mono-benzylphthalate	Plastics
Butylbenzyl phthalate	Plastics
Dioctylphthalate	All plastics, pesticides and cosmetics
Phthalic acid	Phthalate breakdown product
Diheptyl phthalate	Plastics
Aromatic sulphonates	
Naphthalene sulphonates	Azo dyes, detergents, plasticizers
Benzene sulphonates	Azo dyes, detergents, plasticizers
p-Toluenesulphonate	Azo dyes, detergents, plasticizers
Sulphones and sulphonamides	
Diphenylsulphone	Plasticiser and intermediates
N-Butylbenzene sulphonamide	Plasticiser
Phosphonates	
Tributylphosphate	Plasticiser, solvent, antifoaming agent
Triethylphosphate	Plasticiser, solvent, antifoaming agent
Terpenoids	
Terpenoids (general)	Plant by-product, chemical intermediate
Borneol	Chemical, perfume, flavouring intermediates
Camphor	Perfume and incense additive
1,8-Cineole	Flavours and fragrance
Fenchone	Flavouring
Limonene	Flavouring
Menthol	Flavours and fragrance
Pinene	Flavours and fragrance
a-Terpineol	Flavours and fragrance
Tetralins	Flavours and fragrance
Thymol	Flavours and fragrance

(Continued)

Table 6 *(Continued)*

Compounds	Use
Pharmaceuticals	
Ibuprofen	Anti-inflammatory/analgesic-OTC
Propylphenazone	
Phenazone	Analgesic – rarely used today
Clofibric acid	Plant growth reg. and drug intermediate
Pyridines	
Methylpyridine (2–?)	Solvent and substrate for dyes, resins, drugs
Nicotine	Insecticide, tobacco
Cotinine	Formed from oxidation of nicotine
Carboxylic acids	
Benzoic acid	Food preservative, perfumes, creams/drugs, manu.
Phenylacetic acid	Fragrance/flavour, drugs (penicillin)
Benzenetricarboxyl acids	Plastic softeners
Palmitic acid	Food, cosmetics and pharmaceuticals
Stearic acid	Food, cosmetics and pharmaceuticals
Linoleic acid	Food and fragrance
Aliphatics	
n-Tricosane	Plastics and intermediate
n-Triacontane	Intermediate
Alcohols and ethers	
Glycol ethers	Solvent {paint, varnish, inks, pesticides, antifreeze}
General alcohols	Solvents
Diphenylethers	Flame retardant, plasticiser, herbicide
Aldehydes and ketones	
Aldehydes	Solvents {plastics, paints}, stain remover
Ketones	Preservative, resin/dye manu., intermediate
Miscellaneous	
Acetone	Solvent and in manu. of plastics, drugs and fibres
Analines	Ink/dye, resins, drugs, agrochemical intermediate
Benzonitrile	Solvent: dye, drugs, rubber, lacquer manu.
Benzthiazoles	Manu. of drugs, rubber, agrochemicals, etc.
Dibenzofuran	From fossil fuel combustion – incl. Diesel fuel
Caffeine	Food additive, drugs
Esters	Many uses during manufacture
Tetrahydrofuran	Food additive, reagent (drugs, perfumes), solvent
Indane	Fuel and metal cleaning
Indane	Solvent and intermediate
Indoles	Intermediates, food colourant, drugs/hallucinogenics, perfumes, etc.
MTBE	Solvent used as additive in unleaded petrol
Siloxanes	Silicone polymers – varnish, oils/waxes, rubber
Styrene	Naturally occurring, used for plastics/rubber manu.

This indicates that in pollution assessment, both chemical and biological effect measurements need to be addressed simultaneously. This is dealt with in more depth in Chapter 5. However, it is useful to provide a brief summary of approaches that are likely to become more important in future monitoring programmes.

Chemical analyses can provide a definitive measure of the presence of a compound in an environmental matrix. It is not, however, possible to quantify all contaminants.[10] Biological screening can identify whether or not the ecology is being perturbed, but cannot identify the causative agent(s). Also, the effects that are investigated can be specific for the selected organism and not necessarily indicative of the ecosystem as a whole (for example, the presence of herbicides will damage phytoplankton but not necessarily sentinel bivalves which, if selected for monitoring, will not necessarily reveal effects). By combining chemical measurements with a carefully chosen selection of biomarkers of response, both contamination and environmental damage can be appraised.[77] Indeed, the biological effect measures can be utilised to focus more expensive chemical analyses to identify causality. The use of more rapid chemical analyses, such as immunoassays, are also gaining broader acceptance to reduce monitoring costs.[78]

3.9 BIOAVAILABILITY

Lipophilic pollutants in the aquatic environment are largely associated with suspended particulates and colloidal organic carbon,[79–81] and thermodynamic partition coefficients are traditionally used to model particulate – water exchange of the compounds. There is increasing evidence, however, that these are inaccurate owing to the heterogeneity of sorbants, which can alter the kinetics of partition so that equilibrium is not achieved[82–86] (Figure 11). This is further exemplified in the case of tyre abrasion particles. Are the potentially toxic components buried within a rubber matrix biologically available? This is difficult to answer. Measures of contaminants are therefore possibly best made in organisms, although because of the complexity of the matrix they are difficult to analyse. Some recent work, however, has looked into the analysis of biological fluids (*e.g.* crab urine[87,88]), which can identify exposure of the creatures to PAH. This matrix is simple to analyse and can be extracted without harming the organism.

3.10 CAUSALITY OF EFFECTS

The complexity of contaminant mixtures usually renders attribution of cause of effect extremely difficult.[89] If a sewage outfall or a landfill leachate is identified as a cause of environmental damage to the ecology of a river, how

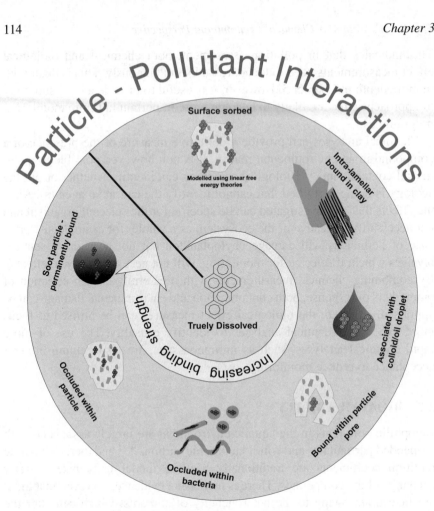

Figure 11 *Diagrammatic representation of binding of benzo(a)pyrene to natural particles*
       (Adapted from Moore *et al.*[79])

can we determine which agents within that outfall are responsible? It could
be simply that severe oxygen demands cause asphyxiation, that ammonia
levels are elevated, or particulate emissions are smothering benthic fauna.
However, as has been noted in preceding sections, a whole host of micro-
contaminants with the potential to induce toxic responses, are also present.

Of the very few instances where apportionment has been clearly demonstrated, perhaps TBT induction of imposex in gastropod molluscs (see previous sections) is the most graphic. The effect is quite specific to the compound
and distributions of imposex correlate well with concentrations of TBT. In
addition, experimental exposures confirmed causality. This is rare and, usually,
it is necessary to try and untangle the complex mixtures of contaminants to
apportion toxic effects. Of the techniques to address this, Toxicity

Table 7 *Substances present in UK estuarine waters, which have been tentatively identified by TIE as contributing to toxic effects (as assessed using the copepod Tisbe battagliaia)*
(Data summarised by Matthiessen and Law[89] and based on Thomas *et al.*[91,92])

Estuary	Substances
Tyne	Chlorobenzene acetonitrile
	4-Chloro-3,5-xylenol
	Methyl acridine
	Methyl pyridine amine
	Monomethyl-trimethyl fluorenes
	Monomethyl-trimethyl naphthalenes
	Naphthalene amine
	Nonyl-phenol
	Pentachlorophenol
	Trichlorophenol
	Tetrachlorophenol
	Triphenylphosphine sulfide
Tees	Atrazine
	Carbophenothion methyl sulfoxide
	4-Chloro-3,5-dimethylphenol
	Diethyl naphthalene carboxamide
	Dimethyl benzoquinone
	Dimethyl naphthalene carboxamide
	Nonyl-phenol
Mersey	Dieldrin
	Dodecylphenol
Milford haven	Bis(2-ethylhexyl)phthalate

Note: Other contributory substances remain to be identified.

Identification and Evaluation (TIE) offers the best option. First introduced by the US Environmental Protection Agency,[90] TIE uses various procedures to fractionate the toxins within a sample. The fractions are then examined using bioassays to investigate which of them are responsible for inducing effects. Further, higher resolution fractionations are then utilised to narrow down the causative agents, and toxicity testing is repeated. Those fractions producing the effects are then analysed using quantitative high resolution GC–MS or LC–MS (as appropriate).

In the section on endocrine disruptors, apportionment of the causative agents was attributed using TIE.[55] Using a bioassay yeast (that had been transfected with the human oestrogen receptor ERα gene linked to a reporter gene system), most oestrogenic activity in water from a sewage outfall could be attributed to 17β-oestradiol, although levels were insufficient to explain observed feminisation of wild flounder in the vicinity. Much higher amounts of oestrogenic activity were, however, associated with extracts from sediments in the locality. While the natural female hormones were considered primarily responsible for the endocrine disruption, nonyl-phenol was also considered to contribute. Thomas *et al.*[91,92] have also applied TIE to investigate more general acute toxicity in UK estuaries and coastal waters employing *T. battagliai* responses as the end-point. Through this, most toxicity was found to be linked to alkyl-phenols, chlorophenols, alkyl-substituted naphthalenes, alkyl-substituted fluorenes, atrazine and dimethyl-benzoquinone. Listings of the compounds identified as toxicants in the various estuaries is given in Table 7, and indicate that effects are estuary specific.

While TIE is shown clearly to provide important information, frequently mass balances of the toxicity identified do not fully explain environmental observations. As described previously, however, analytical constraints do not afford complete resolution of all potential toxicants. TIE does, however, afford excellent potential.

3.11   CONCLUSIONS

Investigating the topic of this chapter using the Internet generates vast numbers of entries from stakeholders with frequently contradicting beliefs on the perceived threats. In dealing with such issues, a balance must be struck between the benefits, which many chemicals bring to society, and protection of the environment. This is difficult in a situation where scientific understanding has many weaknesses. One, in particular, is that adverse environmental effects are frequently attributed to specific individual contaminants, while it is virtually never the case that sources of contamination generate single pollutants. Usually, many contaminants are present and provide stressors for the ecosystem subjected to anthropogenic inputs. The understanding of how these might interact and produce complex threats is very rudimentary, and it would be naïve to consider that this issue can easily be resolved. Clearly, more fundamental research is needed.

REFERENCES

1. B. May and K. Redmond (eds), *AS Level Biology – The Revision Guide*, Coordination Group Publications Ltd, Elanders Hindson Printers, Newcastle upon Tyne, UK, 2003.
2. J.G. Farmer and M.C. Graham, in *Understanding our Environment*, R.M. Harrison (ed), The Royal Society of Chemistry, Cambridge, UK, 1999.
3. B. Crathorne, Y.J. Rees and S. France, in *Pollution, Causes, Effects and Control*, 4th edn, R.M. Harrison (ed), The Royal Society of Chemistry, Cambridge, UK, 2001.

4. Overviews of the European Union activities: Environment. http://europa.eu.int/pol/env/overview_en.htm.

5. Directive Establishing a Framework for Community Action in the Field of Water Policy. Official J., L327, 22 December 2000.

6. EUROPA – Environment DG – Water quality in the EU – WFD timetable. http://europa.eu.int/comm/environment/water/water-framework/timetable.html.

7. EUROPA – Environment DG – Water quality in the EU – Substances on the list. http://europa.eu.int/comm/environment/water/water-framework/priority_substances.htm.

8. White Paper – Strategy for a future Chemicals Policy. Commission of the European Communities COM (2001)88 final. 27 February 2001.

9. Resolution of the European Council of Nice, December 2000 on the precautionary principle which welcomes the Communication from the Commission on the precautionary principle. COM (2000)1, 2 February 2000.

10. C.G. Daughton, *The Critical Role of Analytical Chemistry*. http://www.epa.gov/nerlesd1/chemistry/pharma/critical.htm.

11. J. Kenkel, *Analytical Chemistry Refresher Manual*, Lewis Publishers, Chelsea, MI, 1992.

12. D. Barcelo, *Applications of LC–MS in Environmental Chemistry*, Elsevier, Amsterdam, 1996.

13. *Ocean Planet: Oil Pollution?* http://seawifs.gsfc.nasa.gov/OCEAN_PLANET/HTML/peril_oil_pollution.html.

14. J.W. Readman, S.W. Fowler, J.-P. Villeneuve, C. Cattini, B. Oregioni and L.D. Mee, *Nature*, 1992, **358**, 662.

15. J.W. Readman, J. Bartocci, I. Tolosa, S.W. Fowler, B. Oregioni and M.Y. Abdulraheem. *Mar. Pollut. Bull.*, 1996, **32**(6), 493.

16. M.R. Preston and R. Chester, in *Pollution, Causes Effects and Control*, 4th edn, R.M. Harrison (ed), The Royal Society of Chemistry, Cambridge, UK, 2001.

17. P. Donkin, E.L. Smith and S.J. Rowland, *Environ. Sci. Technol.*, 2003, **37**, 4825.

18. P.J. Squillace, M.J. Moran, W.W. Lapham, C.V. Price, R.M. Clawges and J.S. Zogorski, *Environ. Sci. Technol.*, 1999, **33**, 4176.

19. NAWQA, National Water-Quality Assessment Program. A National Assessment of Volatile Organic Compounds (VOCs) in Water Resources of the United States, 1995–2005, available at http://water.usgs.gov/nawqa/vocs.

20. J.F. Pankow, N.R. Thomson, R.L. Johnson, A.L. Baehr and J.S. Zogorski, *Environ. Sci. Technol.*, 1997, **31**, 2821.

21. D.P. Lince, L.L. Wilson, G.A. Carlson and A. Bucciferro, *Environ. Sci. Technol.*, 2001, **35**, 1050.
22. J.S. Brown, S.M. Bay, D.J. Greenstein and W.R. Ray, *Mar. Pollut. Bull.*, 2001, **42**(10), 957.
23. G.A. Rausina, D.C.L. Wong, W.R. Arnold, E.R. Mancini and A.E. Steen, *Chemosphere*, 2002, **47**, 525.
24. J.F. Pankow, R.E. Rathbun and J.S. Zogorski, *Chemosphere*, 1996, **33**(5), 921.
25. F.J. Squillance, J.F. Pankow, N.E. Korte and J.S. Zogorski, *Environ. Toxicol. Chem.*, 1997, **16**(9), 1836.
26. G. Lethbridge, *Pet. Rev.*, 2000, **54**, 51.
27. T.C. Schmidt, E. Morgenroth, M. Schirmer, M. Effenberger and S.B. Haderlein, in *ACS Symposium Series 799*, American Chemical Society, Washington, DC, 2002.
28. J. Fraile, J.M. Ninerola, L. Olivella, M. Figueras, A. Vilanova and D. Barcelo, *The Scientific World*, **2**, 1235.
29. C. Guitart, J.-M. Bayona and J.W. Readman, *Chemosphere*, 2004, **57**, 429.
30. *Stockholm Convention on Persistent Organic Pollutants*. http://www.pops.int/documents/convtext/convtext_en.pdf.
31. EUROPA–European Commission–Enterprise & Industry–Chemicals–Legislation. http://europa.eu.int/comm/enterprise/chemicals/legislation/markrestr/index_en.htm.
32. US Department of Health and Human Services 11th Report on Carcinogens. http://ntp.niehs.nih.gov/ntp/roc/toc11.html.
33. UK Environment Agency, *Pesticide Use in England & Wales*. http://environment.gov.uk/yourenv/eff/1990084/business_industry/agri/pests.
34. Pesticides Safety Directorate, *Approvals for Pesticides in the UK*. http://www.pesticides.gov.uk/approvals.asp?id=882.
35 J.W. Readman, L. Liong Wee Kwong, L.D. Mee, J. Bartocci, G. Nilve, J.A. Rodriguez-Solano and F. Gonzalez-Farias, *Mar. Pollut. Bull.*, 1992, **24**, 398.
36. C. Alzieu, *Mar. Environ. Res.*, 1991, **32**, 7.
37. I.K. Konstantinou and T.A. Albanis, *Environ. Int.*, 2004, **30**,235.
38. International Coatings Ltd, *Antifoulings – The Legislative Position*, 2004. http://www.yachtpaint.com/superyacht/sy/pdf/antifouling_legislation.pdf.
39. J.W. Readman, L. Liong Wee Kwong, D. Grondin, J. Bartocci, J.-P. Villneuve and L.D. Mee, *Environ. Sci. Technol.*, 1993, **27**, 1940.
40. Assessing Antifouling Agents in Coastal Environments (ACE). http://www.pml.ac.uk/ace.
41. B. Dahl and H. Blanck, *Mar. Pollut. Bull.*, 1996, **32**, 342.

42. J.W. Readman, R.A. Devilla, G. Tarran, C.A. Llewellyn, T.W. Fileman, A. Easton, P.H. Burkill and R.F.C. Mantoura, *Mar. Environ. Res.*, 2004, **58**, 353.

43. I. Watanabe and S. Sakai, *Environ. Int.*, 2003, **29**, 665.

44. C. de Wit, *Chemosphere*, 2003, **29**, 665.

45. R.J. Law, M. Alaee, C.R. Allchin, J.P. Boon, M. Lebeuf, P. Lepom and G.A. Stern, *Environ. Int.*, 2003, **29**, 757.

46. T. Colborn, D. Dumanoski and J. Peterson Myers, *Our Stolen Future*, http://www.ourstolenfuture.org/NewScience/oncompounds/PFOS/2001-04pfosproblems.htm.

47. Organisation for Economic Cooperation and Development (OECD), http://www.oecd.org/dataoecd/23/18/2382880.pdf.

48. K. Kannan, J. Koistinen, K. Beckmen, T. Evans, J.F. Gorzelany, K.J. Hansen, P.D. Jones, E. Helle, M. Nyman and J.P. Giesy, *Environ. Sci. Technol.*, 2001, **35**, 1593.

49. K. Kannan, J.C. Franson, W.W. Bowerman, K.J. Hansen, P.D. Jones and J.P. Giesy, *Environ. Sci. Technol.*, 2001, **35**, 3065.

50. DEFRA, *News releases 2004: UK Acts to Ban Hazardous Chemical*, http://www.defra.gov.uk/news/2004/041019a.htm.

51. Parliamentary Office of Science and Technology (POST), *Technical Report 108*, January, 1998.

52. T. Colborn, D. Dumanoski and J. Peterson Myers, *Our Stolen Future*, http://www.ourstolenfuture.org/Basics/chemlist.htm.

53. EUROPA – Environment – Endocrine disrupters – Strategy, http://europa.eu.int/comm/environment/endocrine/strategy/substances_en.htm.

54. Research Programme on Endocrine Disruption in the Marine Environment (EDMAR), http://www.defra.gov.uk/environment/chemicals/hormone/edmar.htm.

55. K.V. Thomas, M. Hurst, P. Matthiessen and M.J. Waldock, *Environ. Toxicol. Chem.*, 2001, **20**, 2165.

56. O.A. Jones, N. Voulvoulis and J.N. Lester, *Water Res.*, 2002, **36**,5013.

57. EMEA, *Note for Guidance on Environmental Risk Assessment of Medicinal Products for Human Use*, CPMP/SWP/4447/00 draft corr, 2003, available at http://www.emea.eu.int/pdfs/human/swp/444700en.pdf.

58. J.P. Bound and N. Voulvoulis, *Chemosphere*, 2004, **56**, 1143.

59. C.G. Daughton, *The Lancet*, 2002, **360**, 1035.

60. C.G. Daughton and T.A. Ternes, *Environ. Health Perspect.*, 1999, **107** (Supplement 6), available at http://www.ameliaww.com/fpin/ProductsEnv.htm.

61. S. Müller, P. Schmid and C. Schlatter, *Chemosphere*, 1996, **33**(1), 17.

62. E.J. Routledge, J. Parker, J. Odum, J. Ashby and J.P. Sumpter, *Toxicol. Appl. Pharmacol.*, 1998, **153**, 12.

63. T.A. Ternes, M. Stumpf, B. Schuppert and K. Haberer, *Vom Wasser*, 1998, **90**, 295 (cited in Ref. 60).

64. L.M. McMurry, M. Oethinger and S.B. Levy, *Nature*, 1998, **394**, 531.

65. D.E. Latch, J.L. Packer, W.A. Arnolda and K. McNeill, *J. Photochem. Photobiol. Chem.*, 2000, **158**(1), 63.

66. M. Nagtegaal, T.A.Ternes, W. Baumann and R. Nagel, *UWSF-Z für Umweltchem Ökotox*, 1997, **9**, 79 (cited in Ref. 60).

67. J. Hany and R. Nagel, *Deutsche Lebensmittel*, 1995, **91**, 341 (cited in Ref. 60).

68. A. Zacarias-Farah and E. Geyer-Allely, *J. Clean Prod.*, 2003, **11**, 819.

69. R.J. Slack, J.R. Gronow and N. Voulvoulis, *Sci. Total Environ.*, 2005, **337**, 119.

70. US Code, *Solid Waste Disposal Act, as amended 1976 Resource Conservation and Recovery Act: Subtitle D (Solid Waste Program)*, US Code (Acts of Congress) Title 42, Chapter 82, Subchapter 1 (Section 6901), 1976.

71. European Council, Council Directive 91/689/EEC on hazardous waste. *Off. J. Eur. Communities*, L377 (as amended by Council Directive 94/31/EC. Official Journal of the European Communities, L168, 28), 1991, pp. 0020–0027.

72. European Council, Council Directive 99/31/EC on the landfill of waste. *Off. J. Eur. Communities*, 1999, **L182**, 0001–0019.

73. European Council, Council Decision of 19 December 2002 establishing criteria and procedures for the acceptance of waste at landfills pursuant to Article 16 of and Annex II to Directive 1999/31/EC, *Off. J. Eur. Communities*, 2002, **L11**, 27–49.

74. J. Schwarzbauer, S. Heim, S. Brinker and R. Littke, *Water Res.*, 2002, **36**, 2275.

75. T.H. Christensen, P. Kjeldsen, P.L. Bjerg, D.L. Jensen, J.B. Christensen and A. Baun, *Appl. Geochem.*, 2001, **16**, 659.

76. P. Kjeldsen, M.A. Barlaz, A.P. Rooker, A. Baun, A. Ledin and T.H. Christensen, *Crit. Rev. Environ. Sci. Technol.*, 2002,**32**, 297.

77. T.S. Galloway, R.C. Sanger, K.L. Smith, G. Fillmann, J.W. Readman, T.E. Ford and M.H. Depledge, *Environ. Sci. Technol.*, 2002, **36**, 2219.

78. G. Fillmann, T.S. Galloway, R.C. Sanger, M.H. Depledge and J.W. Readman, *Anal. Chim. Acta.*, 2002, **461**, 75.

79. M.N. Moore, M.H. Depledge, J.W. Readman and D.R. Paul Leonard, *Mutation Res.*, 2004, **552**, 247.

80. A. Murdoch, K.L.E. Kaiser, M.E. Comba and M. Neilson, *Sci. Total Environ.*, 1994, **58**, 113.

81. F. Smedes, *Int. J. Environ. Anal. Chem.*, 1994, **57**, 215.
82. J.W. Readman, R.F.C. Mantoura and M.M. Rheas, *Fresenius Z. Anal. Chim.*, 1984, **319**, 126.
83. M.B. Thomson and J.J. Pignatello, *Mechanisms and Effects of Resistant Sorption Processes of Organic Compounds in Natural Particles, in 214th ACS National Meeting of American Chemical Society*, Division of Environmental Chemistry, Preprints of extended abstracts 1997, **37**, 159.
84. D. Thomas, B. Gustafsson and O. Gustafsson, *Environ. Sci. Technol.*, 2000, **34**, 5144.
85. J.L. Zhou, T.W. Fileman, S.V. Evans, P. Donkin, C.A. Llewellyn, J.W. Readman, R.F.C. Mantoura and S.J. Rowland, *Mar. Pollut. Bull.*, 1998, **36**, 597.
86. J.L. Zhou, T.W. Fileman, S. Evans, P. Donkin, J.W. Readman, R.F.C. Mantoura and S.J. Rowland, *Sci. Total Environ.*, 1999, **244**, 305.
87. G. Fillmann, G.M. Watson, E. Francioni, J.W. Readman and M.H. Depledge, *Mar. Environ. Res.*, 2002, **54**, 823.
88. G. Fillmann, G.M. Watson, M. Howsam, E. Francioni, M.H. Depledge and J.W. Readman, *Environ. Sci. Technol.*, 2004, **38**, 2649.
89. P. Matthiessen and R.J. Law, *Environ. Pollut.*, 2002, **120**, 739.
90. D.I. Mount and D.M. Anderson-Carnahan, Methods for aquatic toxicity identification evaluations. Phase 1. Toxicity characterisation procedures (EPA/600/3-88/034). US EPA, Duluth, MN, USA, 1988.
91. K.V. Thomas, R.E. Benstead, J.E. Thain and M.J. Waldock, *Mar. Pollut. Bull.*, 1999, **38**, 925.
92. K.V. Thomas, J.E. Thain and M.J. Waldock, *Environ. Toxicol. Chem.*, 1999, **18**, 401.

Web links verified on 21 March 2006.

CHAPTER 4

Soils and Land Contamination

S.J.T. POLLARD AND M.G. KIBBLEWHITE

Cranfield University, UK

4.1 INTRODUCTION

4.1.1 Soil as a Living System

Soil, with air and water, is one of the three key compartments of the terrestrial environment.[1] It is a living system made up of complex communities of organisms that use carbon from plants as substrate. Soil organisms process approximately 60×10^{18} tonnes of plant carbon globally per year.[2] It is the presence of organic matter and biological activity that distinguishes soil from the regolith (weathered rock).

4.1.2 The Soil Habitat

The biomass in most soils is greater than that in the corresponding above ground habitat. It is of the order of 3 tonnes ha^{-1} in temperate agricultural soils (Table 1). Moreover, it is characterised by extreme diversity, which increases with decreasing spatial scale: at the cm scale, there are only a few mammals; at the mm scale, there are hundreds of larger invertebrates; while at a microscopic scale, there are many tens of thousands of species of microbe.

Table 1 *Abundance of microorganisms and fauna in soil*

Organism	Abundance
Algae	10^4 cells g^{-1}
Fungi	10^6 cells g^{-1}
Nematodes	$10–10^2 g^{-1}$
Protozoa	$10^5–10^6$ cells g^{-1}
Actinomycetes	10^7 cells g^{-1}
Bacteria	10^9 cells g^{-1}

This diversity underpins the multifunctionality of the soil system and its capacity to deliver a wide range of ecosystem services (*e.g.* food and fibre production, the retention of water in catchments) upon which humankind relies.

The "architecture" of the soil, in terms of the physical network and size distribution of pores and the associated behaviour of water and dissolved solutes, has been proposed as a key factor controlling the type and extent of life in soil.[3] The physical and chemical nature of the soil habitat determines the types and condition of biological communities and so the health or quality of the soil system. The spaces between particles and aggregates form a system of pores, which are filled with aqueous soil solution and gases. The extent and size distribution of these pores is critical (Figure 1). In general, a soil that has an open structure that allows a free movement of water, solutes and gases into and out of it will support a richer biological community than one with a massive structure.

Soils are continually subject to physical and chemical stresses. How a soil reacts to stress can be described in terms of response and resilience[5] (Figure 2). Resistance is the change in state for a given level of perturbation, for example, a reduction in the respiration rate arising from compaction. Resilience describes how far the system is able to return to it original state. Restoration of the physical and chemical condition of degraded soil habitats is a pre-requisite for biological recovery, but even with active management, return to fully functioning soil systems may take many years. Methods for assessing the biological health of soil are not yet agreed, although basic measurements of soil biomass and respiration rates are useful, if not always easily interpreted.

4.1.3 Soil Functions

Soil supports a range of ecosystem services that are critical to the quality of human life and the conservation of biodiversity.[6] Soil resources represent

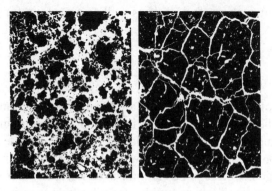

Figure 1 *Soil pore networks*
(Reproduced from Crawford, Ritz and Young (1993))[4]

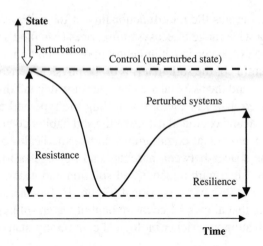

Figure 2 *Graphical representation of trajectory of resistance and resilience to perturbed system* (After Ritz *et al.* (2003)[5])

'natural capital' for current and future generations. Starting from the agricultural revolution of the early nineteenth century, scientific interest in soil has focused on its role as a medium for plant growth and the production of food and fibre. This remains an important focus. Increasingly however, other soil-based services are recognised as being of equal or even more importance.[7] Soil links the atmosphere to surface as well as groundwaters, attenuating and degrading pollutants as well as moderating the hydrology of river basins. Soil conserves biodiversity, both directly below the ground and indirectly above the ground. Soil is a platform for the natural as well as the built terrestrial environment, providing physical foundations for buildings and anchorage for plants and trees. The role of soil in the formation and maintenance of land-scapes, urban as well as rural, is critical to our sense of place and community, while soil is often an artefact itself as well as a medium for archaeological conservation.

4.1.4 Soil Protection

Maintenance of the capacity of soil to deliver this wide range of services requires that the soil system remains intact and its performance is not com-promised by imposed stresses. Particular threats to soil include surface sealing by construction, erosion, contamination and loss of organic matter. Figure 3 shows the range of soil organic carbon across Europe. Other threats that are important in some regions and with particular land uses include increased salinity and compaction.

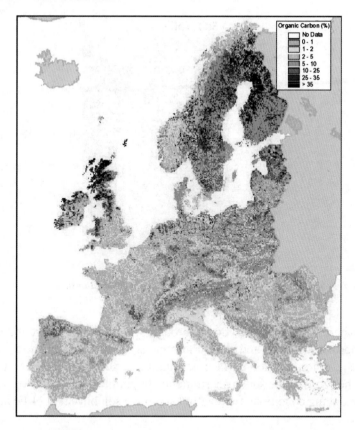

Figure 3 *Organic carbon content (%) in the surface horizon of soils in Europe*
(European Soil Bureau Network)

4.2 SOIL RESOURCES

4.2.1 Soil Formation

There is a great variety of soil types. Different soils form depending on a range of factors, including the available parent material (which may be hard rock but is often superficial deposits), the climate, the landscape topography, the past and present land use and the time over which soil formation has occurred. The result is a three-dimensional continuum of soil properties,[8] with different types of soil prevailing in different locations, each with a set of distinct horizontal layers (horizons) that make up a characteristic soil profile (Figure 4). The process of soil formation is called pedogenesis and combines physical, physicochemical and biological processes.

The soil water regime has a profound influence on soil formation, properties and behaviour. Where precipitation exceeds evapotranspiration (the combined evaporation losses from plants and the soil surface), the tendency

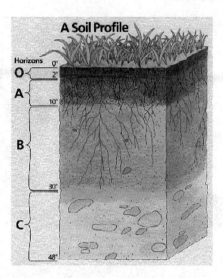

Figure 4 *A stylised soil profile, showing horizons and their relationships*
         (J Turenne, USDA)

is for materials to be leached down through the soil profile to groundwater or
lower horizons. In arid and semi-arid regions, the net movement of water is
upwards, towards the soil surface, leading to an accumulation of salts in the
surface horizons. If the soil is intermittently or permanently water-logged (gley
soils), reducing conditions frequently lead to the solubilisation of Fe and Mn.

4.2.2   Types of Soil

Differences in pedogenesis in space and time, even over short distances,
result in soils becoming highly heterogeneous. Although soil properties tend
to vary continuously in space, it is possible to discern certain soil types with
characteristic soil profiles. For example, podzols form on acid parent material
in temperate regions with higher rainfall. This soil type is characterised by a
surface organic horizon overlying a bleached mineral horizon from which Fe,
Al, Mn and other ions have been leached. Conversely, solonetz soils are found
in arid and semi-arid regions with intermittent rainfall and these have a surface horizon that is saline and base saturated. Within Europe and much of the
world, a system of soil classification called The World Reference Base[9] is
employed (Figure 5). This defines some 30 distinct soil groups and a larger
number of sub-groups. While studying soil in the field or the laboratory, it is
essential to identify the soil type under investigation, because this provides a
basis for the subsequent comparison of results within the appropriate class.

Most soils have been modified by human activity to some extent. The modification is slight and subtle for many soils where the land cover is semi-natural

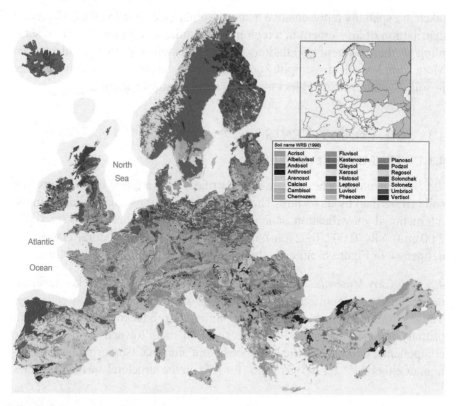

Figure 5 *European soil resources*
(European Soil Bureau, Joint Research Centre, EC)

vegetation and in more remote regions it may be confined to background contamination. Modification is much greater in managed agricultural systems, especially where the water regime has been altered by irrigation or drainage. It is most profound in the built environment, where the original profile is often altered substantially by the removal of surface horizons, the burial of waste or other materials or by mixing due to excavation.

4.2.3 Spatial Variation

Soil exhibits a large degree of spatial variability in physical, chemical and other properties at all scales, from the kilometre to the micrometre. Scientific study and management of soil should take this variability fully into account by careful description of the types of soil being studied and an assessment of the extent to which they are representative of soil at field, landscape and regional scales. For example, decisions about the level of contaminants in a field need to be made on the basis of chemical analysis of a sufficient number of samples taken in a spatially representative manner. Similarly, estimating the background distribution of an element in a region may require a complex pattern of sampling so that different populations of soil types are represented adequately. More complex still are sampling designs that are capable of measuring change in soils, which is also subject to spatial and temporal variation.

4.3 SOIL CONSTITUENTS[10]

4.3.1 The Mineral Fraction

4.3.1.1 Texture. The particle size distribution of mineral components in a soil is described by its texture. Three main size fractions are recognised in the international classification scheme; namely sand (particle diameter 0.02–2.00 mm), silt (0.002–0.02 mm) and clay (<0.002 mm). Textural classes are defined as in Figure 6. Silica is the main component of the sand fraction.

4.3.1.2 Clay Minerals. The clay fraction contains a variety of clay minerals, depending on parent material and the strength and length of time of weathering. These aluminosilicates are secondary minerals produced by weathering of primary minerals in rocks. They are sheet silicates formed from two basic components: tetrahedral silica and octahedral alumina. Isomorphous substitution, of either Al^{3+} for Si^{4+} or Mg^{2+} for Al^{3+} in the structural sheets, generates

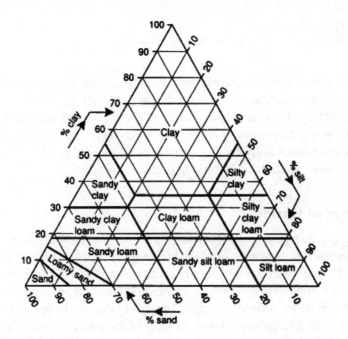

Figure 6 *Soil texture classes*

negative charges on the clay surface. This permanent excess negative charge is independent of the soil pH and contributes to the cation exchange capacity (CEC) of a soil.

Some clay minerals are non-expansive while others shrink and swell with varying soil moisture. This shrinking and swelling has a strong influence on the formation of soil aggregates and structure. The predominant non-expansive clay mineral is kaolinite, which has silica and alumina in a ratio of 1:1 and a relatively low surface area (up to $100 \, m^2 \, g^{-1}$) and a CEC of 2–20 cmol g^{-1}. Smectites are a group of expansive clay minerals with silica and alumina in a ratio of 2:1. They have a large surface area ($700–800 \, m^2 \, g^{-1}$) with access to interlamellar surfaces, which, together with isomorphous substitution, leads to their high CEC (80–120 cmol g^{-1}). Montmorillonite and bentonite are important smectitic clay minerals. Illites have properties which are intermediate between kaolinite and the smectites, while vermiculite has extensive isomorphous substitution of Mg^{2+} for Al^{3+} leading to a high CEC (100–150 cmol g^{-1}).

4.3.1.3 Metal Oxides. Oxides of Fe and Al are ubiquitous in soil and are the ultimate residual weathering products of rocks. The quantities in soil depend on the mineralogy of the soil parent material, the extent of secondary weathering and the prevailing oxidation–reduction (redox) conditions. The prevalence of Fe in its various forms can be presented in a *pe*-pH diagram (Figure 7). Here, the chemical form of Fe is presented by reference to the soil pH and the soil oxidisability (its redox status) represented here by *pe*, the negative common log of the free electron activity. Larger *pe* values favour oxidised species and smaller *pe* values favour reduced species.

The pH-dependent solubility of both Fe and Al leads to their dissolution and precipitation with varying pH. Co-precipitated hydrous oxides of Al and Fe have an amorphous structure. The behaviour of Fe is also controlled by redox conditions, with reducing conditions generating soluble Fe(II) and oxidising ones Fe(III), which is precipitated at higher pH. The ferromagnesian minerals, such as olivine, augite and biotite mica that are found in igneous rocks are a particularly rich source of Fe. Under the oxidising conditions found in freely draining soil environments, Fe(III) oxides are highly stable and give rise to the characteristic brown colouration of many soils. The main forms of Fe oxide are ferrihydrite ($5Fe_2O_3 \cdot 9 \, H_2O$), goethite (alpha-FeOOH) and in warmer soils, haematite (alpha-Fe_2O_3). In poorly drained and waterlogged soils, reducing conditions produce more soluble Fe(II). Commonly, such conditions exist within the core of larger peds (aggregates) even when the overall soil condition is not reducing. Manganese oxide is precipitated as MnO_2 when Mn(II) is oxidised to Mn(IV). If reducing conditions are restored, the reverse reaction is slowed by the solubility of MnO_2.

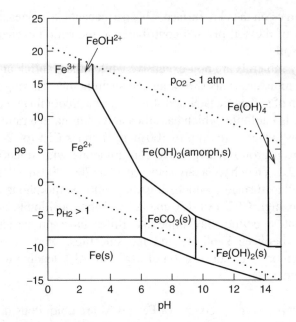

Figure 7 *Phase diagram for the Fe–CO_2–H_2O system in soils*
(G.J.D. Kirk, *The Biogeochemistry of Submerged Soils.* Wiley, Chichester, 2004)

4.3.1.4 Carbonates. Calcium carbonate or calcite ($CaCO_3$) occurs in soils with chalk- or limestone- containing parent material but is also formed by chemical reactions in the soil profile, for example in semi-arid soils. Where free calcite is naturally present in soil, this ensures a soil pH above 7.0. Calcite also contributes to the chemisorptive properties of soil. Liming of soil to modify its pH is a well-established practice in agriculture for controlling soil fertility; it is also used to control the solubility of contaminants that are precipitated at higher pH and would otherwise be phytotoxic or be a source of wider environmental contamination.

4.3.1.5 Sulfides and Sulfates. The forms of sulfur (and also selenium) change with varying redox condition, a process that is mediated biologically. Sulfides occur in anaerobic soils and in soils that are periodically anaerobic due to waterlogging (gley soils). Sulfate moves down or up the soil profile, depending on net water movements. Calcium sulfate (gypsum) is found in the surface horizon of semi-arid and arid soils and in the lower horizons of some temperate soils where groundwater is saturated with calcium.

4.3.1.6 Trace elements. Depending on parent material, past land-use and atmospheric deposition, soil contains varying quantities of a wide range of trace elements. Urban soil often contains much higher concentrations of metals from anthropogenic sources (see Table 2).

Table 2 *Ranges of heavy metal concentrations (mg kg^{-1}) in grassland soils of Austria as compared to public greens and parks in Vienna*[11]

Metal	Grassland soils (mg kg^{-1})	Soils in urban parks
Cd	0.14–0.59	0.33–0.88
Cr	19.8–66.6	16.4–40.3
Cu	11.2–43.2	22.6–81.1
Hg	Upto 0.33	0.14–0.96
Ni	9.1–46.2	22.2–34.5
Pb	12.3–55.7	37.2–143.7
Zn	52.6–145.0	84.2–269.6

4.3.2 Organic Matter

Plant residues and products of their microbial decomposition contribute to the formation of soil organic matter (SOM), which has a highly heterogeneous composition. The levels of organic matter in wetter and colder soils tend to be higher than those in drier and warmer ones. Soils in arid environments typically contain only small amounts of organic matter (<1%), whereas those in humid temperate regions have greater contents (*e.g.* 1–10%). Within a climatic region, soil wetness influences organic matter levels. There is evidence that minimum levels of organic matter, at least in temperate soils, increase with higher clay contents[12], possibly because a combination of strong absorption and occlusion reduces microbial degradation.

Globally, there is much more carbon in soil than in either the atmosphere or above the ground biota, so understanding the dynamics of soil organic carbon is important for modelling of global carbon cycles. The overall dynamics of soil organic carbon reflect changes in different carbon pools that turn over at widely varying rates[13]. Relatively, fresh plant material including roots is quickly colonised by microbes and a small proportion of this is incorporated into SOM (most is respired to the atmosphere). Polysaccharides and other extra-cellular material from microbes are released into soil pores where they may become inaccessible and combined with metal oxides. More recalcitrant plant material such as lignin is degraded more slowly. Except in cold and wet soils, where peat forms, carbon in plant material is either respired to the atmosphere or incorporated into SOM within a few seasons. This newly formed SOM is then repeatedly cycled through the soil biota over lengthening periods, with each cycle releasing carbon to the atmosphere. The most aged soil carbon has half-lives measured in hundreds of years and is in forms that appear to be resistant to biological degradation. This stability may arise from a combination of molecular heterogeneity and strong adsorption within and on clay minerals.

Extensive structural investigation of SOM, lasting more than a century, has provided only incomplete information on its chemical nature. Classical investigations were based on extraction of soil with alkali, followed by acid-ification of the extract to separate an acid-soluble fulvic acid fraction from an acid insoluble humic fraction. Organic matter that could not be extracted from soil by alkali was termed as humin. Attempts to assign formal structures to fulvic and humic acids are probably misleading. These are colloidal mater-ials of indeterminate composition and a wide range of molecular weights, containing varying quantities of co-extracted Al and Fe. Studies of carbon in whole soil using techniques such as ^{13}C NMR and pyrolysis GC–MS have revealed information about the relative proportions of carbon forms in SOM, including the prevalence of aromatic structures[14]. Characterisation of soil extracts has revealed that soil contains most classes of natural organic mol-ecules, including lipids, polysaccharides and polyuronic acids, polyphenols and polycarboxylic acids. The overall nature of SOM can be summarised as being polymeric, heterogeneous and colloidal, with a predominance of carb-oxylic acid and phenolic groups, which support a high, pH-dependent, CEC (up to $200 \, cmol \, kg^{-1}$).

Typically, the nitrogen content of SOM is about one-tenth to one-twentieth of the carbon content. This nitrogen, together with soil phosphate, strongly influences the natural fertility of soil and forms the main reservoir of nitrogen in the terrestrial environment. Surprisingly, given its importance to agriculture and ecosystems in general, the form of this nitrogen is not known completely. About half of it can be accounted for by amino acids, amino sugars, nucleic acids and other identifiable structural components. In some circumstances, a further significant component is adsorbed ammonium.

4.4 SOIL PROPERTIES

4.4.1 Water Retention and Drainage

4.4.1.1 The Soil Pore System. The voids between soil particles and aggregates form a continuous system of pores within the soil profile. The com-plexity of this labyrinthine system reflects the wide range of particle and aggregate sizes and shapes. It is a highly important feature of soil, because it controls the movement of water and air into and from the soil profile, which in turn controls the mass transfer of gases, liquids, sediment and other parti-cles, and dissolved substances.

4.4.1.2 Soil Water. Experimentally, water-retention curves can be deter-mined for soils by applying progressively higher tensions using a porous plate or pressure membrane apparatus and measuring the resulting equilibrium water contents. Water contents are then plotted against applied pressure on a logarith-mic scale (pF) as in Figure 8. The form of the retention curve depends on the distribution of pore sizes, because fine pores hold water more strongly against an applied pressure than do the coarse ones. Figure 8 illustrates retention curves for a clay loam and a fine sandy loam and their relationship to pore size. Pores with diameter greater than 30 μm tend to drain under gravity while water in smaller ones is held by capillary action. When a saturated soil is allowed to drain fully, its water content reduces to "field capacity" and is then largely held in pores of less than 30 μm diameter. The precise water content at which field capacity is reached is not easily determined because the pore structure in soil is not uniform and, within the bulk soil, different volumes have different capacities to hold water.

In addition to the inter-particle pores, larger ones (macropores) are normally present in soil as worm burrows, root channels and cracks opened by shrinking and swelling during seasonal wetting and drying cycles. If the soil is cultivated, macropores may have been introduced by tillage, including sub-soiling. These larger pores allow bulk drainage of water through the soil profile to ground-water, or to field drains.

When a surface load is applied to soil by, for example, vehicles, animals or storage of bulk materials, the soil may become compacted, especially if it is wet and above its plastic limit, which is lower for finer textured soils. Compaction can also be caused by the shear forces set up by the interaction of wet soil and implements during tillage and excavation operations. Soil compaction disrupts

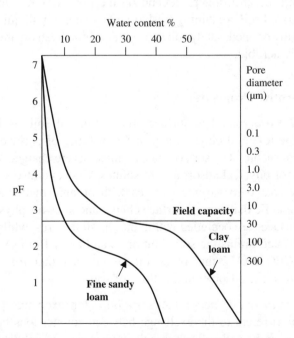

Figure 8 *Moisture retention curves for contrasting soils. Moisture content is plotted against applied pressure on a logarithmic scale (pF)*

the pore system and so water movement. It can lead to surface or lower horizons remaining saturated or at least wetter than soil that is deeper in the profile.

4.4.2 Soil pH and Redox

4.4.2.1 Soil pH. Soil pH is a measurement of the concentration of H^+ ions in the soil solution that are in equilibrium with the negatively charged surfaces of soil particles. Normally, soil pH is measured by equilibrating soil with de-ionised water, although under many field conditions the bulk soil solution contains cations such as Ca^{2+} and, in semi-arid and saline conditions, Na^+.

Soil pH is the most important physico-chemical characteristic of soil because it strongly influences the chemistry of the soil solution, affecting plant growth, microbial ecology and activity and contaminant chemistry. Soil pHs are normally in the range 4–9, although more extreme values occur exceptionally. In general, non-limed soils in humid regions tend to have pH values between 5 and 7 and those in arid regions between 7 and 9. In temperate regions, such as the UK, the optimum pH for arable soils is 6.5 and for grassland it is 6.0 and these values are maintained by liming of non-calcareous soils.

4.4.2.2 Redox Potential. Generally, a high oxygen demand in soil is maintained by aerobic organisms and plant root respiration. Rates of gaseous exchange with the atmosphere depend on the pore structure and its water content. Saturated soil containing fresh organic material will quickly become reducing. A fully drained, sandy soil that contains little fresh organic matter will remain mostly aerobic.

4.4.3 Adsorptive Properties

4.4.3.1 Soil Surfaces. The surface area within all soils is large and is greatest in fine textured clays. Many different types of surface are present and are heterogeneous. The surfaces of clay minerals are charged and therefore provide sites for physical adsorption of solutes. Clay and also sand particles are generally coated to varying degrees with organic matter and hydrous oxides of Al and Fe that carry a surface charge and sites for physical adsorption. Some surfaces are occluded within inaccessible pores, while fluctuating soil water fills and drains the soil solution to and away from surfaces. SOM has a higher CEC at pH 7 than other soil colloids, and therefore plays a very important part in the adsorption reactions of soils.

4.4.3.2 Physical (Non-Specific) Adsorption. Ion exchange (or non-specific adsorption) refers to the exchange between counter-ions balancing the surface charge on the soil colloids and the ions in the soil solution. Generally CEC increases with increase in soil pH, at least up to pH 7. It is the negative charge on soil colloids that is mainly responsible for cation absorption. This negative charge arises in three ways: (a) permanent charges from isomorphous substitution in clay minerals; (b) dissociation of carboxylic and phenolic groups on organic matter; and (c) the pH-dependent charge on oxides of Fe, Al, Mn and Si. Hydrous Fe and Al oxides have relatively high isoelectric points (pH 8) *in vitro* but lower values in the soil system (c. pH 7) and so tend to be positively charged in most soils in which they occur.

A double diffuse layer exists in the volume of soil solution close to a charged soil surface. In the solution volume close to the charged surface, a net opposite charge forms from hydrated counter ions that have diffused a short distance into the soil solution. The thickness of this diffuse double layer depends on both the surface charge density and the soil solution (electrolyte) concentration and composition. The layer contracts with increasing charge density on the surface or in the soil solution. An important consequence of the diffuse double layer is that it acts as a partial barrier to the exchange of ions between the charged surface and the bulk soil solution. When the soil solution is a weak electrolyte the diffuse double layer is thicker and exchange of ions between the surface and the bulk solution requires more energy. In the case of H^+, this means that when the soil solution is saturated with Ca^{2+} as compared to Na^+, the diffuse double layer is more compressed and the effective pool of H^+ available by exchange from the soil surface is greater, and so the apparent pH of the soil solution is decreased. More energy is required to remove an ion with a large hydrated radius from the charged surface and the diffuse double layer, than one with a smaller hydrated radius. Thus cation adsorption strength increases with oxidation state and with increasing ratio of ion charge to radius. For example, although K^+ and Na^+ carry the same charge, Na^+ has a smaller ionic radius and so an increased hydrated radius and will displace K^+. The commonly quoted relative order of replacement of metal ions one by another to a cation exchange surface is as follows:

$$Li^+ = Na^+ > K^+ = NH_4^+ > Mg^{2+} > Ba^{2+} > La^{3+} = H^+$$

The anion adsorption capacity of soils is much less than their CEC. Below pH 7 or 8, Fe and Al hydrous oxides are positively charged and positive charges also occur on 'broken edges' of clay minerals. Simple ions, such as chloride or nitrate are only weakly absorbed by soil. Other ions, such as phosphate and arsenate, are absorbed more strongly, but in these cases the absorption is not purely ionic and non-specific.

4.4.3.3 Chemical (Specific Adsorption) of Metals. Specific adsorption occurs when a chemical bond is formed and therefore it is stronger than non-specific ion exchange. The main sites for specific adsorption are on Al, Fe and Mn oxides and organic matter.

Some oxyanions, such as phosphate and arsenate form strong and insoluble complexes with Al and Fe oxides, effectively removing them from the soil solution. Metals that form hydroxy complexes may also be adsorbed specifically to hydrous metal oxides. The order for increasing strength of specific adsorption is as follows:

$$Cd > Ni > Co > Zn > Cu > Pb > Hg$$

SOM carries a wide range of functional groups that are able to complex metal ions, in particular, carboxylic acid and hydroxyl ones. Normally, a large proportion of the available sites for complex formation are likely to be occupied by Fe and Al. The general pattern of stability of complex formation with SOM might be expected to reflect the order for the replacement of water from aqueous complexes, as follows:

$$Mn^{2+} < Fe^{2+} < Co^{2+} < Ni^{2+} < Cu^{2+} > Zn^{2+}$$

In fact, the general order of stability for complexes of metals is as follows:

$$Cu > Fe = Al > Mn = Co > Zn$$

The strong adsorption of Cu may be enhanced by the presence of amine and other functional groups that adsorb this metal strongly, and the order is also affected by the potentially stronger adsorption when the oxidation state is higher than +2. These relationships and their pH-dependencies are also strongly influenced by the propensities of the different metal ions to form complexes with dissolved organic ligands.

4.4.3.4   Adsorption of Organic Compounds.   Organic contaminants that form anions (*e.g.* the phenoxyalkanoic herbicide, MCPA) are less strongly bound to soil than those that are non-ionic (*e.g.* polyaromatic hydrocarbons (PAH)) or cationic (*e.g.* the bipyridal herbicides, paraquat and diquat). The adsorption to soil of non-ionic and non-polar anthropogenic organic contaminants, which includes many pesticides and persistent organic pollutants (POPs), occurs mainly but not exclusively on organic matter. The mechanisms for adsorption to organic matter are largely non-specific as evidenced by the ability of non-aqueous solvents to quite efficiently extract many of these molecules from soil. Adsorption to clay minerals is similarly non-specific. The strength of adsorption of non-polar organic contaminants by organic matter can be predicted by determining their octanol–water distribution coefficients (K_{ow}). Substances with low K_{ow} values tend to be more hydrophilic, whereas those with higher values are more hydrophobic and adsorbed more strongly by organic matter. For example, the herbicide 2,4-D (2,4-dichloro-phenoxyacetic acid) is relatively soluble in water ($900\,mg\,L^{-1}$) and has a low K_{ow} (log $K_{ow} = 1.57$), whereas the insecticide DDT has a low water solubility ($0.002\,mg\,L^{-1}$), a high K_{ow} (log $K_{ow} = 5.98$) and accumulates in SOM.

Case Study 1: Using Partition Coefficients for Soils and Groundwater

Estimate the dimensionless soil/water partition coefficient for 1,1,1-trichloroethane (log K_{ow} = 2.49) in (i) a silty loam soil (density ρ = 2.66 Mg m^{-3}) containing 19 g SOM per kg and (ii) a peat surface soil (ρ = 0.44 Mg m^{-3}) containing 650 g SOM per kg.

Here we assume that organic matter is 56% w/w organic carbon. We also estimate the organic carbon partition coefficient first knowing the Karickhoff (1981) correlation, which relates K_{oc} to K_{ow}:

The fraction of organic carbon in the soil, $\quad f_{oc}$ = 19/1000 × 0.56
$$= 0.01064$$

From the Karickhoff correlation, $\qquad K_{oc} \approx 0.41 \, K_{ow}$
$$K_{oc} = 0.41 \, (10^{2.49})$$
$$= 126.7 \, \mathrm{L \, kg^{-1}}$$

Assuming for non-polar organics that hydrophobic interaction dominates, the soil water partition coefficient (K_d)

$$K_d = f_{oc} K_{oc}$$
$$= 1.35 \, \mathrm{L \, kg^{-1}}$$

For a dimensionless quantity, we multiply by the density of the soil, remembering that the units Mg m^{-3} are equivalent to kg L^{-1}.

$$\text{dimensionless } K_d = 1.35 \, \mathrm{L \, kg^{-1}} \times 2.66 \, \mathrm{kg \, L^{-1}}$$
$$= 3.59 \text{ for the silty loam; and}$$
$$= 20.29 \text{ for the peat}$$

Assuming a dimensionless K_d for the silty loam of 3.6, estimate the aqueous phase concentration (μg L^{-1}) of 1,1,1,-trichloroethane in the soil solution, assuming the surficial soil is contaminated with 380 ppb 1,1,1,-trichloroethane. Compare the result with a drinking water regulation limit of 30 μg L^{-1} for trichloroethane.

The soil/water partition coefficient is the ratio of the equilibrium concentrations of the chemical in these two media:

$$K_d = C_s / C_w$$
Rearranging, $\qquad C_s = K_d C_w$
$$380 \, \mu\mathrm{g \, kg^{-1}} = 3.6 \, C_w$$

Multiply C_s by soil density:

$$380 \, \mu\mathrm{g \, kg^{-1}} \times 2.66 \, \mathrm{kg \, L^{-1}} = 3.6 \, C_w \, \mu\mathrm{g \, L^{-1}}$$
$$1010.8 = 3.6 \, C_w$$
From which $\qquad\qquad C_w = 280.8 \, \mu\mathrm{g \, L^{-1}}$

This compared to a drinking water criterion of 30 μg L^{-1} for trichloroethane.

4.5 CONTAMINATION THREATS TO SOIL

In their role as a medium for food and fibre production and as an engineering platform for development, soils have been subjected to a number of contamination threats before and since industrialisation.[15] Soil scientists and land managers are faced with a wide range of threats to soil quality, including those from

- potentially toxic elements (pte) and pathogens in sewage sludge applied to agricultural land and forestry;
- industrial sludges (*e.g.* pulp and paper mill wastewater treatment, saline process residues) and canal dredgings applied to land as a means of waste disposal;
- radioactive isotopes released following industrial accidents or from wastes disposal;
- POPs deposited on soils following their aerial release from industrial processes;
- residues from production chemicals used in agriculture (pesticides, fungicides, herbicides);
- spent munitions and explosives following demilitarisation;
- heavy metals (Pb, Hg, As) and organic contaminants (petroleum hydrocarbons, polynuclear aromatic hydrocarbons (PAH)) from historic land contamination, discovered during redevelopment or factory decommissioning;
- nitrogen-rich organic manures and wastes applied to soils in nitrogen vulnerable zones;
- prion materials from wastes removed from the food chain to control bovine spongiform encalopathy (BSE) in cattle; and
- considerable quantities of animal carcass material released to soils during foot and mouth disease (FMD) outbreaks and similar emergencies involving animals reared for human consumption.

While our understanding of how to manage these threats has grown considerably, new threats continue to emerge; for example, those from the application of waste pre-treatment residues and low-grade composts to land, from the disposal of nuclear decommissioning residues to nuclear waste repositories, from new generation agricultural chemicals and from genetically modified organisms (GMOs) released to the soil environment.

The general scientific approach to managing these threats promotes the application of risk assessment and risk management[16] (Figure 9). This is being underpinned by an improved understanding of the environmental capacity of soils and the impacts of hazardous agents (physical, chemical and biological) on soil function.

Central to the management of risk is an understanding of "what specifically is at risk" (the soil function, the soil microbiota, crops, humans) and "what it is at risk from" (loss of buffering capacity, loss of soil structure, reduction in soil quality, loss of future use). Environmental risk assessment has become a discipline in its own right. A straightforward description of risk assessment is provided by considering it as answering four essential questions (Figure 9; Table 3).

Being composed of pore water, pore gas, biota, mineral and organic matter, the soil environment is susceptible to harm from contaminants that might arrive through water, air and direct application routes. For contaminants for which soil is an 'environmental sink', this is of particular concern. Furthermore, with water passing down through soils to groundwaters, laterally through soils to surface waters and with gases rising up from soils to buildings above, contaminated soils can act as sources of contamination for other environmental compartments. More so than any other medium, soils are multifunctional and may act as a *source* of contamination, as the *pathway* by which contamination reaches other environmental compartments and as the *receptor* of contamination in its own right. This complexity can cloud the objectives of soil-quality management.

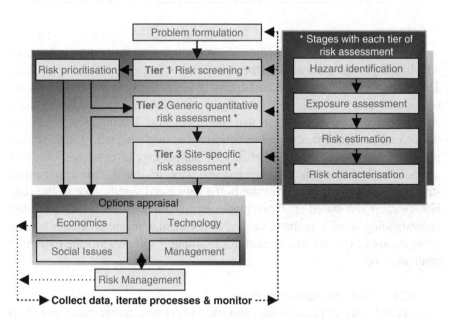

Figure 9 *This framework shows the relationship between good problem definition and a staged approach to risk assessment. Once risks have been assessed, options for the management of risk can be identified and consideration given to the economics, technology required, the social implications and management arrangements*

Table 3 *Essential questions for risk assessment*

Risk assessment question	Stage and description
(1) What hazards are present and what are their properties?	Hazard identification: Identification of the sources of the hazard and assessment of the consequences of the hazard if realised, including the identification of dose–response relationships, where appropriate
(2) How might the receptors become exposed to the hazards and what is the probability and scale of exposure?	Exposure assessment: Evaluating the plausibility of the hazard being realised at the target, and by which mechanisms, allowing an assessment of the probability, magnitude and duration of exposure
(3) Given exposure occurs at the above probability and magnitude, what is the probability and scale of harm?	Risk estimation: Consideration of the consequences of exposure with reference to effects and dose, expressed as a likelihood or probability of the hazardous effects of exposure being realised; and expressed over a range of spatial and temporal fields
(4) How significant is the risk and what are the uncertainties?	Risk characterisation: Evaluating the acceptability and significance of risk with reference to standards, targets, background risks, cost-benefit criteria or risk 'acceptability' and 'tolerability' criteria and commenting on the uncertainties associated with the assessment

In managing risks to or from the soil environment, it is critical to understand which factors contribute most to the risk and how they may be best controlled. The principal value of a risk-based approach is the ability to prioritise between issues and target management strategies and resources towards the more significant risks and those factors that contribute most. Distinguishing between the adverse consequences of soil pollution (what might happen) and their likelihood of occurrence (the probability of happening) offers a discipline of approach and enables policy-makers and regulators to develop risk-management strategies suitable and proportionate to the nature and significance of the risk. However, the risk-based approach places significant demands on our scientific understanding of soils. In the modelling of pollutant transport processes and the application of exposure assessment, for example,[17] soil presents the particular challenges of

– extreme system heterogeneity;
– multiple phases (pore fluids, mineral and organic matter, biota and often 'free product') in intimate contact; and
– analytical difficulties that often hinder exposure model validation.

Risk assessment approaches are proving particularly important where direct exposures are considered and expensive management decisions are being made, for example

- in estimating the likely in-building concentrations of and inhalation exposures from volatile compounds released from soils contaminated with organic compounds;
- in estimating the potential pathogenic load to commercial crops grown where organic wastes have been applied to agricultural land;
- in protecting groundwater resources affected by land contamination; and
- in regulating new substances released to environment at the end of their product cycle.

4.6 LAND CONTAMINATION AND REGENERATION

Soil contamination has been of special concern to developed nations seeking to manage the problems of historic industrialisation. Movement to a post-industrialised society, together with the globalisation of many national companies, has meant that portfolios of former industrial sites have been made available for redevelopment and new uses for the land, such as housing and light industrial use. As companies acquire and divest themselves of their land holdings, former industrial sites have changed hands with the need for liability assessments as new companies acquire the assets of those they have taken over.

House builders developing new homes on former brownfield land, and the more progressive companies seeking to limit their environmental liabilities by cleaning up their own land bank have been faced with land contamination problems involving a wide range of chemical substances[18] (Table 4). Because of the many actors involved in land transactions, the scientific complexities and uncertainties associated with soil and the commercial/legal drivers involved, those involved with managing land contamination have sought to formalise the approach to land management in order to improve clarity and ease decision-making. To be effective, these multiple parties must work together (Case study 2) to reach consensus on how to manage soil contamination problems best.

To help decision-makers navigate the complexities of site investigation, risk assessment and remediation, a number of decision-making frameworks have been developed. This has coincided with an increased research interest in the fate of contaminants in soils, in methods for their analysis and in technologies for their remediation.

Case Study 2: The Importance of Partnerships to Site Regeneration[19]

An example of the value of consensus decisions on historically contaminated land is provided by Pollard and Herbert.[19] The site described was the centre for Scotland's oil shale refining industry at the end of the past century. Initially, shale was subject to distillation at or near the mines to produce crude "retort oil". As shallow mines became exhausted, deep mining of the shale was undertaken and refining was performed at one of three sites. Viscous tarry residues from the refining process were usually disposed of in hollows or ponds near to the sites. From 1948 onwards, the site was also used for detergent manufacture with oil refining eventually ceasing in 1962. The detergent works closed in 1993 when the majority of plant and buildings were demolished.

Post-closure site investigations revealed two principal contaminant groups at the site – surfactants from detergent manufacture and hydrocarbons from oil refining – both with further implications for land and ground-water contamination. Several remedial options were examined including: (i) incineration, (ii) cement kiln, (iii) specialised recycling, (iv) stabilisation, (v) bioremediation (*i.e.* biological removal of contamination), and (vi) land-fill. Of these, only bioremediation and cement grout stabilisation (mixing contaminated material into a cement-based matrix) were shown to be effective.

Tar residues form the oil refining process classed as 'soft' tars and having a relatively low PAH content, were easy to work with and become mobile when warm. These tars were suitable for biological treatment. Trials showed an initial 50–60% w/w reduction in most contaminants due to the bulking process, rising to an 85–95% w/w reduction within 8 weeks of biological treatment. A staged risk assessment, including the derivation of site-specific health-based remediation criteria, was conducted in consultation with the regulator to determine acceptable final concentrations of the key contam-inants, most notably the more carcinogenic PAHs, commensurate with the intended use for the site as an extension to an adjacent golf course ("suitable for use"). From the bioremediation trials, it was estimated that all target concentrations would be achieved in 3–4 months of full-scale treatment.

The hard tars from the refining process, tars form the detergent sul-fonation process and high phenol tars from an adjacent site were not suitable for bioremediation. These materials were treated using *ex-situ* stabilisation. The site was restored to a level suitable for use as a golf course with the bioremediation residues and imported topsoil used as cover for the existing land. Close working in partnership with the regulator throughout proved vital to the successful regeneration of this site.

Table 4 *A prioritised list of 25 chemicals encountered at national priorities list (NPL) sites in the United States Superfund programme, based on a combination of frequency, toxicity and potential for human exposure (after Agency for Toxic Substances and Disease Registry (ATSDR), 2004)*

Chemical substance	Chemical abstracts service (CAS) registry no.
Arsenic	007440-38-2
Lead	007439-92-1
Mercury	007439-97-6
Vinyl chloride	000075-01-4
Polychlorinated biphenyls	001336-36-3
Benzene	000071-43-2
Cadmium	007440-43-9
Polynuclear aromatic hydrocarbons	130498-29-2
Benzo(*a*)pyrene	000050-32-8
Benzo(*b*)fluoranthene	000205-99-2
Chloroform	000067-66-3
DDT, P,P'-	000050-29-3
Aroclor 1254	011097-69-1
Aroclor 1260	011096-82-5
Dibenzo(*a*,*h*)anthracene	000053-70-3
Trichloroethyelene	000079-01-6
Hexavalent chromium	018540-29-9
Dieldrin	000060-57-1
Phosphorus, white	007723-14-0
Chlordane	000057-74-9
DDE, P,P'-	000072-55-9
Hexachlorobutadiene	000087-68-3
Coal tar creosote	008001-58-9
DDD, P,P'-	000072-54-8
Benzidine	000092-87-5

4.6.1 Risk-Based Land Management

Mindful of the potential impacts of soil pollution, governments across the world have sought to intervene at sites that pose the most significant harm to public and environmental health. Since the early 1980s, industrialised nations have progressed national legislation on the assessment and management of land contamination. For many countries and jurisdictions, this has been a 'Pandora's box' and they have uncovered many more historic sites than were initially imagined. With the costs of site assessment and remediation high, policymakers and regulators have to focus their efforts on the most serious sites, using risk assessment as the means of setting national priorities and informing management activities at the site level.

4.6.1.1 Environmental Fate and Transport Properties. The pollution science supporting the risk-based management of contaminated land is founded on the basic physicochemical properties of environmental pollutants in soils.

These properties, shown in Table 5 for priority organic pollutants at sites contaminated with wood-preserving chemicals,[20] determine the propensity of individual compounds to associate with, or partition between the organic matter, mineral matter, the pore gas, the pore water and the wastes itself within a waste-soil matrix. These properties should be used to (i) direct site investigation techniques and the selection of analytical methods by identifying the key

Table 5 *Physicochemical properties of wood-preserving chemicals. Shading indicates physico-chemical properties that dominate the environmental fate of these compounds in soil-water systems (after Pollard et al., 1993)*

Compound	MW $(g\,mol^{-1})$	Aqueous solubility $(g\,m^{-3})$	Vapour pressure (Pa)	K_{oc} (unitless)	Half-life (d)
benzene	78.1	1780	12.7×10^3	65	69
toluene	92.1	515	3800	257	37
phenol	94.1	82000	71	14	1
p-cresol	108.1	1600	15	19	1
2,4,6-trichlorophenol	197.5	900	1	1070	5.3
pentachlorophenol	266.4	14	0.2	50×10^3	20
naphthalene	128.2	32	10	1290	10
benzo(a)pyrene	252.3	3.8×10^{-3}	0.73×10^{-6}	545×10^4	450
carbazole	167.2	1	0.093	1300	0.6
hexachlorodibenzo(p)dioxin	392.9	1.3×10^{-4}	1.93×10^{-6}	26×10^6	500

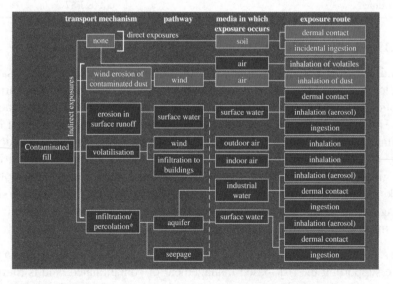

Figure 10 *Generalised exposure model for human exposures to contaminated fill*

phases for sampling and analysis; (ii) identify the key exposure pathways by which contaminants may be transported to receptors; and (iii) inform the selection of remediation technologies, whose principal focus should be the key environmental compartment (soil, water, gas, free organic phase) within which risk-critical compounds are present (Case Study 3).

Case Study 3: Relating Physico-chemical Properties to Site Investigation, Risk Assessment and Remediation

Using Table 5, we can consider the principal environmental medium in which we would expect to find these contaminants in soil systems, propose analytical methods, predict the key exposure pathways and propose approaches to risk management. Consider benzo(*a*)pyrene, for example. This polynuclear aromatic compound has a very low vapour pressure and aqueous solubility and high organic carbon partition coefficient (K_{oc}) and half-life. We therefore expect it not to be soluble or found in air in high concentrations but to be tightly bound to the soil and be persistent, given its poor propensity to be removed (high $t_{1/2}$). We should sample and analyse for it in the soil itself, therefore this leads us to analytical methods that involves its extraction and analysis from soil samples, such as gas chromatography.

   Given the preference of benzo(*a*)pyrene for soil, we expect those exposure pathways involving exposure to soil media to be most important (Figure 10), i.e. dermal contact, incidental ingestion and inhalation of windblown dusts. The relative importance of these exposure routes depends on the end use of the site. Exposure to dust may only be an issue where soils are left exposed at the surface and can become windborne. Ingestion and dermal contact require human contact directly with soils, so for an industrial estate this may be of less concern. For housing, risk management options that address the soil matrix directly will be most important. Bioremediation has limited ability to degrade persistent compounds and where residues persist above concentrations of concerns, the soils themselves may have to be removed from the site.

4.6.1.2 Using Risk Assessment Tools. Assessing the risks to human health, ecosystems and other receptors has become central to making sound decisions at contaminated sites. International approaches adopted in the US, the Netherlands, Australia and the UK have been devised, in general, by considering a notion of an 'acceptable risk' below which further remediation has marginal benefit, and developing toxicologically-based site assessment criteria as triggers for detailed site assessment. These site assessment criteria are often presented as 'soil screening levels'. When individual contaminants exceed these levels, site owners are guided to consider detailed investigations that will help them better understand the nature and magnitude of the risk.

To assist decision-makers, many of the standard exposure pathways encountered at contaminated sites (Figure 10) have been coded into desktop risk assessment software packages. For potential risks to human health, these allow the concentrations of soil contaminants identified on site to be translated into environmental concentrations at an off-site exposure point and into indicative doses (mg contaminant per kg body weight per day) that can be compared with acceptable daily intakes or national reference doses. To facilitate a quick initial screening, most models adopt generic exposure scenarios (conditions of exposure and specific land uses) that standardise many of the numerous assumptions made in assessing the site risk. Using such models, land managers can identify the key pollutants, exposure pathways and receptor types that contribute most to the site risk. Risk assessments can then be used to inform and help prioritise the management of risk.

Probabilistic models, that account for the myriad of uncertainties in a stochastic fashion, are now widely used. They are powerful in identifying the principal sources of variance in estimates of exposure[21] (Figure 11) and in providing more credible assessments of risk. This is important because the costs of remediation are high and contamination must be managed to ensure cost-effective protection of the environment and public health.

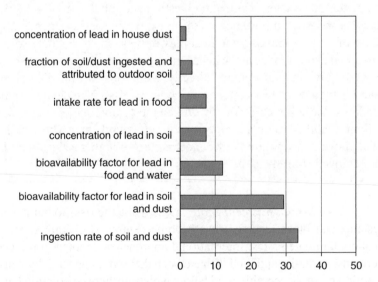

Figure 11 *Contribution to % variance in total lead uptake from key parameters in a*
probabilistic blood lead exposure model
(Redrawn from Lee *et al.* (1995))

4.6.1.3 Scientific Rigour in Site Investigation and Environmental Analysis. The confidence with which decisions can be taken on the remediation of contaminated sites relies heavily on the quality of the site investigation data. Risk assessments conducted in the absence of quality data provide a false sense of security in their estimate of the magnitude of the risk. Using risk as the basis for prioritising remediation across a 'portfolio' of sites requires managers to compare the nature, extent, magnitude and depth of contamination. Databases and decision-support systems that can store and process vast amounts of site data are improving practice in this area.

Central to many pollution issues is the characterisation of the source of pollution – often called the 'source term'. Understanding the source, its chemical, physical and microbiological characteristics allows the nature of the hazard to be identified – benzene, for example is a known human carcinogen by inhalation. It has a considerable vapour pressure and a slight aqueous solubility; hence it volatilises from soils and also dissolves in groundwater.

Because many risk estimates rely on modelled exposures rather than measured data at the point of exposure, high-quality data on the source term is critical. The interpretation of site investigation data, especially for large sites, often employs statistical analysis. Many risk assessment tools require contaminant distributions, or at least the use of 'worst-case' and mean concentrations within deterministic exposure assessment models. A central concern is how to best present 'hot spots' (highly contaminated areas for immediate treatment), total contaminated areas (of value for liability assessment) and randomised site data in statistical risk assessment packages. Where these different needs of the site data are combined in the same site investigation, the overlay of sampling points and separation of data sets to avoid the introduction of bias will be required.

Examinations of site risk frequently reveal the significant geochemical and analytical uncertainties associated with land contamination. Despite substantial improvements in environmental diagnostics and instrumental analysis over the past 25 years, risk estimates remain constrained by analytical uncertainties and an overriding variance in contaminant concentrations in soils.[22] Analytical techniques, even for well understood contaminants such as arsenic, harbour significant methodological uncertainties. Figure 12 illustrates the performance of a range of laboratories (individual bars) and analytical methods (grouped together as methods 131–135) about a grand median value of arsenic in a reference soil sample of 33 mg kg^{-1}. The scale of this variance must be considered by regulators in evaluating the significance of soil contamination; here for example, in the context of a soil guideline value for arsenic of 20 mg kg^{-1}. Case study 4 illustrates the number of assumptions involved in estimating exposures from a single exposure route, the incidental ingestion of soil by children.

Case Study 4: Quantifying Exposure

Exposure assessment is used to quantify chemical doses from contaminated soils to humans. Estimates of exposure rely, in principle, on straightforward equations, such as that for the incidental ingestion of soils. However, there is considerable debate about the amounts of soil ingested by infants, for example, as well as the inherent variability associated with the analysis of contaminants in soil (Figure 12). This just for two of the parameters involved. For carcinogens, the lifetime average daily dose (LADD):

$$\text{LADD} = \frac{C \; IR \; B \; ED}{(bw \; AT)}$$

where C is the mean exposure concentration (mg kg^{-1}), IR the soil ingestion rate ($\text{kg}_{soil} \, \text{d}^{-1}$); B the bioavailability (unitless), ED the exposure duration (d), bw the body weight (kg), and AT the averaging time (a lifetime, 25,550 days = 70 years).

Say we wish to estimate the LADD for an infant exposed every day to soil containing $19 \, \text{mg kg}^{-1}$ arsenic for 4 years of its early life. We assume that the infant ingests, on average, $100 \, \text{mg}_{soil} \, \text{d}^{-1}$, though this may vary between 20 and 1000 or even more.

$$\text{LADD} = 19 \, \text{mg}_{As} \, \text{kg}_{soil}^{-1} \times 100 \times 10^{-6} \text{kg}_{soil} \, \text{d}^{-1} \times 1.0$$
$$\times (4 \times 365 \, \text{d})/(15 \, \text{kg} \times 25{,}550 \, \text{d})$$
$$\approx 7 \times 10^{-6} \text{mg}_{As} \, \text{kg}_{bw}^{-1} \text{d}^{-1} \; (0.000007 \text{ milligrams of arsenic}$$

per kg of body weight per day, averaged over a lifetime).

Risk assessments that use single data points are referred to as 'deterministic'. These provide single numerical estimates of risk using a combination of assumptions populated by individual data points, some gathered by site investigation and analysis, some modelled and some assumed from the literature. A deterministic risk estimate is an increased likelihood of a specific adverse event occurring over a specified time period; for example, an above-background theoretical cancer risk attributable to specific exposures from carcinogens present at a contaminated site. For most sites, these incremental risks are usually very small indeed (10^{-5}, 10^{-6}, 10^{-8}). Because they are estimated using deterministic tools, deterministic risk estimates do not reflect the considerable uncertainties in the assumptions and data that they rely on. Managing these uncertainties has become one of the key difficulties in applying the risk-based approach to situations such as the buying and selling of land where uncertainties can undermine the confidence of the parties involved. For decision-makers, focusing on the relative magnitude of the risks to

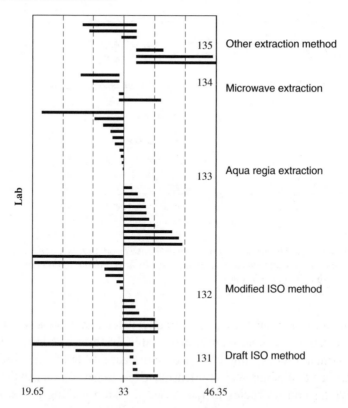

Figure 12 *Distribution of analytical data for arsenic in a sample of homogenised soil*

inform management of the land rather than the absolute value of risk estimates is one approach to overcoming these difficulties.

4.6.1.4 Bioavailability. Risk analysts are aware that risk estimates based on doses received at exposure points are only surrogates for the toxicological risk, because they do not account for processes that occur inside the body. Here, what counts is the availability of the species to the biological system that it interacts with. The bioavailability of compounds *in vivo* is an important concept therefore with respect to the toxicological response[23] (Figure 13).

4.6.1.5 Communicating Risk, Environmental Justice and Participation in Decision-Making. The regeneration of contaminated land has always required a multi- and trans-disciplinary approach. Increasingly, scientists, engineers, planners and lawyers are turning to the social sciences for advice. Researchers and practitioners are now gaining valuable insights into how we might involve others in decision-making and into issues of equity and the perceptions and reporting of risk. The interpretation and communication of an environmental risk assessment often plays a key role in the decisions that are

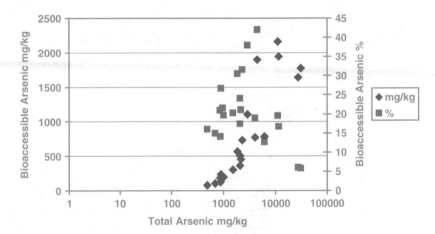

Figure 13 *Bioaccessible arsenic sourced from Great Devon Consoles. Different soils with similar concentrations of total arsenic may have exhibit different bioavailabilities* (After Bragg *et al.* (2001))

made on land contamination. Historically a technical preserve, the new agenda on risk-based land management is requiring us to consider how the risk assessment process might be made more accessible to non-expert audiences. Recognition of this social dimension to land regeneration and the need to engage wider constituencies in risk-based decision-making has stimulated research on participation in risk assessment.[24] The following aspects are important:

- providing information, consulting with others and involving stakeholders in decisions are critical so that decisions about land contamination remain credible;
- consultation that allows options to be developed and explored, rather than offering single solutions are preferred;
- decision processes that have clearly defined stages help;
- taking time to clearly identify and define the problem under consideration is important; and
- involving a wide range of people in decisions also needs to be sensitive to the timescales within which regulatory decisions must be made.

4.7 CONCLUSION – MANAGING SOIL FOR SUSTAINABLE COMMUNITIES

This chapter has summarised the pollution science that supports our understanding of soils in the natural and built environment. It is clear that soil can only be protected as a resource where there is respect for its functions. The risks that contaminated soils pose need to be managed cost-effectively and in consultation with the communities whose health and well-being is at the heart of land regeneration strategies.

QUESTIONS

1. Describe the main functions of soil and how these are supported by the soil system. How does this system respond to pressures, including pollution from the atmosphere and wastes?
2. How do soil-forming factors, such as climate and parent material affect soil type and variability? What are the consequences of soil variability for soil protection?
3. Describe the main variables controlling the chemistry of Fe in soil systems.
4. How does SOM interact with geochemical cycles? Consider the cycling of metals and other elements as well as carbon.
5. What are the properties of soil that affect its ability to absorb pollutants and attenuate their release to groundwater?
6. What are the benefits of a risk-based approach to land management? What arguments would you advance that this was a cost-effective means of managing land contamination?
7. Describe the environmental fate of phenol. How would you devise an analytical strategy for phenol at a creosote-contaminated site where phenol was present? What might you expect the key exposure routes to be?
8. Distinguish between deterministic and probabilistic risk assessment. What advantages does probabilistic risk assessment offer site remediation engineers?
9. Summarise an approach to investigate a former gas works site. What would you do first? What information would you require to make a judgement on the significance of the risks posed?

REFERENCES

1. N.C. Brady and R.R. Weil, *The Nature and Properties of Soils*, 13th edn, Prentice-Hall, Upper Saddle River, NJ, 2002.
2. P. Smith, *Soil Use Manage.*, 2004, **20**, 212–218.
3. N. Nunan, K. Wu, I.M. Young, J.W. Crawford and K. Ritz, *FEMS Microbio. Ecol.*, 2003, **44**, 203–215.
4. J.W. Crawford, K. Ritz and I.M. Young, *Geoderma*, 1993, **56**, 157–172.
5. K. Ritz, M. McHugh and J.A. Harris, *OECD Expert Meeting on Soil Erosion and Biodiversity Indicators*, Rome, 2003.

6. R.S. De Groot, M. Wilson and R.M.J. Boumans, *Ecol. Econ.*, 2002, **41**, 393–408.
7. W.E.H. Blum, in *Soil and Environment*, Vol 1, H.J.P. Eijsackers and T. Hamers (eds), Kluwer Academic Publisher, Dordrecht, 1993.
8. E.A. Fitzpatrick, *Soils: Their Formation, Classification and Distribution*, 2nd edn, Longman, London, 1986.
9. FAO, *World Reference Base for Soil Resources*, FAO, Rome, 1988.
10. D.L. Sparks, *Environmental Soil Chemistry*, Academic Press, San Diego, 1995.
11. L. Van-Camp, B. Bujarrabal, A.R. Gentile, R.A. Jones, L. Montaranella, C. Olazabal and S.-K. Selvaradjou, *Reports of the Technical Working Groups Established under the Thematic Strategy for Soil Protection*. EUR 21319 EN/4. Office for Official Publications of the European Communities, Luxembourg, 2004, 872pp.
12. J. Webb, P. Bellamy, P.J. Loveland and G. Goodlass, *Soil Sci. Soc. Am. J.*, 2003, **67**, 928–936.
13. D.S. Jenkinson and J.H. Rayner, *Soil Sci.*, 1977, **123**, 298–305.
14. M.H.B. Hayes, P. MacCarthy, R.L. Malcolm and R.S. Swift (eds), *Humic Substances: II. in Search of Structure*, Wiley, Chichester, 1989.
15. B.J. Alloway, in *Understanding Our Environment*, R.M. Harrison (ed), Royal Society of Chemistry, Cambridge, 1999, 199.
16. Department of Environment, Transport and the Regions, Environment Agency and Institute for Environment and Health, *Guidelines for Environmental Risk Assessment and Management, Revised Departmental Guidance*, DETR, London, 2000, 88pp.
17. S.J.T. Pollard and R. Duarte-Davidson, in *Forecasting the Environmental Fate and Effects of Chemicals*, P.S. Rainbow, S.P. Hopkin and M. Crane (eds), Wiley, Chichester, 2001, 55.
18. Agency for Toxic Substances and Disease Registry, *2003 CERCLA Priority List of Hazardous Substances*, 2004, available at http://atsdr1.atsdr.cdc.gov:8080/clist.html.
19. S.J.T. Pollard and S.M. Herbert, in *Contaminated Soil '98, Proceedings 6th International FZK/TNO Conference on Contaminated Soil*, Edinburgh, Thomas Telford, London, Vol 1, May 17–21, 1998, 33.
20. S.J.T. Pollard, R.E. Hoffmann and S.E. Hrudey, *Can. J. Civ. Eng.*, 1993, **20**, 787.
21. R.C. Lee, J.R. Fricke, W. Wright and W. Haerer, *Environ. Geochem. Health*, 1995, **17**, 169.

22. S.J.T. Pollard, M. Lythgo and R. Duarte-Davidson in *Assessment and Reclamation of Contaminated Land*, R. Hester and R.M. Harrison (eds), Royal Society of Chemistry, Cambridge, 2001, 1.
23. S.E. Hrudey, W. Chen and C. Rousseaux, *Bioavailability in Environmental Risk Assessment*, Lewis Publishers, CRC Press, Boca Raton, FL, 1996, 294pp.
24. J. Petts, S.J.T. Pollard, A.-J. Gray, P. Orr, J. Homan and P. Delbridge, Proceedings of 8th International FZK/TNO Conference on Contaminated Land, Ghent, 2003, 2969.

CHAPTER 5

Investigating the Environment

C. NICHOLAS HEWITT[a] AND ROB ALLOTT[b]

[a]Lancaster Environment Centre, Lancaster University, Lancaster,
LA1 4YQ, UK
[b]Environment Agency, Preston, PR3 8BX, UK

5.1 MONITORING: A SYSTEMATIC INVESTIGATION OF THE ENVIRONMENT

The gathering of information on the existence and concentration of substances in the environment, either naturally occurring or from anthropogenic sources, is achieved by measurement of the substance or phenomenon of interest. However, single measurements of this type made in isolation are virtually worthless, since temporal and spatial variations cannot be deduced. Rather, a systematic investigation of the environment requires the parameter of interest to be *monitored* by repeated measurements made over time and space, with sufficient sample density, temporally and spatially, that a realistic assessment of variations and trends may be made. Hence, this chapter focuses on the principles and practices of environmental monitoring.

Monitoring of the environment may be undertaken for a number of reasons and it is important that these be defined before sampling takes place. The generalization that "monitoring is done in order to gain information about the present levels of harmful or potentially harmful pollutants in discharges to the environment, within the environment itself, or in living creatures (including ourselves) that may be affected by these pollutants"[1] may be expanded as follows:

(a) Monitoring may be carried out to assess pollution effects on man and his environment, and so to identify any possible cause and effect relationships between pollutant concentrations and, for example, health effects, or environmental changes.

(b) Monitoring may be carried out in order to study and evaluate pollutant interactions and patterns. For example, source apportionment[2] and pollutant pathway studies usually rely on environmental monitoring.

(c) Monitoring may be carried out to assess the need for legislative controls on emissions of pollutants and to ensure compliance with emission standards. An assessment of the effectiveness of pollution legislation and control techniques also depends upon subsequent monitoring.

(d) In areas prone to acute pollution episodes, monitoring may be carried out in order to activate emergency procedures.

(e) Monitoring may be carried out in order to obtain a historical record of environmental quality and so provide a database for future use, for example, in epidemiological studies.

(f) Monitoring may also be necessary to ensure the suitability of water supply for a proposed use (industrial or domestic) or to ensure the suitability of land for a proposed use (for example, for housing).

A basic problem in the design of a monitoring programme is that each of the above reasons for carrying out monitoring demands different answers to a number of questions. For example, the number and location of sampling sites, the duration of the survey, and the time resolution of sampling will all vary according to the use to which the collected data are to be put. Decisions on what to monitor, when and where to monitor, and how to monitor are often made much easier once the purpose of monitoring is clearly defined. Therefore, it is most important that the first step in the design of a monitoring programme should be to set out the objectives of the study. Once this has been done, the programme may be designed by consideration of a number of steps in a systematic way (see Figure 1) such that the generated data are suitable for the intended use. It is important also that the data produced by a monitoring programme should be continuously appraised in the light of these objectives. In this way, limitations in the design, organization or execution of the survey may be identified at an early stage.

The aim of this chapter is to present and discuss the most important and relevant considerations that must be taken into account in the design and organization of a monitoring exercise. It is not intended to be a manual or practical guide to monitoring; rather it is hoped that it highlights the types of approaches that may be used and some of the problems likely to be encountered. The inclusion of case studies and references direct the reader to the more specific practical information that is available elsewhere.

5.2 TYPES OF MONITORING

The Earth's surface is comprised of four distinct media; the atmosphere, the hydrosphere, the biosphere and the land. Pollutants can occur in either the

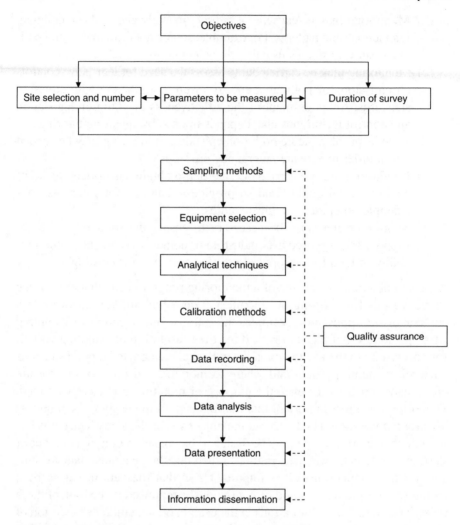

Figure 1 *Steps in the design of a monitoring programme*

solid, liquid or gaseous phases. However, the environment is not a simple system and consequently each of the four media may contain pollutants in each of the three phases. Monitoring may therefore be necessary for a particular pollutant in a specific phase in a particular environmental compartment (*e.g.* sulfur dioxide in air) or it may encompass two or more phases and/or media (*e.g.* dissolved and particulate phase metals in water). Pollutants in the environment originate from a multitude of different types of sources and the identification of these is a necessary prerequisite to the design of a monitoring programme.

First, pollutant sources may be classified by their spatial distribution as point sources, line sources or area sources. Point sources include industrial stacks and chimneys, liquid waste discharge pipes and localized toxic waste dumps on land. Line sources may include highways, airline routes and runoff into rivers from agricultural land, while area emissions may arise from extensive urban or industrial complexes. Sources may also be classified as either stationary or mobile, motor vehicles being the obvious example of the latter. Classification may also be made for air pollutant sources on the basis of the height of discharge, *i.e.* at street level, building level, stack level or above the atmospheric boundary layer level. A further important distinction may be made between "planned", "fugitive" and "accidental" emissions to the environment

(a) Planned emissions arise when (as is invariably the case) it is economically or technically impossible to completely remove all the contaminants in a discharge and hence the process operation allows pollutants to be discharged to the environment at known and controlled rates. Obvious examples of planned emissions include sulfur dioxide from power generation plants and low-level radioactive effluent during nuclear fuel reprocessing.

(b) Fugitive emissions arise when pollutants are released in an unplanned way, normally without first passing through the entire process. They therefore occur at a point sooner in the process than the stack or duct designed for "planned" emissions. They generally originate from operations that are uneconomic or impractical to control, have poor physical arrangements for effluent control or are poorly maintained or managed. An example is the escape of heavy metal contaminated dust from a factory on vehicle tyres, arising from poor dust control and wheel washing arrangements.

(c) Accidental emissions result from plant failure, such as a burst filter bag or faulty valve or from an accident involving either equipment or operator error (*e.g.* the Chernobyl nuclear reactor accident or Bhopal chemical plant accident). Accidental emissions can give rise to very high mass emission rates and ambient concentrations but they normally occur only infrequently.

Classification of the sources of pollutants in this way allows the distinction of two differing approaches to their monitoring. On the one hand, samples may be taken of the effluent before discharge to, and dispersion in, the environment (source monitoring). Alternatively, samples may be taken of the ambient environment into which discharges occur, for example, of the air or receiving waters, without consideration of source strengths and rates. Obviously neither one of these approaches alone can necessarily provide all the data required to resolve a particular problem and often it is desirable to complement one with the other.

5.2.1 Source Monitoring

5.2.1.1 General Objectives.   Source monitoring may be carried out for a number of reasons.

(a) Determination of the mass emission rates of pollutants from a particular source, and assessment of how these are affected by process variations.
(b) Evaluation of the effectiveness of control devices for pollution abatement.
(c) Evaluation of compliance with statutory limitations on emissions from individual sources.

5.2.1.2 Stationary Source Sampling for Gaseous Emissions.   A common feature of many industrial processes is that effluent output rates exhibit cyclical patterns. These may be related to working shift arrangements or be a function of the operations involved, but both require that source testing or monitoring be planned accordingly. Process operations should be reviewed so that discharges during the period of sampling are representative of the plant output in order to ensure that the samples themselves are representative of the effluent, and that the final pollutant analysis will be a representative measure of the entire output.

There are detailed guidance and standards available for source monitoring for gaseous emissions.[3,4] The US Environmental Protection Agency (EPA) has published standard test methods for emission monitoring. The UK Environment Agency has established a monitoring certification scheme (MCERTS) to provide a framework within which quality environmental measurements can be made. MCERTS provides for the product certification of monitoring systems (for example, instruments, meters, analysers and equipment), the competency certification of personnel and the accreditation of laboratories under the requirements of European and international standards.

Two requirements may be specified for valid source monitoring. First, the sample should accurately reflect the true magnitude of the pollutant emission at a specific point in the stack at a specific instant of time. This requirement is met by adequate sampling instrument design. Second, enough measurements should be obtained over time and space so that their combined result will accurately represent the entire source emission. This requires consideration of the emissions both in time and in space, across the entire cross-section of the duct or stack (the sample plane). The sampling plane should be situated in a length of straight duct (preferably vertical) with constant shape and cross-sectional area. Where possible, the sampling plane should be as far down-stream and upstream from any disturbance, which could produce a change in the direction of flow (*e.g.* bend, fan or a partially closed damper). This sample plane criterion is usually met in sections of duct with five hydraulic diameters of straight duct upstream of the sampling plane, two hydraulic diameters downstream (so long as stack outlet is no nearer than five hydraulic diameters downstream). The hydraulic diameter is defined as four times the cross-sectional area of the sample plane divided by the length of the sample plane perimeter.

In a circular flue, sampling at the centroids of equal-area annular segments will ensure that emission variations across the sample plane are quantified. In a rectangular flue sample points should be located at the centroids of smaller equal-area rectangles. For larger ducts and stacks there should be a minimum of 12 sampling points across the sample plane and at least four sampling points per m^2 of the sample plane.[3] When particulate pollutants are sampled within the stack, it is important that an isokinetic sampling regime is maintained.

5.2.1.3 Mobile Source Sampling for Gaseous Effluents. Vehicle and air-craft emissions are heavily dependent upon the engine operating mode (*i.e.* idling, accelerating, cruising or decelerating) and the results obtained by sampling must be considered specific to the type of operating cycle used during the test. Emission tests are usually performed with the vehicle on a dynamometer equipped with inertia flywheels to represent the vehicle weight and brake loading on a level road.

5.2.1.4 Source Monitoring for Liquid Effluents. Liquid wastes and efflu-ents often tend, like gaseous effluents, to be inhomogeneous and care is needed in selecting sampling positions. Having considered the location of the site in relation to the plant operation (*e.g.* should the site be before or after a particular stage of the process or treatment) it is desirable that a region of high turbulence and/or good mixing be chosen. As for gaseous emissions, several samples may have to be taken across the cross-section of a pipe or channel. Sampling from vertical pipes is less liable to be affected by deposition of solids than sampling from horizontal pipes, and a distance of approximately 25 pipe diameters downstream from the last inflow should ensure that mixing of the two streams is essentially complete.[5] If suitable homogeneous regions for sampling cannot be found, particularly where suspended materials are present, samples may have to be taken from several positions along the efflu-ent stream.

Where the composition of a liquid effluent is known to vary with time, grab samples may be collected at set intervals, either manually or by use of an automatic sampler. An alternative approach is to sample at intervals varying with the flow rate so that a more representative composite may be obtained (*i.e.* flow porportionate sample).

5.2.1.5 Source Monitoring for Solid Effluents. Solid effluents may arise from a number of different processes, including sludge after sewage treatment, ash residue from municipal waste incinerators or low-grade gypsum from desulfurization plants attached to coal fired power stations. In general, solid wastes are even less homogeneous than either liquid or gaseous effluents. Therefore, great effort must be made to ensure that samples are representative of the bulk waste (see Section 5.3.3). Monitoring of sewage sludge is particularly common due to the sludge acting as an efficient sorption material for heavy metals. Typically, 80–100% of the input lead in a sewage treatment plant is incorporated into the sludge, which historically resulted in sludge lead concentrations of 100–3000 $\mu g\,g^{-1}$, although the concentrations are now generally $<500\,\mu g\,g^{-1}$. Consideration must therefore be given to the concentrations of pollutants in the material before it is used as a fertilizer, incinerated or used as landfill. The determination of the metal balance of a sewage treatment work may be necessary when considering the fate of the treated effluent and solid waste.

In most countries, guidelines exist to control the disposal of sewage sludges to land, usually based primarily upon the zinc, copper and nickel content of the sludge. Hence, considerable quantities of other metals, including lead, may be added to land over a normal 30-year disposal period. In the UK, the disposal of lead-rich sewage sludges to land is controlled where direct ingestion by animals of contaminated grass or soil can occur.

Until fairly recently, most trace metal analysis of environmental samples was designed to give a measure of the total elemental concentration in the sample as it was felt that this gave an adequate measure of the pollution load of that metal. However, in the past three decades it has become apparent that total metal concentrations are often not sufficient and that information based upon some form of physicochemical speciation scheme is required. This may include, for example, solubility of the pollutant in acids of different strengths, the size distribution of particles and the association with organic compounds. This is because the physical, chemical and biological responses to a pollutant will vary according to its physical and chemical speciation. One disadvantage of this type of analysis is that it is complicated and time consuming compared with total metal determinations. Thus, speciation studies are invariably limited to a few samples where many (tens or even hundreds) would be taken in a total metal study.

> *Case Study 1: Speciation of Mercury Emissions from a Landfill Site*[6]
>
> Landfill sites are widely recognized as sources of toxic, radiatively active (*i.e.* greenhouse gas) and explosive substances to both the atmosphere and to groundwater, and the aftercare and monitoring of such sites are essential. The contamination of groundwater by toxic organic chemicals from landfill leachate is of major environmental concern, as are methane emissions, especially when dwellings are built in close proximity to, or even on, a disused site. Landfills often contain numerous sources of mercury and, because of the presence of anaerobic bacteria, this mercury can be methylated to produce highly toxic organic compounds of the metal. In this study[6], municipal landfill sites in Florida, Minnesota, Delaware and California were investigated and total gaseous mercury concentrations in the $\mu g\,m^{-3}$ range, and methylated mercury compounds in the $ng\,m^{-3}$ range were found in the emitted landfill gas. Although landfill gas flaring should ensure that only inorganic Hg is released, many landfill sites around the world still do not use this technology, and hence landfill sites remain a significant source of methylated mercury compounds to the atmosphere.

5.2.2 Ambient Environment Monitoring

5.2.2.1 General Objectives. Monitoring the environment may be carried out for a number of reasons, as outlined above in Section 5.1. However, whatever the purpose of the survey the overriding consideration when designing a programme is to ensure that the samples obtained provide adequate data for the purpose intended. Invariably, this means that samples should be representative of conditions prevailing in the environment at the time and place of collection. Thus, not only must the sampling location be carefully chosen but also the sampling position at the chosen location.

The selection of a specific monitoring site requires consideration of four steps: identification of the purpose to be served by monitoring; identification of the monitoring site type(s) that will best serve the purpose; identification of the general location where the sites should be placed; and finally, identification of specific monitoring sites.

5.2.2.2 Ambient Air Monitoring. Air pollution problems vary widely from area to area and from pollutant to pollutant. Differences in meteorology, topography, source characteristics, pollutant behaviour and legal and administrative constraints mean that monitoring programmes will vary in scope, content and duration, and the types of station chosen will also vary. However, ambient monitoring sites may be divided into several categories

(a) Source-orientated sites for monitoring individual or small groups of emitters as part of a local survey (*e.g.* one particular factory).
(b) Sites in a more extensive survey, which may be located in areas of highest expected pollutant concentrations or high population density, or in rural areas to give a complete nation wide or regional coverage.
(c) Baseline stations to obtain background concentrations, usually in remote or rural areas with no anticipated changes in land use.

Location of Source-Orientated Monitors. Occasionally, the effects and impact of a specific pollutant source are of sufficient interest or importance to warrant a special survey. This will usually include a site at the point of antici- pated maximum ground-level concentration and also a nearby site to character- ize the "background" conditions in the area. Examination of meteorological records will usually be necessary in order to choose suitable locations for the sites. Several computerized models are available for determining the areas of maximum average impact from a point source. Ground-level concentrations rise rapidly with distance from the source to a maximum and then fall grad- ually beyond the maximum,[7] as shown in Figure 2.[8] This is for meteorologi- cal conditions of neutral stability and different heights of emission (H metre). The ordinates in this graph represent concentration (kilogram per cubic metre) normalized for emission rate (Q kilogram per second) and wind speed (U metre per second) and the various curves are for different source heights (H) and different limits to vertical dispersion (L metre). It is prudent, therefore, to locate the monitoring site somewhat beyond the distance where the maximum concentration is predicted. This allows some margin for error by placing the monitor in a region of relatively small concentration gradients. If monitoring data are coupled with meteorological data to produce pollution roses (as dis- cussed in Section 5.6.2), then it may be possible to confirm the source of the emission by triangulation of the directions of high air concentrations.

In some cases, pollutants are emitted to the atmosphere from a single source but in a more diffuse manner than from a single stack. Calculations of mass emission rates and distance of maximum ground-level concentration are more difficult to make for such diffuse or fugitive emissions, which in some cases, may have significant impacts on the local air quality.

Location of Monitors in Larger-Scale Surveys. Often it is important to know the geographical extent of atmospheric pollution, and to have localized

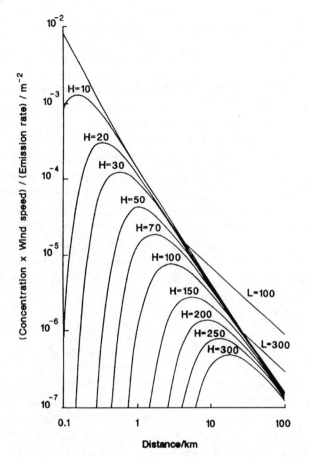

Figure 2 *Normalized ground-level concentrations from an elevated source for neutral stability. The effective stack height (H) is the sum of the release height (e.g. chimney height) and the height gained by the plume due to momentum and buoyancy*

information on source strengths or ground-level concentrations within a plume is not sufficient. For example, Member States of the European Union are obliged to carry out large scale monitoring of a number of air pollutants for a number of reasons. These include the following

(a) To understand air quality problems in order that cost-effective policies and solutions can be developed.
(b) To assess how far UK and European standards and targets are being achieved.
(c) To provide public information on current and forecast air quality.
(d) To assist the assessment of personal exposure to air pollutants.

Case Study 2: National Air Quality Monitoring in the UK[9]

Different member states of the European Union approach their obligations to air quality monitoring, incumbent upon them under the terms of
the Framework Directive on air quality management and assessment, in
different ways. In the UK, this obligation is discharged by central government by monitoring at over 1300 national air quality monitoring sites
across the UK, organized into several automatic and non-automatic networks. Each of these networks has a different scope and coverage, with
different clearly defined objectives which have been used to optimize network design, select priority pollutants and appropriate measurement
methods, and to determine the required level of quality assurance/control
and data management. In total, 125 sites (16 in London, 82 in other urban
areas and 22 rural) constitute the automatic urban and rural network
(AURN). Together with five sites in the hydrocarbon network, these operate automatically, and provide high-resolution hourly information on a
range of pollutants that is communicated rapidly to the public. The pollutants measured vary from site to site, but include NO_2, O_3, SO_2, CO and
particles (PM_{10}). The non-automatic sites measure average concentrations
over a specified sampling period (typically from a day to a month) instead
of instantaneous concentrations, but still provide invaluable data for
assessing levels and impacts of pollution across the country as a whole.
These include "smoke", SO_2, NO_2, lead and other metals, ammonia and
nitric acid. There are other networks in which acid deposition, metal deposition, benzene and 1,3-butadiene, polycyclic aromatic hydrocarbons
and toxic organic micro-pollutants are determined. Considerable detail of
these networks, the pollutants measured and the methods used, together
with a full archive of data, is available (http://www.airquality.co.uk).

One important feature of monitoring networks of this complexity is obviously the organizational and management structures used. In the case of the
AURN network in the UK, a highly devolved structure is used, with a central management unit, an equipment support unit, a large number of local site
operators, an independent QA/QC unit and a public information service.
With current information technology more or less real time data are available
to the public online, with many years of archived data from all sites (both
those currently used and those now redundant) also retrievable.

As well as the national air quality monitoring networks outlined above, a
considerable amount of other monitoring is carried out at the local level by
local authorities and other bodies, including the environment agency, and the

data obtained by them may, in some cases, be used to supplement that obtained from the national networks, and vice versa.

Location of Regional-Scale Survey Monitors.   On the regional or global scale, monitoring is usually concerned with long-term changes in background concentrations of pollutants and so the principal siting requirement is that truly representative baseline or background levels may be measured over a long time period without interference from local sources. It has been suggested[7] that baseline stations should be located in areas where no significant changes in land use practices are anticipated for at least 50 years within 100 km of the station, and should be away from population centres, major highways and air routes. Such locations are hard to find, but Cape Grim in Tasmania is one example.

The OECD "long-range transport of air pollutants" project, which measured chemical components in precipitation and SO_2 and particulate sulfate in air, is an example of a network established to monitor distant sources for regional or global effects. The long-term measurement of carbon dioxide concentrations in air at Mauna Loa since 1956 is probably the best example of "baseline" monitoring (http://cdiac.esd.ornl.gov).

Case Study 3: National and Regional Radon surveys in the USA[10] and UK[11]

An increasing awareness of the importance of human exposure to the naturally occurring radioactive gas radon has resulted in national surveys being commissioned for many countries, including the USA and UK. ^{222}Rn, commonly known as radon, is a colourless odourless gas, which results from the decay of ^{238}U and in turn decays to a series of radioactive daughters. Two other isotopes of radon exist, ^{219}Rn and ^{220}Rn (thoron) within the ^{232}Th and ^{235}U radioactive decay series. However, it is ^{222}Rn which has the greatest health significance, since it has the largest proportion of alpha emitting, high activity progeny. The inhalation of radon and particularly its daughters is believed to be a major contributor to the overall annual dose received from ionizing radiation.

Several methods exist for the measurement of radon in air and may be divided into active or passive techniques. The active techniques are generally only used for research or special survey purposes and normally consist of a pump that draws air through a filter trapping the radon decay products. The alpha radiation from these daughters is measured by scintillators or semi-conductor detectors. The most common method for monitoring radon is the passive alpha track detector and these have been extensively used in the USA[10] and UK.[12] Radon is allowed to diffuse into a small container, but radon daughters present in the ambient air are excluded. Inside the pot is placed a piece of polycarbonate, cellulose nitrate or allyl diglycol carbonate film. Alpha particles, formed by the decay of radon, damage chemical bonds on the surface of the film. After a period of exposure, the surface of the film is etched with NaOH and tracks appear. The number density of these tracks is proportional to the average radon concentration and may be either counted using a computerized image analysing system under a microscope, or the tracks filled with a scintillant (ZnS – Ag) which fluoresces when exposed to an alpha source in proportion to the number of tracks present. The detectors may be calibrated by exposure to known radon concentrations for known periods of time. Exposure times of weeks to months are required with this method.

The US survey found an arithmetic annual average radon concentration in homes of $46\,Bq\,m^{-3}$, with about 6% of homes having radon levels greater than the US EPA's action level for mitigation of $148\,Bq\,m^{-3}$. The EPA has used this survey data along with information on geology, aerial radio-activity, soil permeability and foundation type to assess the potential radon concentrations in homes and to present this information as a map of radon zones (www.epa.gov).

The primary influence of bedrock geology on relative radon concentrations throughout the UK is exhibited in Figure 3. High levels of radon in homes (reaching $8000\,Bq\,m^{-3}$) were found in areas where the bedrock contained high concentrations of uranium. These included granites and areas of mineralization in SW England and Scotland (up to 2000 ppm U) and some shales and limestones in north and central England (about 800 ppm U).

Other factors that influence radon levels were also identified, including season, time of day, meteorological conditions, ventilation (*e.g.* existence of double glazing), usage and other occupancy habits. These can result in a factor of two differences between summer and winter concentrations. Furthermore, radon concentrations may vary between rooms, with the highest radon levels in basements and, on average, first floor bedrooms have concentrations two-thirds those of living rooms. Indoor radon concentrations can be remediated by either preventing the gas entering the building (*e.g.* by sealing floors), increasing the ventilation rate (*e.g.* by extractor fan) or removing the radon decay products from the air (*e.g.* by electrostatic precipitator).

Monitoring networks may also be established to detect accidental releases. Following the Chernobyl nuclear reactor accident, the UK Government set up a national response plan for dealing with overseas nuclear accidents, which included establishment of a national radiation monitoring network and nuclear emergency response system (radioactive incident monitoring network (RIMNET)).

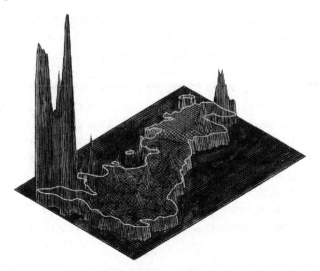

Figure 3 *Relative radon concentrations in homes throughout the UK*
(Reproduced with permission from the National Radiological Protection Board)

5.2.2.3 Environmental Water Monitoring. Pollutants enter the aquatic environment from the air (by dry deposition or in precipitation occurring either directly onto the water surface or elsewhere within the catchment area), from the land (either in surface runoff or *via* sub-surface waters) and directly through effluent discharges (either domestic, industrial or agricultural). The undesirable effects of pollutants in natural water may be due to

(a) stimulation of water plant growth – eutrophication – which ultimately leads to deoxygenation of the water and major ecological change;
(b) their direct or indirect toxic effects on aquatic life; and
(c) the loss of amenity and practical value of the water body, particularly as a source of water for public supply.

Apart from the monitoring of sources of pollutants in liquid effluents (Section 5.2.1.4 above) sampling may be carried out

(a) in rivers, lakes, estuaries and the sea in order to obtain an overall indication of water quality;
(b) for rainwater, groundwater and runoff water (particularly in the urban environment) to assess the influence of pollutant sources;
(c) at points where water is taken for supply, to check its suitability for a particular use; and
(d) using sediments and biological samples in order to assess the accumulation of pollutants and as indicators of pollution.

The use of *in situ* biomonitors or "biomarkers", *i.e.* organisms either collected directly from contaminated sites for direct chemical analysis or transplanted to a contaminated site for analysis after a period of pollutant accumulation, has grown in recent years. Often such work is accompanied by laboratory-based toxicity testing. Biomonitoring in aquatic systems might involve, for example, the collection of mussels and their subsequent analysis for heavy metals. This method has also been applied to terrestrial ecosystems, for example, using lichens as accumulators of metals. However, the measurement of chemical and physical parameters as well as the quantitative or qualitative assessment of aquatic flora and fauna is often used to give an indication of the presence or absence of pollution, and well-recognized relationships exist between the abundance and diversity of species and the degree of pollution. This is often used to assess the cleanliness of natural fresh waters and is known as biological or ecological monitoring.

Location of Sampling Sites. There are two main causes of heterogeneous distribution of quality in a water body. These are

(a) if the system is composed of two or more waters which are not fully mixed (such as in thermally stratified lakes or just below an effluent discharge in a river) and

(b) if the pollutant distributes non-uniformly in a homogeneous water body (for example, oil which tends to float, and suspended solids which tend to settle out of the water). Also chemical and/or biological reactions may occur non-uniformly in different parts of the system, so resulting in heterogeneous pollutant concentrations. When the degree of mixing is unknown, it is advisable to conduct a preliminary survey before deciding on sampling locations. Rapidly obtainable data of water temperature, pH, dissolved oxygen or electrical conductance may be used in this respect.

Sampling locations should generally be at points as representative of the bulk of the water body as possible, *e.g.* away from river or lake banks or the walls of channels or pipes, but often it will be desirable (and necessary) to take samples from several locations in order to obtain the required information.

When sampling from rivers and streams downstream of effluent discharges longitudinal, transverse and vertical sampling arrays may be necessary to ensure that truly representative data are obtained. Studies of some pollutants require sampling at considerable distances downstream of effluent inputs, *e.g.* in investigating the sag in dissolved oxygen content. When a temporally varying effluent discharge is under study, it may be desirable to sample as close to the point of discharge as mixing allows in order to monitor short-term variations in concentration. However, if long-term average water quality is of interest

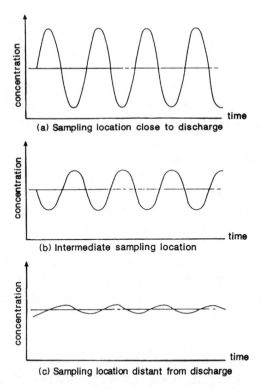

(a) Sampling location close to discharge

(b) Intermediate sampling location

(c) Sampling location distant from discharge

Figure 4 *Schematic diagram of the dependence of pollutant concentration on the distance downstream from a cyclically varying waste discharge*

then sampling should be carried out further downstream where longitudinal dispersion and mixing will have smoothed out the short-term variations (see Figure 4).

Sampling in estuaries presents special problems as great spatial and temporal variability may be exhibited. The appropriate locations for sampling will vary from estuary to estuary and will depend on the parameters of interest. If one considers a compound which has an input at only one end of an estuary and which is not removed from or added to solution during its lifetime in the estuary, then the concentration of that compound in the estuary will be solely dependent upon the dilution ratio between the river water and the seawater. Thus, for example, the concentration of chloride ion or salinity is dependent only upon the mixing of the fresh and the saline water bodies. This concept of "conservative" behaviour, is an important one which must be taken into account when monitoring estuarine concentrations. If a graph is drawn of the concentrations of the element of interest in an estuary against salinity then the data points will fall on a straight line (the theoretical dilution line) if physical mixing is the only process controlling the concentration of that element in the water. However, if the element of interest is added to or removed from the solution during mixing, then the data will not plot on a straight line. In the case of lakes and reservoirs, vertical stratification of pollutants may be very pronounced due to a reduction in dissolved oxygen from the surface downwards. A minimum of three samples is then probably necessary, at 1 m below the surface, 1 m above the bottom and at an intermediate point.

When water is abstracted from a river, lake, reservoir or from an aquifer, samples should be regularly taken at the point of abstraction and at the point where the water enters the distribution system. Several excellent handbooks with full descriptions of water sampling and analytical methods are available.[13–15]

The determination of concentrations of trace metals in natural waters is a fundamental stage in the calculation of their budgets or cycles, but is subject to the same problems of sample contamination that occur for atmospheric samples from remote areas. All stages of the analysis, from sampling collection, storage and filtration to actual laboratory manipulation require care to prevent contamination from occurring. Indeed for many years, measured levels of many trace elements in seawater were purely an artefact of contamination during sampling and analysis.

Case Study 4: Estuarine Trace Metal Distributions

A study[16] was carried out in the Trinity river estuary of Galveston Bay, Gulf of Mexico to examine phase speciation for a number of trace metals (*e.g.* Cd, Cu, Co, Fe, Ni, Pb, Zn). The aim was to investigate the size distribution of colloidal metals as well as their scavenging and production processes in an estuarine environment. Surface water samples were collected over a salinity range from 0 to 35 parts per thousand. Samples were collected using a peristaltic pump system and an ultra-clean sampling protocol to prevent sample contamination. Water samples were filtered through an acid-cleaned 0.45 μm pore size inline cartridge filter, placed on ice and stored in the dark. At the laboratory, some water samples were ultrafiltered to a 1 kDa nominal molecular weight cut off. Trace metals in the $\leqslant$1 kDa phase were classed as truly dissolved, whilst those in the range 1 kDa–0.45 μm were classed as colloidal.

Figure 5 shows the concentrations of copper and lead during two surveys. In the first survey, copper demonstrated conservative behaviour with concentrations falling on a theoretical mixing line between the freshwater and seawater end members. However, this behaviour was not observed for the second survey, with excesses of copper in the lower salinity ranges. Lead showed non-conservative behaviour for both surveys, with a removal process at lower salinities. All metals showed significant colloidal fractions and in most cases

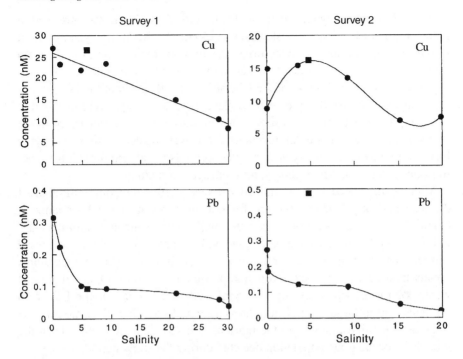

Figure 5 *Relationship of estuarine concentrations of copper and lead (<0.45 μm) with salinity in the Galveston Bay for two survey periods*
(Reproduced from Ref. 16, with permission from Elsevier)

colloidal metal concentrations correlated significantly with colloidal organic matter. Overall, the study revealed a dynamic system, resulting in both conservative and non-conservative mixing behaviour, perhaps as a result of time-dependent benthic fluxes and/or biological activity in the water column.

Case Study 5: Algal Blooms in Lakes

Thick surface accumulations or "blooms" of cyanobacteria in freshwater ecosystems are primarily attributed to nutrient, particularly phosphorus, enrichment. Other environmental factors that have been reported to favour cyanobacteria include high water temperatures, as stable water column, low light availability, high pH, low dissolved carbon dioxide and low total nitrogen to total phosphorus ratios. Cyanobacteria blooms often create unsightly surface scums, decreased water column transparency, unpalatable drinking water and noxious odours. Some species of cyanobacteria also produce toxic compounds ("cyanotoxins") that have been implicated in livestock, wildlife and pet fatalities as well as human poisonings worldwide.

In 1997, the toxic cyanobacterium *Microcystis aeruginosa* was observed in Lake Sammamish (western Washington, USA) for the first time. Cyanobacterial activity and environmental conditions that may promote toxic cyanobacteria were investigated in this study.[17] Water samples of 2 L were collected at 3–5 m depth intervals to the bottom of the lake at six sites throughout the lake (maximum depth of 25 m). Samples were collected weekly. Traps were installed at two sampling locations to investigate the migration of cyanobacteria from the sediments into the water column. Microcystins were detected using an enzyme linked immunosorbent assay and samples were also analysed for soluble reactive phosphorus, total phosphorus, total nitrogen and nitrate.

Microcystin concentrations ranged between 0.19–3.8 μg dm^{-3} throughout the lake and at all depths (except for one high result at the boat launch). Comparison of the conditions associated with toxic cyanobacterium episodes indicated that Microcystis was associated with a stable water column, increased surface total phosphorus concentrations ($>10\,\mu$g dm^{-3}), surface temperatures greater than 22°C, high total nitrogen to phosphorus ratio (>30) and increased water column transparency (up to 5.5 m). External loading of the lake with nutrients due to large rainfall may have triggered the 1997 algal bloom. The migration of Microcystis from the nutrient rich sediments may have been the inoculum for the toxic population detected during the study period.

5.2.2.4 Sediment and Soil Monitoring. Soils and sediments may become polluted by a number of routes, including the disposal of industrial and domestic solid wastes, wet and dry deposition from the atmosphere and infiltration by contaminated waters.

The main pollution hazards on land have been identified as follows:[1]

(a) Harmful substances may get into the soil or plants and so into the food supply.
(b) Substances may wash from the land and so pollute water supplies.
(c) Contaminants may be resuspended and subsequently inhaled.
(d) Substances polluting the land may make it potentially dangerous or unsuitable for future use (*e.g.* for housing or agriculture).
(e) Ecological systems may be damaged, with consequent loss to conservation and amenity.

Some potentially harmful substances, such as mercury or lead, are naturally present in soils but at concentrations which are not normally deleterious. Some activities, however, can cause elevated levels of these compounds. For example, mining may cause soils to be contaminated by metals, and the dumping of solid wastes on land will invariably introduce a wide variety of pollutants to the soil. On the other hand, there are compounds that do not occur naturally, and their presence in soils and sediments is due entirely to man's activities. These substances include pesticides, (particularly the organochlorine compounds such as DDT, toxaphene, aldrin, dieldrin) and artifical radionuclides (*e.g.* ^{137}Cs, ^{239}Pu).

A great number of studies have been conducted on dust and soil contamination with heavy metals, in particular lead, due to its predominant source as an antiknock agent added to petrol. These studies have established the magnitude of different sources of lead in the environment and some studies[18] have investigated the relative contribution of these sources of lead to a child's total lead intake.

As with the other types of media discussed above, it is important that the background levels of pollutants be established in soils, sediments and vegetation. One example of a large-scale investigation of this type is the geochemical survey of stream-bed sediments carried out in England and Wales.[19] Stream sediments are considered to represent a close approximation to a composite sample of the weathering and erosion products of rock and soil upstream of the sampling point and in the absence of pollution provide information on the regional distribution of the elements.

A second type of monitoring programme is required to establish actual levels of contamination in land or sediments known or believed to be affected by pollutants. In this case, a much more specific and localized monitoring may be required in order to quantify the degree of contamination. The contamination of sites often arises from their previous uses, particularly as coal gas manufacturing plants, sewage works, smelters, waste disposal sites, chemical plants and scrap yards. A typical contaminated site may be found to contain variable concentrations of toxic elements and organic compounds, phenols, coal-tars and oils, combustible material from undecomposed refuse, acidic or alkaline waste sludges and sometimes methane accumulations. In addition, contaminated land is often formed by waste tipping and so may be poorly compacted and very inhomogeneous.

Some sites may contain underground pipe work and structures from their previous uses and so present formidable sampling problems. Site investigation of this type is very expensive, involving bore holes or trial pits and, in cases where a detailed site history is unavailable, determination of a large number of pollutants. However, it is most important to be sure that the investigation is sufficiently rigorous as remedial measures are extremely costly and must be based upon adequate data. Common problems which have been identified are

- an inadequate number of samples,
- an inadequate range of determinants,

- bulking of samples when individual samples from specific locations are preferable,
- inappropriate analytical methods,
- inadequate referencing of sample locations,
- inadequate descriptions of samples,
- inadequate descriptions of trial pit strata, and
- an ignorance of the nature of the required information.

Monitoring should be carried out following the application of sewage sludge or waste waters to agricultural land. Samples of surface water, groundwater, site soil, vegetation and the sludge applied would normally be tested for faecal coliform, nutrients, heavy metals and pH. Details of the necessary chemical and physical methods may be found elsewhere.[13] The results from the monitoring exercise may be compared to predicted levels derived from the application rates of sludges to land, soil type, nitrogen, phosphorus and heavy-metal contents of the waste and the nutrient-uptake characteristics of the cover crop.

When monitoring background levels and more specific pollution on land or in the sediments of a water body, measurements will often be made of levels in the plants or organisms that the soil or sediments support. In many cases, flora or fauna provide excellent indicators of the degree of pollution as they may act as bioconcentrators (*e.g.* of heavy metals from suspended material in shellfish). Furthermore, it is obviously important to monitor pollution levels in food and through the food chain. The simultaneous measurements of pollutant levels in soils and plants as well as in water, sediments and aquatic biota are therefore often carried out. However, the relationships between levels in these various media are often not simple and sampling and analysis of one of these is no substitute for a comprehensive monitoring programme. For example, laboratory experiments indicate that metals are readily absorbed by some plants but measurements of metals in soil, rainwater and plants often reveal a lack of correlation between the corresponding sets of concentrations.

Monitoring of concentrations of trace metals in crop plants grown on sewage sludge-amended soils indicates that the levels found will vary with crop species and properties of the soil substrate on which they have been grown. More recent studies have concentrated on the physicochemical speciation of the metals in the applied material and the receiving soil and have demonstrated the importance of organic complexes in reducing free metal activity.

The dioxin and furan congener profile (fraction of each congener present) was used to define a "signature" or "fingerprint" for the incinerator based on the previous stack emission tests. This signature was found to be present in the incinerator ash, ambient air and soil samples. There was a clear gradient in the soil results, with declining distance as a function of distance from the incinerator (Figure 6). The average soil concentrations ranged from 458 pg g^{-1}

Case Study 6: Dioxins in Soils around an Incinerator

Emissions from incinerators may be a source of dioxins and furans, particularly for older plant with poor emission abatement systems. One study[20] focused on the municipal waste-to-energy facility in Columbus, Ohio, USA, which operated from June 1983 to December 1994. During that period it was estimated to have released nearly 1000 g of dioxin toxic equivalents (TEQ) per year. TEQ is a toxicity weighted sum of dioxin congeners. This compares to a total release of dioxins from all sources within the USA, at the time of operation of the incinerator, of 9300 g TEQ per year.

Because of the magnitude of the emissions from this single source, a study was undertaken to evaluate the impacts on air and soil near the incinerator. Since the incinerator had ceased operation at the time of the study, ambient air monitoring was no longer an option and it was decided that monitoring of soil was the best means to evaluate long-term impacts. However, the results of some previous monitoring of stack gas emissions, ambient air and incinerator ash, undertaken whilst the incinerator was operational, were assessed as part of the study.

The design of the soil monitoring programme involved selecting sampling locations on the site of the incinerator; at distances of up to 8 km away (within the city of Columbus); and at a background rural location 45 km away. Sampling locations were randomly selected within four major quadrants around the incinerator (northeast, southwest, *etc.*). The following conditions were sought during site selection: (a) level, undisturbed soils, (b) away from trees, (c) not adjacent to roads (d) not near pressure treated wood and (e) not known for soil to have known or suspected high dioxin concentrations for any other reason.

TEQ (15747 pg g^{-1} total dioxin) for the on-site samples to 1.4 pg g^{-1} TEQ (235 pg g^{-1} total dioxin) for the rural background samples. The urban background for soil samples more than about 1 km from the site was 4–10 pg g^{-1} TEQ. It was postulated that the soil concentrations of total dioxin (sum of the dioxin and furan homologue C14–C18 group concentrations) would decrease exponentially reaching an urban background level and it was possible to fit such a model curve to the data (Figure 6).

Despite the high emissions of dioxins and furans from this incinerator, the overall air and soil concentrations in the Columbus urban area did not appear to be elevated in comparison to urban air and soil from other studies around the World. For example, concentrations in urban soil of 26 pg g^{-1} have been reported for England and 10–30 pg g^{-1} for urban areas in Germany.

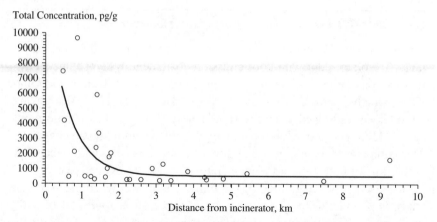

Total Concentration, pg/g

Figure 6 *Relationship of soil concentrations of dioxins with distance from the Columbus municipal waste-to-energy facility*
(Reproduced from Ref. 20, with permission from Elsevier)

5.2.2.5 Biological Monitoring. Biological monitoring may often take the form of examining the number and types of species or taxonomic families at a particular location and comparing these statistics to that which would be expected from an unpolluted reference location. When assessing the biological quality of rivers, it is common to use macro-invertebrate communities as the basis of monitoring. Macro-invertebrates include insects such as mayflies and caddis flies, together with snails, shrimps, worms and many other organisms. They are widely used for assessing river water quality because they are found in virtually all fresh waters, they tend to stay in one location and they respond to pollutants in the water as well as to physical damage to their habitat. They can be affected by pollutants that occur infrequently or in very low concentrations and which may be missed by chemical sampling.

Macro-invertebrates are sampled in rivers by actively disturbing the bed sediment for 3 min and collecting the organisms in a 900 μm mesh pond net. Care is taken to sample all in-stream habitats within a sampling location in proportion to their occurrence. Also, a 1 min visual inspection may be made to remove any organisms living on the water surface or attached to rocks, logs or vegetation. Samples will be returned to the laboratory for a detailed identification and classification.

In the UK, the occurrence of about 80 different taxonomic groups of freshwater macro-invertebrates at a sampling location are compared to what would be expected based on the river invertebrate prediction and classification system (RIVPACS) (www.ceh.ac.uk). In the USA, a multimetric approach has been the most commonly used technique in monitoring water quality. This involves the measurement of an array of metrics or indices that each provide particular information on the state of the site, and, when integrated to a single value, also provide an overall indication of the site's ecological condition. The use of biomarkers is becoming more common to measure the effect pollutants have on organisms (*e.g.* change in certain biochemicals).

Case Study 7: Multibiomarker Approach to Environmental Assessment

The use of ecologically relevant biomarkers has been promoted as a means of more directly measuring the impact of environmental pollution. In a study[21] of the effects of pollution at sites within Southampton Water, UK, a suite of biomarkers was used to measure molecular damage, developmental abnormality and physiological impairment to organisms living in the estuary. The aim of the study was to determine the viability of combining chemical analyses of sediments with biomarkers to characterize the relationship between anthropogenic pollution and the response of biota. Southampton Water was chosen as it is subject to a high level of shipping activity and receives discharges from a refinery and another industry.

Invertebrate species, including the common shore crab (*Carcinus maenas*) and edible cockle (*Cerastoderma edule*), were chosen for the study as they are abundant in most aquatic environments and representatives of different trophic levels can be found. Organisms and sediments were sampled from Cracknore Hard at the head of the estuary down to Fawley, Calshot and then Hillhead at the mouth of the estuary.

Biomarker measurements were made for cellular effects (cell viability, lysosomal integrity), physiological status (heart rate, haemolymph protein), immunotoxicity (phagocytosis), genotoxity (micronucleus formation), endocrine disruption (intersex), metal exposure (metallothionein induction), pesticide exposure (inhibition of esterase activity) and PAH exposure (PAH metabolites). Sediments were analysed for metals (*e.g.* Fe, Zn, Cu, Cd) and PAH.

Evaluation of the results of the biomarker measurements revealed that overall a detrimental effect to biota was apparent at sites towards the head of the estuary, whereas sites towards the mouth of the estuary appeared unaffected. This is consistent with the relative distribution of metals and PAH in sediments. Multivariate analysis was performed on the results to enable patterns of variation and the inter-relationships in the whole data set to be synthesized and summarized simultaneously. This showed a clear difference in the combined response of biomarkers for the common shore crab, for the Cracknore Hard and Calshot sites compared to the Fawley and Hillhead sites.

5.3   SAMPLING METHODS

Monitoring of pollutants will often involve the collection of samples for subsequent analysis at a laboratory. It may also involve the use of instruments, in particular for monitoring air and more recently, water. Instrumentation monitoring is discussed in Section 5.5.1.

5.3.1   Air Sampling Methods

Sampling systems for airborne pollutants usually consist of four component parts, the intake component, the collection or sensing component, the flow measuring component and the air moving device. All these must be constructed of materials that are chemically and physically inert to the sampled air (other than the collector or sensor itself).

5.3.1.1   Intake Design.   The nature of the intake is determined by the type and objective of the sampling technique, and may vary from a vertical opening for the passive collection of dustfall in a deposit gauge to a thin-walled probe used for source sampling of aerosols. Common problems that may require consideration are the non-reproducible collection of the sample portion from the air mass due to poor inlet design, adhesion of aerosols to tube walls, loss or change of analyte by chemical reaction with inlet materials, adsorption of gaseous components on inlet materials, and condensation of volatile components within the transfer lines.

The sampling of aerosols presents particular difficulties in inlet design. A basic requirement is that the velocity of the sample entering the system intake should be the same as the velocity of the gas being sampled. This is necessary because, if the streamlines of the sampled gas are disturbed by the intake probe, particles travelling in the gas flow and possessing inertia directed along the stream lines will continue into the probe while the "carrier gas" will be diverted away from (if the probe intake velocity is too low) or into (if the intake velocity is too high) the inlet (Figure 7). Thus, either a greater number of particles per unit volume of gas than that exists in the actual gas flow, or a fewer number, will be collected. Only when the intake velocity at the face of the probe is equal to the approach velocity of the gas stream will the streamline pattern remain unaltered and the correct number of particles per unit volume of gas enter the probe. This is known as isokinetic sampling.

Sampling of ambient air masses is seldom made isokinetically as sophisticated equipment is required to maintain the inlet facing into the wind and to adjust the sampling velocity to match changes in wind speed. This is feasible, but what is difficult is then to interpret the analysis of the collected

(a)

(b)

(c)

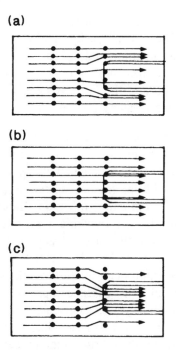

Figure 7 *Schematic diagram of (a) under-sampling of suspended particles, (b) isokinetic sampling and (c) oversampling of suspended particles*

sample as the flow rate is temporally variable. However, isokinetic sampling is usually practicable, and indeed necessary, when sampling flue gases.

5.3.1.2 Sample Collection. The methods most commonly used for the collection of atmospheric particulate samples are

- filtration;
- impingement: wet or dry impingers, cascade impactors;
- sedimentation: by gravity in stagnant air, thermal precipitators; and
- centrifugal force, cyclones;

and for gaseous samples are

- adsorption;
- absorption;
- condensation; and
- grab sampling.

Filtration. This is by far the most common technique. The type of filter medium chosen will depend upon a number of factors. These include the collection efficiency for a given particle size,[22] pressure drop and flow characteristics of the filter type,[23] background concentrations of trace constituents within the filter medium and the chemical and physical suitability of the filter with regard to the sampling environment. Where concentrations of pollutants are low then it will be necessary to use high volume air samplers with a sampling rate greater than $1 \, m^3 \, min^{-1}$.

Impingement. Impingers consist of a small jet through which an air stream is forced, so increasing the velocity and momentum of suspended particles, followed by an obstructing surface on which the particles will tend to collect. Wet impingers operate with the jet and collection surface under liquid and require high flow rates for optimum collection efficiency.

Cascade impactors use the aerodynamic impaction properties of particles to separate the sample into different size fractions by use of sequential jets and collection surfaces. Increasing jet velocity and/or decreasing gaps between the jet and collection plate fractionates the sample. Figure 8 shows the principle of a commonly used cascade impactor. This consists of up to 15 stages backed by a membrane filter, each stage containing accurately drilled holes that align over a solid portion of the adjacent plates. The holes in each successive stage are smaller than those in the preceding plate, and since air is drawn through the instrument at a constant flow rate, the effective velocity at each stage increases. The largest particles are impacted on the first stage and the smallest are collected on the back-up filter. The range of particle diameters collected on each stage may be determined by laboratory calibration or by theoretical calculations. However, impaction sampling at normal flow rates (about $0.01–0.04 \, m^3 \, min^{-1}$ or $0.6–1.0 \, m^3 \, min^{-1}$ for Hi-Vol cascade impactors) is only efficient for particles with aerodynamic diameters $>0.3 \, \mu m$. Also the collection efficiency of each stage will vary according to particle

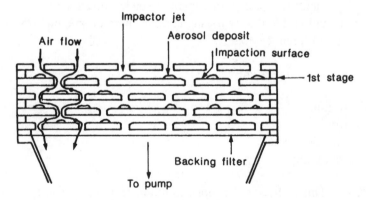

Figure 8 *Schematic representation of a cascade impactor*

type, some being very "sticky", others liable to bounce off. Other problems of cascade impactor sampling include wall losses and the aggregation of particles and the mechanical breaking of agglomerates, which result in inaccurate size distribution measurements.

Sedimentation. The collection of particulate material by allowing it to deposit into a collection vessel is the simplest of all air pollution measurement techniques. However, the presence of the bowl or cylinder in the path of the falling particles will change their flow pattern and it is not clear whether the collected material is truly representative of actual conditions. The methods are not as widely used now as previously. The British standard deposit gauge consists of a collection bowl connected to a bottle and supported by a galvanized steel stand. During wet weather, dust is washed down from the bowl, but during dry weather, high winds may blow dust out of or into the bowl, so producing erroneous dust loadings. At the end of the sampling period (usually 1 month) a measured volume of water is used to wash any dust in the bowl into the collection bottle, and the pH, total particulate mass and water volume determined. Other non-standard methods may be used, including frisbee-type collectors with better aerodynamic characteristics than the British standard gauge.

Adsorption. The adsorption of gases is a surface phenomenon. Gas molecules become bound by intermolecular attraction to the surface of a collection phase and so become concentrated. Under equilibrium conditions at constant temperature, the volume of gas adsorbed on the collection phase is proportional to a positive power of the partial pressure of the gas, and is also dependent upon the relative surface area of the adsorbent. Materials commonly used as adsorbents include activated carbon, silica gel, alumina and various porous polymers.

When selecting a suitable adsorbent, the relative affinity for polar or non-polar compounds must be considered. For example, activated carbon is non-polar and therefore will adsorb non-polar organic gases, but exclude polar compounds such as water vapour. The wide range of gas chromatographic supports available vary in their degree of polarity and so allow selection of the appropriate type.

The adsorbent used must not react chemically with the collected sample unless chemisorption is used intentionally. Also the analytes must not react with other constituents of the sampled air, either during collection or storage. It has been found, for example, that some hydrocarbons may decompose on polymers by reaction with the atmospheric ozone. This may be prevented by the use of a selective prefilter which removes the oxidant from the air stream but allows the analytes to pass through.

It is important to determine the retention volume of the adsorbent (*i.e.* the volume of air which may be passed without breakthrough of the analyte) with respect to the species being collected. This should be high enough to allow sufficient of the analytes to be collected for analysis. The desorption properties of the material are also important to ensure a quantitative recovery of the sample, preferably with regeneration of the adsorbent for subsequent use. Activated carbon is a very efficient adsorber, so making quantitative desorption difficult. Steam stripping may result in hydrolysis reactions with the analytes and vacuum distillation and solvent extraction are not without problems. Compounds on support bonded porous polymers may conveniently be thermally desorbed by flushing with an inert carrier gas. In the case of reactive hydrocarbons on polymer, a two-stage thermal desorption system utilizing an intermediate cryogenic trap cooled with liquid N_2 followed by flash heating allows quantitative recovery as well as direct injection of the sample in a very small volume of gas onto the gas-chromatograph column.

Since adsorption is temperature dependent, collection efficiency and an increase in retention volume may be achieved by cooling the adsorbent. However, problems with blockages by ice may then occur. With the increase in sophistication of detection systems in recent years, particularly by the interfacing of chromatographic separation techniques with mass selective detectors, the use of adsorbents as preconcentrators is also increasing. However, care must always be exercised to avoid non-quantitative collection, break-through effects due to exceeding the retention volume of the system, decomposition and non-quantitative recovery of the sample.

Absorption. Gases may be collected by being dissolved in a liquid collection phase or by chemical reaction with the absorbent. The simple Dreschel bottle may be used or may be modified by the inclusion of a fritted diffuser to create small bubbles and so enhance the collection efficiency.

An example of a simple absorption technique that allows an estimate of NO_2 concentration to be made with relatively little capital outlay is the use of triethanolamine diffusion tubes.[24] A small acrylic tube is fitted with a fixed cap at one end. A fine wire mesh coated in triethanolamine is placed in the closed end of the tube and absorbs NO_2 as it diffuses from the open end. The NO_2 is determined spectrophotometrically at the end of the sampling period. These passive samplers are very cheap to construct and analyse and have been used as an effective primary survey technique before embarking upon a more expensive monitoring exercise based upon the standard chemiluminescent technique.

Condensation. By cooling an air stream to temperatures below the boiling point of the substance of interest it is possible to condense gases from the air and so concentrate them. However, a limitation of the method is that water vapour present in the air will also freeze and so progressively block the trap. This may be overcome by using a first trap of large volume designed to collect water and a second trap at a sufficiently low temperature to collect the analytes. Care needs to be taken with coolants of temperatures $-183°C$ or lower (*e.g.* liquid N_2) as they may condense atmospheric oxygen and so may result in a combustion hazard.

Grab Sampling. Rather than utilizing a concentration technique in the field, samples may be collected in an impermeable container and returned to the laboratory for analysis. Grab samples of this type have been collected in FEP-Teflon bags or specially treated stainless steel cans for hydrocarbon determination by GC, for example. In the bag technique, the deflated container is housed within a rigid box which is slowly evacuated. Air is thus drawn into the flexible bag that may be sealed when inflated. Samples can then be drawn at a later stage from the bag by hypodermic gas-tight syringe. Pumps constructed of inert material may be used to fill rigid cans to a high pressure, allowing a large volume of air to be sampled.

Case Study 8: Hi-Vol Method for Suspended Particulates

This method is the current US EPA reference method for total suspended particulates (TSP) and is used in the US National Air Sampling Network,[25] although it should be noted that the majority of sampling for suspended particulate material now focuses on the smaller particles known as PM_{10} and $PM_{2.5}$ (see below) which are believed to carry the highest risks to health. A high flow rate blower draws the air sample into a covered housing and through a 20×25 cm rectangular glass fibre filter at $1.1–1.7\,m^3\,min^{-1}$. The mass of particles collected on the filter is determined gravimetrically and extraction techniques may then be used to remove the material for chemical analysis. However, when glass fibre filters are used, reaction with acidic components may result in an artifact formation, for example, sulfate from gaseous SO_2. Although 24 h sampling periods are commonly used, timer devices are available to switch on and off the blower at predetermined intervals. However, passive sampling by the settlement of particulates onto the filter during periods when the pump is not operating may cause a positive error in the determination. Recent modifications include size-selective inlets, which will exclude particles of greater than a given size. For example, inlets which excludes particles of aerodynamic diameter $>10\,\mu m$ (the PM_{10} instrument) or $>2.5\,\mu m$ ($PM_{2.5}$) are now being used with an increasing frequency as it become more and more apparent that human health effects are due to smaller, inhalable particles, rather than all particles present in ambient air. However, the cut-off efficiency is usually dependent upon wind speed and may not be sufficiently selective.

5.3.1.3 Flow Measurement and Air Moving Devices.   In order to measure the concentration of an airborne constituent, it is necessary to know the volume of air sampled. This may be achieved by measuring the rate of flow with a rate meter or by directly measuring the volume of air passed with a dry or wet test gas meter, a cycloid gas meter or by the use of a mass flow controller. All these devices require calibration and regular checking for leaks. Calibration should be undertaken following purchase of a sampler, after motor change or repair, following adjustment or repair of the rate meter and at least every 6 months. As air volume is dependent upon both temperature and pressure, the measurement of these two parameters at the meter inlet is essential so that volumes may be expressed at standard temperature and pressure.

Many different types of pump are available for air sampling, both mains and battery operated, but the precise type chosen will depend upon the required flow rate, availability of power and whether continuous or intermittent flow is required.

5.3.2   Water Sampling Methods

For many applications no special water sampling system is required as an appropriate sample container immersed in the water may be adequate. The main requirement is that a portion of the material under investigation small enough in volume to be transported and handled conveniently but still accurately representing the bulk material should be collected. Typically a 0.5–2 dm^3 volume is sufficient. When samples are required from depth, two types of collection vessel may be used.[13] The first consists of a cylinder with hinged lids at both ends. The container is lowered into the water with both lids opened and at the desired depth, a messenger weight is sent down the wire that closes them. This type is not suitable for trace metal work as contaminated surface waters may result in contamination of the vessel and the messenger may scour metallic particles from the wire. The second type consists of a sealed container filled with air, which is lowered to the required depth. A messenger is again sent down to open, and another to close, the lid. Alternatively, pressure sensors may activate the lid.

Automatic sequential samplers are available which will collect a given volume of water into an array of bottles. They have been used, for example, in collecting storm water runoff from roads and the sampling sequence may be triggered when the flow in a flume reaches a certain height.

Adsorption or filtration media may also be used to concentrate the species of interest *in situ*. Using a completely self-contained sealed unit of inert material housing a peristaltic pump and power supply with only the adsorption tube inlet and outlet open to the water, contamination of the sample can be completely avoided. Very large volumes of water may be processed. Another on-going development is the use of passive permeation devices that may be immersed in water for long periods of time, so giving time-averaged concentrations.

When considering sampling methods for use on inland waters or in coastal waters and estuaries, the sophisticated techniques developed for use at sea may be found to be impracticable due to their need for heavy lifting gear on the sampling vessel. Monitoring work must often be undertaken on such waters using small boats without such equipment. One ingenious method of collecting water samples at different depths using very limited resources on a small boat is to lower a weighted plastic tube to the desired depth and to use a small peristaltic pump to draw water up and into a collection bottle. In this way, completely uncontaminated samples may easily be obtained from depths of 30 m or more. Whether the particulate fraction of material present in the water can be quantitatively collected in this manner is not clear.

Whichever method of collecting samples is used, care must be taken to ensure that neither the sample storage containers nor any collecting vessels used contaminate or alter the sample. This may occur by

(a) leaching of contaminants from the surface of imperfectly cleaned containers,
(b) leaching of organic substances from plastics or silica and sodium or other metals from glass,
(c) adsorption of trace metals onto glass surfaces or organics onto plastic surfaces. In the case of metals this may be avoided by prior acidification of the container, but this may in turn exacerbate problem (a),
(d) reaction of the sample with the container material, *e.g.* fluoride may react with glass, and
(e) change in equilibrium between pollutants in particulate and solution phases.

If a solvent extraction technique is used to concentrate the analytes prior to analysis, care must be taken to ensure that the reagents and containers used are themselves sufficiently clean. Some commonly used materials and techniques have been shown to cause severely elevated metal levels in water.

Different determinants require different methods of preservation in order to prevent significant changes between the time of sampling and of analysis. Generally acidification to pH 2 and refrigeration to 4°C will be adequate although complete stability of every constituent can never be guaranteed and, at best, chemical, physical and biological processes affecting the sample can only be slowed down. Samples may be filtered directly after collection in the field to separate the particulate and solution phases. The solution phase may then be acidified to prevent adsorption of the pollutant to the container wall.

5.3.3   Soil and Sediment Sampling Methods

Soils and sediments are typically very inhomogeneous media and large lateral and vertical variations in texture, bulk composition, water content and pollutant content may be expected. For this reason, large numbers of samples may be required to characterize a relatively small area. Although surface scrapings may be taken, it is often necessary to obtain cores so that vertical profiles of the determinants may be obtained or cumulative deposition estimated. Plastic or chromium plated steel tubing of ~2.5 cm internal diameter is often suitable, and if the samples are sealed into the tubes and air excluded, they may be satisfactorily stored at low temperatures until required. Otherwise, they may be extruded in the field and stored in plastic bags. Various core-sampling devices are available for obtaining cores of bottom sediments from lakes *etc.* (*e.g.* the Jenkin corer).

Grab samples of soils are easily obtained manually and stored in precleaned plastic bags. Sometimes composite samples formed by the bulking together of a number of individual samples may be sufficient, but generally analyses of individual samples are to be preferred. In the case of sediments, grab samplers are available for operation at considerable depths, examples being the Ponar, Orange-peel and Peterson grabs. Alternatively, a dredge may be used to obtain a composite sample along a strip of the sediment surface.

Wet soils and sediments which are to be analysed while still wet should not be collected or stored in bags, but in rigid containers. The vessel should be filled as completely as possible leaving no airspace at the top and a bung inserted so as to displace excess water without admitting air.

Some determinants in soils and sediments are liable to change during storage and require the use of preservation techniques. For example, nitrate in soil can be extracted into potassium chloride solution and preserved with toluene. Usually, however, air-dried soils and sediments may be disaggregated, sub-sampled by coning and quartering (see Figure 9) and stored in

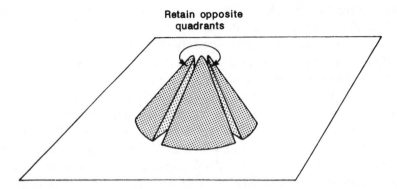

**Retain opposite
quadrants**

Figure 9 *Schematic diagram of the sub-sampling of dried soil or sediment using the technique of coning and quartering*

suitable containers, but as always sample contamination must be avoided at each stage. When sub-sampling by coning and quartering, it is usual to form the soil or sediment into a cone, then flatten until an equal shallow depth of material is obtained. The sample is divided into four equal portions using a fine edged rod and two diagonally opposite portions are removed for further coning and quartering. The other two portions are discarded.

There are important effects associated with grain size, which should be considered in the analysis of soils or sediments. First, many pollutants are associated with particle surfaces and therefore occur in highest concentrations in the smaller grain sized material. Second, sub-sampling from a bulk sample may be very difficult due to size segregation effects and it may be necessary to grind the sample in ball mill to a very fine powder to ensure homogeneity prior to division of the sample.

5.4 MODELLING OF ENVIRONMENTAL DISPERSION

A characteristic feature of environmental monitoring studies is that substances may be found over very large ranges of concentrations, and therefore the analytical techniques employed must be extremely flexible. Some typical concentrations of substances in polluted environmental media are given in Table 1. Not only will large differences of concentrations be found from area to area but even small temporal and lateral changes can result in large changes in pollutant concentrations.

In order to appreciate the variability of pollutant levels, and hence appreciate the complexity of designing an adequate monitoring programme, it is necessary to have some understanding of environmental dispersal, mixing and sink processes and of the time scales on which these processes act. Rather than discussing in detail the physical, chemical and biological processes responsible for changes in pollutant concentrations after discharge to the environment, which have been extensively presented elsewhere, the salient features of some of the techniques by which these changes can be anticipated will be shown. A summary of some of the computer codes that are currently available for modelling environmental dispersion is provided in Table 2.

Table 1 *Typical concentrations of substances in a polluted environmental media*

Pollutant	Medium	Typical ranges
Cadmium	Air	0.1–10 ng m^{-3} (daily average)
Lead	Seawater	2–200 pmol dm^{-3}
Lead	Soil	5–5000 mg kg^{-1}
Lead	Air	1–100 ng m^{-3} (daily average)
Sulfur dioxide	Air	0.5–200 ppb (10^{-9} v v^{-1}) (hourly average)
Sulfate	Air	0.1–20 μg m^{-3} (daily average)
Benzo(*a*)pyrene	Freshwater	0.1–10 ng dm^{-3}
Carbon monoxide	Air	0.1–50 ppm (10^{-6} v v^{-1}) (minute average)

Table 2 *Examples of environmental dispersion modelling tools*

Environmental media	Code	Overview	Availability
Air	ADMS	ADMS is a PC-based model for atmospheric dispersion of passive, buoyant or slightly dense releases from single or multiple sources	Cambridge Environmental Research Consultants (www.cerc.co.uk)
	AERMOD	Similar to ADMS	US EPA (www.epa.gov)
Freshwater	WASP6	WASP6 is an enhanced version of the water quality analysis simulation program (WASP). This version runs more quickly than previous versions of WASP, and allows for graphical presentation of results. This version includes kinetic algorithms for (1) eutrophication/conventional pollutants, (2) organic chemicals/metals, (3) mercury and (4) temperature, fecal coliform and conservative pollutants	US EPA (www.epa.gov)
	QUAL2K	QUAL2K is a stream water quality model designed primarily to simulate conventional constituents (*e.g.* nutrients, algae, dissolved oxygen) under steady-state conditions, both with respect to flow and input waste loads	US EPA (www.epa.gov)
Multi	MMSOILS	The MMSOILS model is a methodology for estimating the human exposure and health risk associated with releases of contamination from hazardous waste sites	US EPA (www.epa.gov)
	MULTIMED	The multimedia exposure assessment model (MULTIMED) simulates the movement of contaminants leaching from a waste disposal facility or contaminated soils	US EPA (www.epa.gov)

5.4.1 Atmospheric Dispersal

Material discharged into the atmosphere is carried along by the wind and mixed into the surrounding air by turbulent diffusion. In the vertical plane, the dispersion continues until the turbulent boundary layer is uniformly filled whilst in the horizontal plane, dispersion is theoretically unlimited and usually proceeds more rapidly than in the vertical plane. The degree of turbulence, and hence of mixing, is dependent upon the amount of incoming solar radiation, wind speed and cloud cover, but surface roughness is also important in producing turbulence, especially in the case of large buildings or topographic features.

Factors that need to be taken into account when modelling pollutant concentrations following a release to atmosphere are

- wind direction and wind speed;
- solar radiation;
- release height of plume and plume buoyancy;
- topographic features (*e.g.* hills, valleys);
- surface obstructions (*e.g.* buildings);
- loss of pollutants by deposition from the plume at the ground surface; and
- chemical reactions of pollutants in the plume.

Probably, the most simple model of plume dispersion[26,27] is that described by a Gaussian distribution of pollutant concentrations. At short distances downwind, with steady winds, neutral or stable weather conditions, uniform terrain and no local obstructions the Gaussian plume model may be expected to give reasonable estimates (within a factor of three) of short-term air concentrations. The Gaussian plume model will also give good estimates of long-term air concentrations (*e.g.* annual average). However, in non-ideal conditions these soon become order-of-magnitude estimates only and it may be necessary to use new generation air modelling codes (*e.g.* AERMOD, ADMS) to obtain more reliable predictions of concentrations. The way in which the new-generation models treat the vertical profile of turbulence is a significant improvement upon Gaussian models.[28]

All attempts to model atmospheric diffusion and mixing processes are liable to be, at best, only good estimates of the real situation and care must be taken to

(a) understand the physical, chemical and mathematical limitations of the model, and
(b) avoid treating the output from models as providing definite answers.

Although accurate measurements may be made of wind speed it is in the determination of the turbulence characteristics of the atmosphere that uncertainties arise, which in turn lead to uncertainties in the model.

5.4.2   Aquatic Mixing

The physical transfer and transport of pollutants in the aquatic environment is determined by the same two processes that determine the mixing of pollutants in the atmosphere. These are:

(a) advection, caused by the large-scale movement of water, and
(b) mixing or diffusion, due to small-scale random movements that give rise to a local exchange of the pollutant without causing any net transport of water.

The combined effect of advection and diffusion is known as dispersion. These two processes occur over a very wide range of scales, both of spatial extent and frequency, which necessitates the use of averaging procedures when defining their role in pollutant transfer. As in the lower atmosphere, there is usually a constraint on the vertical dispersion component, induced either by water depth, or in deeper waters by thermal or density stratification. On a large scale in the oceans, there will be a vertical circulation driven by density differences but on a small scale, vertical motion due to turbulence or eddy diffusion will tend to be suppressed by stratification.[29] Similar restrictions on vertical mixing occur in rivers due to limited depth, but here horizontal mixing in the crosschannel direction is also constrained.

A large number of models have been developed to describe the movement of pollutants in the aquatic environment, many being analogous to those used in air pollution studies. The fact that such models are necessary is due to limitations in our detailed knowledge of the velocity field, and this in turn may lead to uncertainties in the prediction of dispersion patterns.

The dispersion of pollutants in rivers has attracted a great deal of study, particularly in the context of effluent discharges and the ability of rivers to dilute them. Unlike the oceans which have traditionally, but unreasonably, been considered to have an infinite capacity for dilution, the deterioration in water quality due to pollutant discharges is often manifestly apparent in rivers.

Various models have been applied to the problem of modelling river dispersion with varying levels of complexity. An extremely simple model of water concentration (C/ $kg\,m^{-3}$), which is adequate for estimating long-term average concentrations at the point of discharge from a continuous release into a river is

$$C = \frac{q}{v(1 + k_{d}S)}$$

where q is the rate of input of a substance into the river ($kg\,s^{-1}$), v the flow rate of the river ($m^3\,s^{-1}$), k_d the partition coefficient between sediment and water for the substance ($m^3\,kg^{-1}$) and S the suspended solid load ($kg\,m^{-3}$).

A simple model to predict the change in water quality downstream from a discharge into a river would be to assume that an exponential decay in concentration occurs,

i.e.
$$C_x = C_o\, e^{-kt}$$

where C_x is the concentration at point x, C_o the concentration at the point of discharge, k the decay rate and t the time taken for flow from the point of discharge to point x.

An alternative approach to river modelling is to consider the river to be divided into a number of reaches, with each reach being considered as a continuously stirred tank reactor, and between each reach an appropriate time delay is placed. It is assumed that the major dispersive mechanism can be explained by the aggregated effect of all "dead-zone" phenomena in the river between the two sampling points. This "aggregated dead-zone" model is thus a combination of the continuously stirred tank reactor and a factor to account for the advection component of dispersion. Two approaches to modelling river dispersion represented by the Fickian diffusion analysis and the aggregated dead zone (ADZ) model have been compared with data obtained from a tracer experiment.[30] Figure 10 compares the monitored concentrations with those predicted by the two models, and in this example the ADZ model is able to explain the water concentrations arising from an instantaneous release better than the conventional diffusion model.

More recent models[31] take account of the various physiochemical processes a pollutant may be subjected to following a spill into a river (*e.g.* sorption, volatilization, hydrolysis, photolysis, oxidation and biological degradation). The effects of weirs on pollutant removal may also be modelled.

5.4.3 Variability in Soil and Sediment Pollutant Levels

Obviously, the same mechanisms of dispersion that operate in fluid or gaseous media do not occur in soils and sediments. Physical mixing may occur, such as during agricultural practices, dredging of estuaries or bioturbation by burrowing organisms, but usually only on a fairly limited scale. The level of contamination in a soil or sediment will depend upon the deposition rate of the pollutant and its subsequent rate of movement through the soil or sediment column. The rate of movement of a contaminant through these solid media is dictated by the degree of adsorption to or leaching from the particles and the flux rate of pore water, transferring the pollutant to deeper horizons. The physicochemical properties of the water, soil or sediment particles and the pollutant, all influence the rate of adsorption or leaching. For example, conditions that favour the adsorption of radiocaesium and lead in soils include low rainfall and high clay (particularly illite) content, whereas mercury and

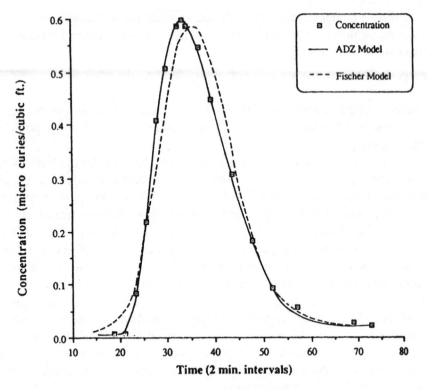

Figure 10 *Modelling of Copper Creek dispersion data: comparison of ADZ model fit with the results obtained by Fischer using the conventional Fickian diffusion model* (Reproduced with permission from Ref. 30)

copper tend to accumulate in soils with a high organic content. Contaminant concentrations are generally higher in soils or sediments with finer grain sizes, due to the increased total surface area available for adsorption. These effects are illustrated in Figure 11 for trace metal concentrations in sewage sludge in treated and control soil plots. Copper concentrations in sludge-treated plots show a strong correlation with organic carbon content, while zinc concentrations in both treated and control plots increase with the decreasing particle size, and increase with the increasing organic carbon content. All these factors lead to a great variability in pollutant concentrations, which simply emphasizes the need for carefully designed monitoring programmes.

5.5 DURATION AND EXTENT OF SURVEY

5.5.1 Duration of Survey and Frequency of Sampling

The duration of a pollution monitoring programme is entirely dependent upon the purpose of the study, and can vary from the time taken to collect a

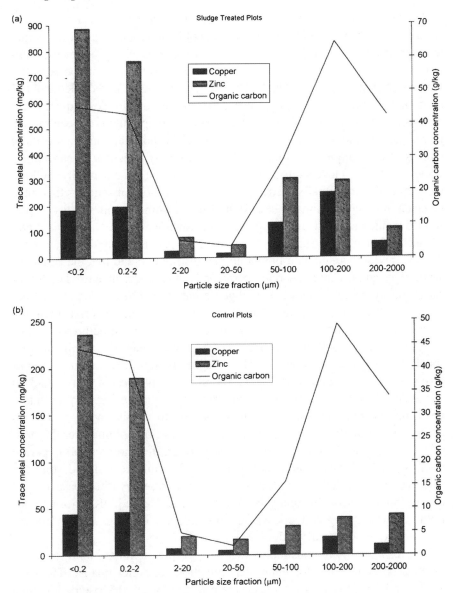

Figure 11 *The distribution of copper and zinc concentrations in (a) sludge-treated soil plots and (b) control soil plots*
(After J. Ducaroir and I. Lamy, Analyst, 1995, **120**, 741–745)

limited number of, for example, street dust samples in an urban area to tens of years for long-term surveillance projects such as the US EPA ambient air monitoring programme. The choice of the frequency of sampling, *i.e.* the duration of each sample period and the interval between successive measurements, is also dependent upon the objectives of the study. Pollutant concentrations in air and water fluctuate with varying degrees of rapidity and in order to characterize their behaviour, it is necessary to measure these changing levels; long-term mean data may be sufficient for some purposes but will not be adequate where information of short-term high-level episodes is required. Generally, if random sampling techniques are used, the number of samples required will increase as the standard geometric deviation of the pollutant concentrations increases, *i.e.* the greater the fluctuation of the pollutant level, the more numerous the samples that must be taken accurately to assess the variations. If the variations in levels during the period of interest are essentially random, independent and normally distributed, then the number of samples that must be taken in order to estimate the period of mean within certain limits and prescribed confidence limits may be calculated. However, in order to do this, a reasonable estimate of the standard deviation of the data is required: it is assumed that the random variations follow a normal distribution and that the results of successive samples are not serially correlated but are independent. These criteria are rarely met. As a more general guideline, it may be assumed to be necessary to have a sampling interval at least ten times shorter than the fluctuation cycle time. For example, the entire variance structure of a diurnally fluctuating pollutant concentration profile may be obtained from roughly 12 samples, each with a 2 h averaging period. If the annual trend of levels is required, then probably 12 monthly samples each year would be adequate. Thus, although a continuous and instantaneous record of pollutant level may be required for some purposes (*e.g.* to monitor very short-term changes in air quality due to the oscillating passage of a plume over the sampling station), it is not always necessary. Further, if a short sampling period is chosen it rapidly becomes necessary to efficiently record and store the large volumes of data generated, which will themselves often be averaged over a longer period for analysis. A case in point is the UK automatic air pollution-monitoring network where 1 min average readings are themselves averaged to give 1-hourly values.

Fast-response continuous monitors are now available for the more common gaseous pollutants and for many determinations of water quality. The term "fast-response" implies a response time measured as a 90% rise of less than about 2 min, but in most cases is of the order of seconds. Thus, very rapid temporal variations are measurable and with the advent of online microprocessor data handling and reduction systems, the large amounts of data produced are more easily handled.

A proper calibration of air sampling equipment is essential. Fast-response continuous monitors do not generally have a predictable response and cannot be calibrated solely in terms of chemical stoichiometry and hence need calibration with a standard atmosphere or solution. For automated instruments, *e.g.* those used to measure ozone, sulfur dioxide or oxides of nitrogen, a fortnightly or monthly visit by the operator with a span gas and zero air calibration may be appropriate. Other equipment may need more frequent checks and calibration. Some of the commonly employed fast-response methods of gaseous air pollutant analysis are shown in Table 3, and some of the methods used for water analysis are shown in Table 4.

Table 3 *Summary of commonly employed methods for measurement of air pollutants*

| | | Sample collection period | |
| | | Response time* (continuous techniques) | Minimum concentrations |
Pollutant	*Measurement technique*		
Total hydrocarbons	NDIR	5 s	1 ppm (as hexane)
	Flame ionization analyser	0.5 s	10 ppb (as methane)
Specific hydrocarbons	Gas chromatography[†]		1 ppb
Carbon monoxide	NDIR	5 s	0.5 ppm
	Catalytic methanation/FID[‡]		10 ppb
	Electrochemical cell	25 s	1 ppm
Sulfur dioxide	Fluorescent analyser	1 min	0.5 ppb
Oxides of nitrogen	Chemiluminescence reaction with ozone	30 s	0.5 ppb
Ozone	UV absorption	30 s	3 ppb
Peroxyacetyl nitrate	GC/electron capture detection[‡]		1 ppb
Particles	Tapered element oscillating microbalance	30 min–24 h	$1 \mu g\ m^{-3}$

*Time taken for a 90% response to an instantaneous concentration change.
[†]Grab samples of air collected in an inert container or on an adsorbent and concentrated prior to the analysis.
[‡]Instantaneous concentrations measured on a cyclic basis by flushing the contents of a sample loop into the instrument.

Table 4 *Summary of some methods of analysis of water*

Pollutant or determinant	*Measurement technique*	*Response time*
pH	Electrometry	10 s
Biochemical oxygen demand	Dilution/incubation	5 days
Chemical oxygen demand	Dichromate oxidation	2 h
Metals	Atomic absorption spectroscopy	–
Organometallics	Gas chromatography – atomic absorption spectroscopy	–
Nitrate	Colourimetry	5 min
Nitrate	UV spectrophotometry	1 min
Formaldehyde	Photometry	6 min
Phenols	Gas chromatography	30 min

One further consideration when deciding on sampling frequency is that if the measurements are being made in order to assess whether a given environmental quality standard is being satisfied, then the data resolution must be sufficient for that purpose. As an example, the US ambient air quality standard for sulfur dioxide specifies that 3- and 24-hourly values are required. The method used must therefore be capable of giving concentrations averaged over these time periods. One advantage of using a method with an apparently higher than necessary time resolution is that instrument failure of other problems can potentially be identified rapidly, minimising data loss. This is particularly advantageous in automated monitoring networks.

5.5.2   Methods of Reducing Sampling Frequency

Once the desired sampling frequency has been selected, it may be found to be impracticable with the resources available and some means of reducing the number of samples will then be required. This may be done by

(a) reducing the number of sampling locations,
(b) reducing the sampling frequency, or
(c) reducing the number of determinants.

It has already been shown above that there are several methods available for the rationalization of an existing monitoring network where the quality at one location is correlated sufficiently well with that at another location, and these methods of analysis may be applied to all types of monitoring networks. Similarly, statistical analysis of past data may show that one determinant is sufficiently well correlated with one or more others, such that it may be used as an indicator of quality.

When sampling water, soils, sediment and flora and fauna, the use of composite samples may be of value in reducing sample numbers. Composite samples are formed by mixing together individual samples to give an indication of the average quality over an area or during a sampling period. They may also be formed by continuous or intermittent collection of samples into one container over a given period, as for example, the collection of atmospheric particulate material on a filter. Individual samples collected at different locations may be mixed together in proportion to the volumes of the sampled bodies, as in the case of a non-homogeneous water body, and so give a better indication of an average quality.

In the case of atmospheric pollutants it is often desirable to estimate the likely daily maximum as well as the daily mean concentration, and several methods have been proposed which allow this to be done on the basis of a few discrete samples of short duration. For example, the same statistical concentration frequency distribution of SO_2 levels as is provided by continuously recorded data can be obtained from a limited number of randomly collected short-term samples.

5.5.3 Number of Sampling Sites

The choice of the number of sampling sites to be used in a particular survey is very dependent, as are so many other design parameters, on the objectives of the study. In the simplest case of source-orientated monitoring of atmospheric emissions from a single stack or site then one or two sampling locations might be considered sufficient, provided they are located at a suitable distance from the source and that monitoring continues for an adequate period of time. This may be thought to be too few but operational constraints may prevent this number being increased. In the case of the UK national survey of air pollution it was previously thought necessary to obtain daily data on sulfur dioxide and "smoke" concentrations from about 1200 sites, although this number has now been greatly reduced.

When too few sampling sites are used for too short a time period in a source-orientated monitoring programme, it is possible that atmospheric (or aqueous) emissions may pass between them without being detected at all. The probability of a fixed number of sample stations detecting a release is a function of the quantity released, the number of samplers, the distance of the samplers from the source, the plume dimensions, the height of release, and its duration. This type of analysis has been applied to the environmental monitoring of atmospheric releases.

Hand-held global positioning systems (GPS) are now often used during sampling programmes to accurately determine the longitude and latitude of a sampling point on the Earth's surface. This is particularly useful when it is necessary to return to the same point for further sampling. In some cases, where monitoring using instruments is being undertaken the instrument readings may be logged with GPS data to enable maps of the measurements to be produced.

5.6 PREREQUISITES FOR MONITORING

Before monitoring begins certain information, techniques and methodologies must be available in order for the survey to be successfully carried out. As has already been stressed, a prime requirement is that the objectives of the study should be defined, but along with this, the following need consideration:

- definition of a monitoring protocol;
- availability of meteorological or hydrological data;
- availability of emission data;
- likely pollutant concentrations to be expected;

- availability of suitable monitoring equipment;
- availability of sensitive and specific analytical techniques; and
- definition of suitable environmental quality standards.

5.6.1 Monitoring Protocol

Monitoring is a complex task and carefully planned and documented procedures are necessary to ensure that reliable and comparable results are obtained. The documented procedure by which this is achieved is known as a monitoring protocol. The main components of a monitoring protocol are as follows:

(a) Reference measurement methods for sampling and analysis – standard methods may not always be available, so some reference methods may actually be non-standard.

(b) A methodology that sets out
- (i) location of sampling (including background);
- (ii) the frequency of sampling;
- (iii) the duration of sampling and the number of samples to be taken;
- (iv) the pollutants to be measured;
- (v) variability of the pollutants under investigation (perhaps supported by modelling);
- (vi) the accuracy required (to some extent this is dictated by the pollutant and the sampling method used);
- (vii) the analytical resolution (*e.g.* detection limit);
- (viii) the general principles of sampling (*e.g.* position, isokinetic);
- (ix) plant operating conditions (for source monitoring);
- (x) meteorological or hydrological conditions;
- (xi) health and safety considerations of sampling staff (and plant operators for source monitoring);
- (xii) access, power supply and security requirements; and
- (xiii) procedures to be adopted when meaningful measurements by standard procedures are not possible.

(c) Quality control procedures which define requirements for the
- (i) calibration of measurement devices;
- (ii) maintenance of instruments;
- (iii) sample storage and transport to ensure that the sample is identifiable throughout the sampling, sample preparation and analysis and to ensure the sample integrity is maintained; and
- (iv) data handling and reporting.

(d) Quality assurance programme to ensure that
- (i) measurements are made in accordance with a standard methodology;
- (ii) the correct quality control procedures are in place;

(iii) the quality control procedures are being adhered to;

(iv) sample identification and routing procedure is well documented; and

(v) the correct reporting procedures are used.

It is most important to have a documented monitoring protocol when source monitoring is being carried out for comparison against emission limits or environmental quality standards. Regulatory authorities may wish to agree to such a protocol prior to the monitoring being carried out. Since each monitoring programme is different both in the way in which it is operated and in the pollutants that are monitored, separate protocols will be needed for each programme. Measurement teams, operators and contractors should be instructed in the use of the protocols.

5.6.2 Meteorological and Hydrological Data

When carrying out air pollution measurements, it is desirable, and often essential, to have an access to meteorological data. Care must be taken to ensure that the data used are meaningful and representative of the area of study, wind data in particular being very susceptible to local interference. Light, robust anemometers (*e.g.* hot-wire) and wind vanes are now generally available and when mounted on a dismountable 10 m mast may be used in the field, thus obviating the need to rely on data obtained from another, possibly a less representative, source. Less easily obtained parameters that may be required are the lapse rate and the height of any atmospheric temperature inversions. These are rather difficult to measure, requiring accurate measurements of temperature at increasing heights or acoustic radar observations, but such data are usually obtainable from large meteorological stations.

Measurements of low-level atmospheric turbulence are made using a bivane and anemometer on a 10 m mast. This instrument measures the horizontal and vertical fluctuations of the wind; siting of the mast is obviously critical and a data logging system is required to cope with the large amounts of data generated.

Meteorological data may be required for several reasons. First, when fast response measurements are unavailable it may be desirable to construct time-weighted pollution roses that show how pollutant concentrations vary with surface wind directions. For this the hourly wind direction at 10 m height is required, the record divided into sixteen 22.5° sectors and the duration in each sector is tabulated. The rose may then be calculated using

$$(\text{TWMC})_n = \frac{\sum_{i=1}^{m} (t_{i,n} C_i)}{\sum_{i=1}^{m} t_{i,n}}$$

where $(TWMC)_n$ is the time-weighted mean concentration of the pollutant for the *n*th sector, $t_{i,n}$ the number of hours for which the wind was in sector *n* during the *i*th sampling period, c_i the concentration of the pollutant for the *i*th period, *m* the number of sampling periods and *n* takes values from 1 to 16 (sector $n = 1$ being 0°–22.5°, sector 2 being 22.5°–45°, *etc.*).

Alternatively, the trajectory of a parcel of air over synoptic-scale distances may be required for source apportionment or dispersion studies. For this the surface pressure field over a very large area is needed and the geostrophic wind vector estimated from the isobar spacing and alignment.[32]

The analogous hydrographic data (*e.g.* tidal range) may be required for dispersion studies in the sea, and flow data for rivers and lakes may be needed although it may not be so readily available as meteorological data.

5.6.3   Source Inventory

One, often very cost-effective method of identifying the likely occurrence of pollutants prior to actual monitoring is by collecting and collating detailed information on the pollution emissions in a given area or to a particular river. Such an emission or source inventory should contain as much information as possible on the types of source as well as the composition of emissions and the rates of discharge of individual pollutants. Supplementary information describing the raw materials, processes and control techniques used should also be collected. Detailed discussion on the test procedures to adopt in compiling an emission inventory is available,[33] and the applications of the completed inventory were identified by these authors to be

(a) guiding emission – reduction efforts;
(b) helping to locate monitoring stations and alerting networks;
(c) indicating the seasonal and geographic distribution of the pollution burden;
(d) assisting in the development of implementation strategies;
(e) pointing out the priority of air or water quality problems;
(f) aiding regional planning and zoning;
(g) air and water quality diffusion modelling;
(h) predicting future air and water quality trends;
(i) determining cost-benefit ratios for air and water pollution control; and
(j) community education and information programmes.

Although a very useful tool, the emission inventory is no substitute for actual measurements of pollutant levels and should be considered as a complementary technique to monitoring, not as an alternative.

5.6.4 Suitability of Analytical Techniques

As was shown in Table 1, pollutants may be found in environmental media over very wide concentration ranges. Often there are several procedures available by which a pollutant may be analysed, but they may have widely differing sensitivity and specificity.

Some of the considerations that will affect the choice of an analytical method may be summarized

- sensitivity (depends on detection limit and pollutant levels);
- specificity (to allow unequivocal determinations);
- response – time;
- response – range (particularly linearity of continuous monitors);
- ease of operation;
- ease of calibration;
- cost and reliability; and
- precision and accuracy.

Some of the more commonly used instrumental methods are shown in Table 5 and described comprehensively elsewhere.[34]

Figure 12 highlights the difference between precision and accuracy for an analytical technique. Precision may be defined as the reproducibility of analyses, whereas accuracy is a true measure of the determinant present. In order

Table 5 *Instrumental analytical methods*

Method	Sample*	Specificity	Sensitivity†
Gravimetric	SLG	Good	$>1\,\mu g$
Titrimetric	SLG	Good	$>10^{-7}$ M in solution
Visible spectroscopy	SL	Fair	>0.005 ppm in solution
Ultraviolet spectroscopy	SLG	Fair	>0.005 ppm in solution
Flame emission spectroscopy	SL	Good	>0.001 ppm in solution
Atomic absorption spectroscopy	SL	Excellent	>0.001 ppm in solution
Gas chromatography	LG	Excellent	>1 ppm for volatile organic compounds
Liquid chromatography	SL	Good	>0.001 ppm
Polarography	L	Good	>0.1 ppm
Anodic stripping voltammetry	L	Good	>0.001 ppm
Spectrofluorimetry	SL	Good	>0.001 ppm
Emission spectroscopy	SL	Excellent	>0.1 ppm
X-ray fluorescence	SL	Good	>10 ppm in solid samples
Neutron activation	SL	Excellent	>0.001 ppm
Mass spectrometry	SLG	Good	$>0.1\,\mu g$

*S = solid, L = liquid, and G = gas.
†Approximate only, depending upon the particular element being analysed.

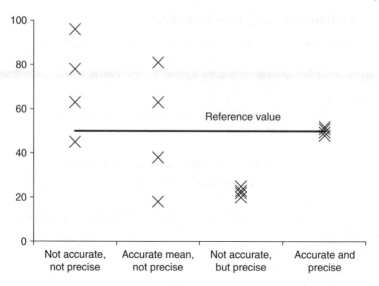

Figure 12 *Schematic illustration of precision and accuracy*

to assess the precision of an analytical method it is necessary to analyse separate representative sub-samples of the sample under investigation. These sub-samples should be included at intervals during the analysis. In this way, the drift in the precision of a technique may also be detected. It is not possible to know how accurate a series of determinations have been, even if they are precise. However, a best estimate of the accuracy of an analytical technique may be found from the regular analysis of national and international standard reference materials. These have undergone inter-laboratory analyses, using different techniques and calibration standards, for a variety of species.

5.6.5 Environmental Quality Standards

Of the six possible reasons for carrying out a monitoring programme outlined above, four rely upon the prior formulation of a standard of environmental quality. Only in the case of source apportionment and pollutant interaction and pathway studies or when monitoring is carried out with the intention of obtaining a historical record of environmental quality is this not a prior requirement. There is little point in monitoring in order to pinpoint pollutant health effects or to assess the need for legislative controls on emissions, for example, unless a certain pollutant level has been defined as being undesirable or likely to cause damage. Environmental quality standards have been devised and adopted for many atmospheric and water pollutants, some of which are shown in Tables 6, 7 and 8.

Table 6 *US ambient air quality standards*

Pollutant	Measurement period	Standard
PM_{10}	24 h average	$150 \mu g\ m^{-3}$
	Annual arithmetic mean	$50 \mu g\ m^{-3}$
$PM_{2.5}$	24 h average	$65 \mu g\ m^{-3}$
	Annual arithmetic mean	$15 \mu g\ m^{-3}$
Sulfur dioxide	Annual arithmetic mean	0.03 ppm
	24 h average	0.14 ppm
	3 h average (secondary standard)	0.5 ppm
Carbon monoxide	8 h average	$10\ mg\ m^{-3}$
	1 h average	$40\ mg\ m^{-3}$
Ozone	8 h average	0.08 ppm
		0.12 ppm
Nitrogen dioxide	Annual arithmetic mean	$100 \mu g\ m^{-3}$
Lead	Quarterly average	$1.5 \mu g\ m^{-3}$

Note: The number of exceedances allowable vary from pollutant to pollutant; check www.epa.gov for details.

Table 7 *European commission air quality standards (selected examples)*

Pollutant	Measurement period	Concentration ($\mu g\ m^{-3}$)
Particles (PM_{10})	24 hour mean*‡	50
	Annual mean	40
Sulfur dioxide	8 hour mean‡	12
Ozone	8 h running mean‖	120
	1 h mean§	180
	1 h mean‖	240
Nitrogen dioxide	3 h mean¶	400
	Annual mean†	30
	Annual mean‡	40
	1 h mean‡,**	200

*Maximum of 35 exceedances a year.
†Vegetation protection thresholds.
‡Health protection thresholds.
§Population information threshold, public to be informed in the event of exceedance.
¶Population warning threshold, public to be informed in the event of exceedance.
‖Maximum of 20 exceedances per year.
**Maximum of 18 exceedances per year.
Note: European Union limit values for a wider range of pollutants appear in Chapter 2.

Having adopted a standard for environmental quality there may be great difficulty in ensuring compliance. In the case of, for example, lead in drinking water, various reduction strategies are possible, culminating in the wholesale removal of lead pipes (although this may not entirely solve the problem when lead-based solder is used with copper pipe) and in the case of primary air pollutants similar "simple" remedies are possible. The difficulties arise in the case of secondary pollutants (*i.e.* those formed within the atmosphere itself) or for pollutants with both primary and secondary origins.

Table 8 *EC water quality standards for inland waters (e.g. rivers, lakes)*

Pollutant	Standard (annual mean)
Endrin, isodrin	$0.01 \mu g\ L^{-1}$
Aldrin, dieldrin	$0.005 \mu g\ L^{-1}$
Total 'drins	$0.03 \mu g\ L^{-1}$
Cadmium and its compounds	$5 \mu g\ L^{-1}$ (total)
Carbon tetrachloride	$12 \mu g\ L^{-1}$
Chloroform	$12 \mu g\ L^{-1}$
DDT (all isomers)	$0.025 \mu g\ L^{-1}$
p,p'-DDT	$0.01 \mu g\ L^{-1}$
Hexachlorobenzene	$0.03 \mu g\ L^{-1}$
Hexachlorobutadiene	$0.1 \mu g\ L^{-1}$
Hexachlorocyclohexane (all isomers)	$0.1 \mu g\ L^{-1}$
Mercury and its compounds	$1 \mu g\ L^{-1}$ (total)
Pentachlorophenol and its compounds	$2 \mu g\ L^{-1}$

In the case of atmospheric suspended particles both primary and secondary sources are important. Primary emissions have in recent years been greatly reduced by the use of efficient control techniques on industrial sources and substantial change in domestic fuel usage away from coal towards cleaner fuels. However, secondary particles, produced in the atmosphere by formation of the ammonium salts of strong acids from industrial emissions of SO_2, NO_x and HCl, together comprise a substantial proportion of the atmospheric aerosol. Thus, reduction of primary emissions of particles does not necessarily ensure a reduction in atmospheric concentrations or compliance with a standard. One secondary air pollutant that is likely to prove difficult to adequately control in the next decade or so is ozone. This is formed in the lower atmosphere by reactions involving several primary pollutants, and our present understanding of the chemistry of the atmosphere is probably insufficient to accurately predict the effect of control strategies.

5.7   REMOTE SENSING OF POLLUTANTS

Many highly sophisticated techniques are now available for the remote sensing of atmospheric and water pollutants. However, their use is almost exclusively restricted to specialized monitoring exercises due to the very considerable capital cost of the instrumentation. Probably, the cheapest and most widely used methods are those of aerial photography, including infra-red sensing and optical correlation spectrometry. Uses of aerial photography include the monitoring of liquid effluent dispersion using dye tracers and conventional colour film. Infra-red photography has been used for monitoring the condition of crops and forests. Airborne heat-sensing infra-red line-scanning equipment has been routinely used to monitor thermal plumes in waters receiving industrial effluents and have also been used for the detection and mapping of oil spills at sea using thermal infra-red data from satellites.

The most common use of the correlation spectrometer in air pollution analysis is for the determination of sulfur dioxide and nitrogen dioxide concentrations in plumes from tall stacks, and provides a good technique for studying the transport and dispersion of a plume. The instrumentation may be ground based in a mobile laboratory or airborne, in which case the plume is viewed from above.

The use of tunable lasers allows long-path absorption measurements of a range of gaseous pollutants such as SO, NO_2, SO_2, CO and O_3 and minor reactive species such as HO. Reliable measurements of this latter species are of great importance due to its dominant role in the chemistry of the troposphere. In long-path laser absorption methods, a detector is used to monitor absorption of specific wavelengths in the light path. In lidar techniques, however, the backscattered radiation from a laser is monitored. By using a pulsed system, the time taken for receipt of backscatter can be related to the distance of travel, allowing spatial resolution of pollutant concentration data within the light path. By monitoring backscatter intensity at two close wavelengths, one strongly absorbed by the species of interest and one unabsorbed, the species' concentration may be inferred as well as its spatial distribution. Care is required to avoid spectral interferences but this method has been successfully used for measurement of sulfur dioxide up to a range of *ca.* 2 km. Further significant developments of laser methods using the Raman backscatter, which is highly characteristic of the scattering molecule, are likely. A relatively frequently used technique is that of differential optical absorption spectroscopy (DOAS), which relates the quantity of light absorbed to the number of gas molecules in the light path. This technology is used in instruments that can measure a number of different pollutants along a single light beam, which may be up to 1 km long. It has been used, for example, to observe gaseous emissions from a motorway.[35]

Case Study 9: Detection of Asian Dust Aerosol Using Satellite Data[36]

Wind-blown dust plays an important role in modifying climate and providing nutrients to the surface in some areas of the globe. The transport and dispersion of wind-generated dust from the deserts of China to the Pacific Ocean in the vicinity of Japan during the period 2000–2002 has been investigated using data from the meteorological satellites NOAA/AVHRR and GMS-5/VISSR. The brightness temperature difference between the 11 and 12 μm bands was used and found to be very effective for identifying dust regions in dry air which can be distinguished from water vapour and water droplets. Validation of the method was achieved by using measurements of suspended particulate material in the air, and visual observations from the ground recorded by camera.

5.8   ANALYSIS AND PRESENTATION OF DATA

A monitoring programme, particularly one incorporating automatic or fast-response systems, can generate a very large amount of information in a short time. In order for the data to be assimilated and understood, some means of organizing the information and summarizing its most essential characteristics is required so that changes, trends or patterns in behaviour over time and space may be apparent. For this, the methods of descriptive statistics are required. Another group of statistical methods, those of inferential statistics, are used when information of the relationships and processes operating between measurements is required. Details of these methods are available from standard texts and from books devoted to the environmental sciences. [37–39]

Many statistical tests depend upon having data that are normally distributed, but often environmental analytical data do not satisfy this criterion. In a normal distribution the arithmetic mean and median are the same, but in log – normally distributed data, the *geometric* mean and median are the same. This is the situation that applies to many environmental data sets and comes about from a few results having high values whilst the majority of results are closely grouped together. If such data are treated as belonging to a normal distribution, too much weight will be applied to the outlying values and the wrong deductions may be made. In this case, some method of transforming the data is required before statistical analysis is carried out. For example, it might be appropriate to use the logarithms of the data, or the square or cube roots. Similarly, it is often better to quote the 95 percentile range of values, excluding the extreme 5%, again in order to avoid giving prominence to a few outliers.

Monitoring data may be incorporated in geographical information systems (GIS), which are specialized database management systems for handling geographical data – *i.e.* data whose key characteristic is "place". These manage data on land use, industry, roads, habitations, and so on. Data includes both their positions and their various related properties (for example, information about a road may include its traffic capacity, traffic flows and traffic speeds). In addition to simply storing such data, GIS can manipulate and analyse the data – for example, calculating contours of NO_x emissions around road networks – and include visualization tools.

Many graphical methods of representing data are available to illustrate the changes or emphasize differences or similarities in the results (*e.g.* Figure 13). These may take the form of bar charts/histograms (Figure 13a), line graphs (Figure 13b), *XY* plots (Figures 13c and d), pie charts (Figure 13e), stacked bar charts (Figure 13f), geographical presentations (Figures 13g and h) or indeed combinations of these. Figure 13c illustrates the inclusion of error bars (or standard deviations) in a graphical presentation and also a regression line. Data exhibiting an exponential decay will plot as a straight line or a log-normal graph (Figure 13d) and regression analysis may be performed with the

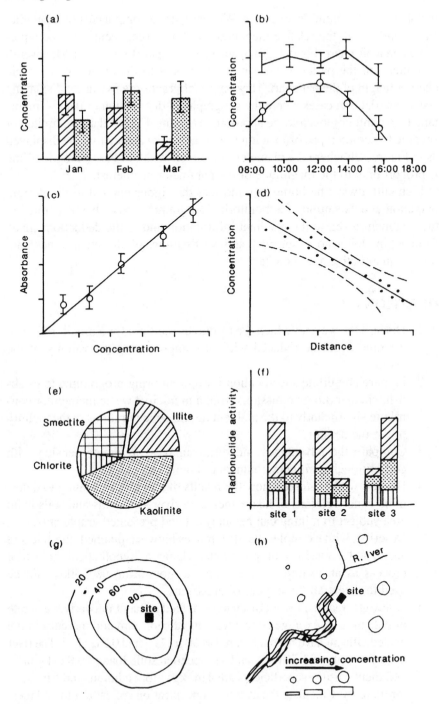

Figure 13 *Examples of different approaches to presenting data*

inclusion of 95% confidence limits. Where separate components of a parameter are measured, the relative magnitude of those components may be represented on a pie chart (Figure 13e) or stacked bar chart (Figure 13f). The overall magnitude of the parameter can be proportional to the diameter of the pie chart or height of the bar chart. These types of graphs are useful for displaying speciation data. In cases where the geographical distribution of data is important, then contour plots may be appropriate (Figure 13g). If data are limited or specific to certain types of location (*e.g.* road or river) they may be displayed as in Figure 13h where the diameter of a circle, width of bar or length of line is proportional to the magnitude of the parameter of interest.

It should always be borne in mind that the rigour applied to the design, operation and execution of a monitoring programme must also be applied to the treatment given to the resultant information and to the deductions made from it, in doing so ensuring that investigations of the environment are robust, quantitative and reliable.

QUESTIONS

1. Discuss the ways in which the resources required for a monitoring programme could be reduced without compromising the validity of the results.
2. Prepare an outline specification for a monitoring programme to establish whether dioxin emissions from a municipal waste incinerator contribute significantly to the ambient dioxin concentrations in the vicinity of the incinerator.
3. Compare the advantages and disadvantages of remote sensing with conventional *in-situ* environmental monitoring.
4. Discuss with examples, how the results of a monitoring study to determine concentrations of heavy metals in river sediments and soils in an area subject to mining can be analysed and presented graphically.
5. Describe, with examples, the differences between planned, fugitive and accidental emissions of pollutants. Using a hypothetical industrial process as an example, outline how a monitoring network could be used to detect all three types of emissions.
6. A sewage treatment work discharges treated effluent into a river at a rate of $0.23 \, \text{m}^3 \, \text{s}^{-1}$. Average concentrations of cadmium and mercury in the treated effluent have been measured as $60 \, \mu\text{g} \, \text{L}^{-1}$ and $0.5 \, \mu\text{g} \, \text{L}^{-1}$. The river has a flow rate of $3.7 \, \text{m}^3 \, \text{s}^{-1}$ and a suspended solid load of $0.025 \, \text{kg} \, \text{m}^{-3}$. Sediment partition coefficients are $4 \, \text{m}^3 \, \text{kg}^{-1}$ for cadmium and $5 \, \text{m}^3 \, \text{kg}^{-1}$ for mercury. Calculate the average concentrations of cadmium and mercury in the river at the point of discharge of treated effluent and determine whether environmental quality standards are likely to be exceeded.

REFERENCES

1. Department of the Environment, *The Monitoring of the Environment in the United Kingdom*, Report by the Central Unit on Environmental Pollution, HMSO, London, 1974, 66pp.
2. P.K. Hopke, *Receptor Modelling in Environmental Chemistry*, Wiley, New York, 1985.
3. Environment Agency, *Technical Guidance Document (Monitoring) M1: Sampling Requirements for Monitoring Stack Emissions to Air from Industrial Installations*, Preston UK, 2002.
4. Environment Agency, *Technical Guidance Note (Monitoring) M2: Monitoring of Stack Emissions to Air*, Preston UK, 2004.
5. A.L. Wilson, Design of sampling programmes, in *Examination of Water for Pollution Control*, Vol 1, M.J. Suess (ed), Pergamon, London, 1982.
6. S.E. Lindberg, G. Southworth, E.M. Prestbo, D. Wallschleger, M.A. Bogle and J. Price, Gaseous methyl- and inorganic mercury in landfill gas from landfills in Florida, Minnesota, Delaware and California, *Atmos. Environ.*, 2005, **39**, 249–258.
7. World Meteorological Organization, *WMO Operations Manual for Sampling and Analysis Techniques for Chemical Constituents in Air and Precipitation*, WMO, No 299, Geneva, 1971, 22pp.
8. D.B. Turner, Workbook of atmospheric dispersion estimates, *Nat. Air Pollut. Cont. Adm.*, 1969, 84pp.
9. G. Davison and C.N. Hewitt, *Air Pollution in the United Kingdom*, Royal Society of Chemistry, London, 1997.
10. F. Marcinowski, R.M. Lucas and W.M. Yeager, National and regional distributions of airborne radon concentrations in US homes, *Health Phys.*, 1994, **66**, 699–706.
11. B.M.R. Green, J.C.H. Miles, E.J. Bradley and D.M. Rees, Radon Atlas of England and Wales, NRPB–W26, HMSO, London, 2002.
12. National Radiological Protection Board, *Exposure to Radon in UK Dwellings* NRPB-R272, G.M. Kendall, J.C.H. Miles, K.D. Cliff, B.M.R. Green, C.R. Muirhead, D.W. Dixon, P.R. Lomas and S.M. Goodridge, HMSO, London, 1994.
13. APHA, *Standard Methods for the Examination of Water and Wastewater*, 15th edn American Public Health Association, New York, 1980.
14. M.J. Suess, *Examination of Water for Pollution Control – A Reference Handbook*, 3 Vols, Pergamon Press, Oxford, 1982.
15. L.G. Hutton, *Field Testing of Water in Developing Countries*, Water Research Centre, Wallingford, 1983, 120pp.

16. L.-S. Wen, P. Santschi, G. Gill and C. Paternostro, Estuarine trace metal distributions in Galveston Bay: importance of colloidal forms in the speciation of the dissolved phase, *Mar. Chem.*, 1999, **63**, 185–212.

17. B.R. Johnston and J.M. Jacoby, Cyanobacterial toxicity and migration in a mesotrophic lake in Washington, USA, *Hydrobiologia*, 2003, **495**, 79–91.

18. D.J.A. Davies, I. Thornton, J.M. Watt, E.B. Culbard, P.G. Harvey, H.T. Delves, J.C. Sherlock, G.A. Smart, J.F.A. Thomas and M.J. Quinn, *Sci. Total Environ.*, 1990, **90**, 13–29.

19. J.S. Webb, I. Thornton, M. Thompson, R.J. Howarth and P.L. Lowenstein, *The Wolfson Geochemical Atlas of England and Wales*, Oxford University Press, Oxford, 1978.

20. M. Lorber, P. Pinsky, P. Gehring, C. Braverman, D. Winters and W. Sovocool, Relationships between dioxins in soil, air, ash, and emissions from a municipal solid waste incinerator emitting large amounts of dioxins, *Chemosphere*, 1998, **37**, 2173–2197.

21. T.S. Galloway, R.J. Brown, M.A. Browne, A. Dissanayke, D. Lowe, M.B. Jones and M.H. Depledge, *Environ. Sci. Technol.*, 2004, **38**, 1723–1731.

22. M. Katz, *Measurement of Air Pollutants, Guide to the Selection of Methods*, WHO, Geneva, 1969.

23. B.Y.H. Liu, D.Y.H. Pui and K.L. Rubow, in *Aerosol in the Mining and Industrial Work Environments*, Vol 3, V.A. Marple and B.Y.H. Liu (eds), Ann Arbour Science, Ann Arbor, MI, 1983, 898–1038.

24. M.R. Heal and J.N. Cape, *Atmos. Environ.*, 1997, **31**, 1911–1923.

25. EPA, *Ambient Air Monitoring Reference and Equivalent Methods: Designation of One New Reference Method for PM10, Four New Equivalent Methods for PM2.5, and One New Reference Method for NO₂ US Federal Register, April 2, 2002* (Volume 67, Number 63).

26. F. Pasquill, *Atmospheric Diffusion*, Ellis Horwood, Chichester, 1974.

27. R.H. Clarke, *A Model for Short and Medium Range Dispersion of Radionuclides Related to the Atmosphere*, National Radiological Protection Board Report NRPB-R91, Harwell, 1979.

28. Defra, Scottish Executive, National Assembly for Wales and Department of the Environment in Northern Ireland, Part IV of the Environment Act 1995, Local Air Quality Management Technical Guidance, LAQM TG(03), 2003.

29. G. Kullenberg, *Physical Processes in Pollutant Transfer and Transport in the Sea*, Vol 1, G. Kullenberg (ed), CRC Press, Boca Raton, FL, 1982.

30. P.C. Young, Quantitative systems methods in the evaluation of environmental pollution problems, in *Pollution: Causes, Effects and*

Control, 2nd edn, R.M. Harrison (ed), Royal Society of Chemistry, London, 1990.

31. A. Watson, L.S. Fryer and J.W. Clark, *Aqueous Pollution Modelling – Approaches used by PRAIRIE™*, AEA Technology, AEAT-0843, 1996.
32. R.I. Sykes and L. Hatton, *Atmos. Environ.*, 1976, **10**, 925–934.
33. A.T. Rossano and T.A. Rolander, The preparation of an air pollution source inventory, in *Manual on Urban Air Quality Management*, World Health Organisation, Copenhagen, 1976.
34. C.N. Hewitt (ed) *Instrumental Analysis of Pollutants*, Elsevier, London, 1991.
35. C.v. Friedeburg, I. Pundt, K.-U. Mettendorf, T. Wagner and U. Platt, *Atmos. Environ.*, 2005, **39**, 977–985.
36. N. Iino, K. Kinoshita, A.C. Tupper and T. Yano, *Atmos. Environ.*, 2004, **38**, 6999–7008.
37. A. Pentecost, *Analysing Environmental Data*, Longman, London, 1999, 214pp.
38. W. Ott, *Environmental Statistics and Data Analysis*, Lewis Publishers, Boca Raton, FL, 1995, 313pp.
39. G.P. Patil and C.R. Rao (eds) *Multivariate Environmental Statistics*, North Holland, Amsterdam, 1993, 596pp.

CHAPTER 6

Ecological and Health Effects of Chemical Pollution

STEVE SMITH

Department of Biochemistry, King's College London, Franklin-Wilkins Building, 150 Stamford Street, London SE1 9NH, UK

6.1 INTRODUCTION

The chapter is concerned with the damaging effects or the hazards of chemical pollution to human health and ecological systems. An evaluation of these requires knowledge of how chemicals exert their effect, the concentrations which produce effects, and the likelihood of hazardous concentrations occurring in the environment. The fundamental aim in tackling or preventing chemical pollution hazards is to protect man and the myriad species in ecological systems from the adverse effects of chemicals. In practice, this means defining a level or concentration in the environment, which to the best of our knowledge can be regarded as "safe" or of minimal risk to health; *i.e.*, not associated with any known damaging effects. The whole process is put under the umbrella of environmental risk assessment (ERA) and it is designed for tackling ongoing chemical pollution problems as well as preventing new ones from arising. A broad distinction is that a chemical which is detectable in the environment but at a concentration that has no known effects is considered as a contaminant, whereas one that is present at levels associated with harmful effects is a pollutant. The distinction between the two can easily become blurred and advances in understanding of the toxicity of chemicals may mean that that a "no-effect" level requires revision.

The number of chemical substances produced by man and which enter the environment is very large indeed; equally there are countless numbers of species in all the different types of ecosystems that may come into contact with these chemicals. This leads to the questions of what is a hazardous or dangerous level and how do we assess the hazards of such a large number of chemicals to a vast number of species that inhabit diverse habitats? A good starting point is to accept a basic maxim of toxicology, which recognizes that all substances are dangerous or toxic above a certain level. It follows that organisms may be able to cope with a small amount of a very toxic substance, while a less toxic one or an otherwise essential one may reach high levels and adversely affect organisms. Not all species are equally affected by the same level of pollution, with some more sensitive than others.

The impact of chemicals is dependent on how much enters the environment and what happens to them in the environment. In the process of dispersion in air, water, or soil, pollutants may undergo transformations to more innocuous forms or to ones that are more hazardous. In the latter case the impact is due to secondary forms of pollution. Dispersion in the environment leads to dilution of pollution, although certain environmental processes can concentrate pollutants and where this occurs, the impact may be greater. On the other hand, they may become bound up or immobilized in some way and in this form they are less available to biota and may be considered less hazardous. The fate of pollutants in the environment is covered in some detail in the preceding chapters but an important point here is that the total measurable concentration of a chemical in the environment is not necessarily the same as that which is available for uptake by an organism. Characterizing the so-called bioavailable fraction of a chemical is an important part of pollution studies.

Chemical pollutants exert their effect on individuals, but recognition of an effect is a subjective decision of the observer and it may be studied at an individual, population, or community level. The rate of supply or the amount of pollutant reaching an organism, population, or community is referred to as the *exposure* and an *effect* is a biological change, usually damaging, caused by the exposure. A damaging effect is referred to as the toxicity of a substance and to exert a toxic effect, a chemical must gain access to the exterior membranes and/or the interior of tissues and cells of organisms. The amount that is taken into an organism is referred to as the *dose*; essentially it is a function of concentration and the period of exposure, although increasingly dose is associated with the concentration that interacts at the biomolecular level to produce the basic toxic effect. Measuring dose of all manner of different organisms in the ambient environment can be a difficult task, and so measuring exposure has become the common means of assessing how much is received by organisms. A fundamental goal of pollution studies is to establish the relationship between *exposure and effect*, so that a level at which no effect occurs can be identified and this can form the basis of setting objectives for controlling pollution or preventing pollution.

An exposure–effect relationship for a chemical in one species will not necessarily be the same for other species. In recent years, there has been a growing recognition that differences in sensitivity to chemical pollution can be explained by differences in the processes which govern uptake and elimination, metabolic processes involved in transformation of chemicals to less toxic forms and "storage" of chemicals in "inert" forms. For example, some species may be more efficient than others in the same niche in eliminating a chemical or in transforming it to a product, which is more easily excreted from the organism. In which case, such species would accumulate less of a chemical and as a result there would be less in the body to exert a toxic effect. These processes are described under the heading of the biodynamics of chemical pollutants.

An exposure may occur as a large pulse (acute exposure) or over extended periods (chronic exposures) or even as a series of pulses. Acute exposures typically lead to immediate effects and they are rarely reversible and are very often fatal. Chronic effects or damage follow a period of prolonged exposure, which results in biochemical or physiological disturbances. Outwardly they may be manifested as visible or clinical symptoms of damage, *e.g.* chlorotic regions on plant leaves, and the inability to maintain homeostatic balance, co-ordinate activities, or breed. These symptoms are often quite diffuse and not specific to a particular pollutant. Following cessation of the exposure, chronic damage is frequently reversible, although continued exposure may prove fatal. However, it may be argued that any severe disturbance will impair the efficiency of an individual and so shorten its life. With recent advancements in molecular biology techniques, more and more emphasis is directed at the fundamental mechanisms of toxicity and how these are expressed at the whole organism level. One paradigm that is emerging is that chemicals with similar structural characteristics and/or with similar classes of reactivity act toxicologically in the same way. A second paradigm is that the differences in sensitivity to chemicals can be explained by the differences in biodynamic processes both at a species level and in individuals making up a population; these concern rates of uptake and elimination, storage forms and capacity, and metabolism of chemicals. A third paradigm recognizes a link between biomolecular interactions, cellular and physiological responses, whole organism effects, and population and community changes. Figure 1 is still very relevant in this respect, even though it was originally produced in 1979.[1] It presents the levels of biological organization as a series of linked reservoirs with each one having a built-in capacity to withstand a certain input of pollution, if this capacity is exceeded, then damaging effects appear at that level of organization.

6.2 HISTORICAL PERSPECTIVE

Chemical pollution of the environment emerged as an important issue in the 1950s and 1960s, when the effects of organochlorine pesticides and methylmercury first came to light. DDT and related substances had devastating effects on predator bird populations across large swathes of the world where these

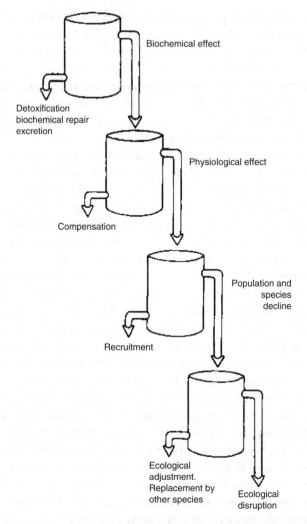

Biochemical effect

Detoxification
biochemical repair
excretion

Physiological effect

Compensation

Population and
species
decline

Recruitment

Ecological
adjustment.
Replacement by
other species

Ecological
disruption

Figure 1 *An idealized model of 'cascading' effect of pollution depicting successive levels of organization as a series of reservoirs. Effects do not spill over from one level to another unless inputs exceed the capacity of relieving systems shown on the left of each reservoir. Variation between species and individuals of the species can be thought of in terms of variation in capacity of these systems*
(Reproduced with permission from Ref. 1)

pesticides were used intensively after the Second World War. Methylmercury fatally affected many villagers and caused permanent damage to the nervous systems of many others in Minamata in Japan as a consequence of ingesting the substance from locally caught seafood that was later shown to have been contaminated by discharges from a nearby factory. These examples stimulated the development and rapid growth of environmental chemistry, environmental toxicology, and a new discipline of ecotoxicology. In these early stages, new

concepts emerged which are just as important now; the first and most important is that organochlorine pesticides and methylmercury have the capability
of being transferred through food chains and of becoming more concentrated
during the transfer. The second is that DDT and related compounds persist in
the environment for a long time, often for many years and even decades.
Persistence and accumulation in food chains are characteristic of a number of
other chemical pollutants and they are fundamental in explaining their adverse
effects in animals and humans. Their recognition as important factors influencing the fate of chemicals in the environment caused a major shift in thinking about the environment, as hitherto air, water, and soil were assumed to have
an infinite capacity to absorb and dilute man's waste products.

Looking back, other major chemical pollution problems have influenced
our understanding of how chemicals interact with the environment and affect
humans and wildlife generally. They include air pollutants such as sulfur
dioxide, ozone and particulates, acid aerosols, lead and other metals such as
cadmium, and oil spillages. Other organochlorine compounds such as polychlorinated biphenyls (PCBs) have caused adverse effects in wildlife and
more recently other types of organic compounds, which are stable and therefore persistent in the environment have emerged as pollution issues. They
include the brominated flame retardants. For the last decade or so the major
chemical pollution issue has been the recognition that natural and synthetic
steroidal hormones can be detected in the environment and that they are associated with abnormal sexual development in wildlife populations, and possibly
in humans as well. The issue is even more serious as many organic chemicals
such as DDT, PCBs, and phenolic substances released from the breakdown
of detergents have been shown to interfere with the endocrine systems of animals. The whole area is known as endocrine disruption and the scale of the
problem and its root cause remains a matter of debate. It is possible that the
reproductive and development effects in wildlife population associated with
organochlorine substances that have been catalogued since the 1950s are
actually examples of endocrine disruption.

As a result of each pollution event, lessons have been learnt and action taken
to reduce, control, or eliminate the problem. In a number of instances, significant reductions in the environmental levels of chemical pollution have occurred
and there are many reports of recovery of populations and whole communities. Despite these improvements, society is far from achieving the ultimate
goal of preventing chemical hazards in the first place.

6.3   DIVERSITY OF POLLUTANTS

Pollutants encompass a broad range of chemical and physical properties, which
strongly influence both their fates in the environment and their effects on

biological systems. Particularly during the past century vast quantities of many different chemical substances have been released into the environment; the majority of these substances are waste products generated by industry and society consuming manufactured goods. Synthetic fabrics and fibres, pharmaceuticals, fertilizers, pesticides, paints, and building materials, as well as chemicals for industrial processes, are just some of the products of the chemical industry that are integral to almost every aspect of modern living. Many such substances are natural constituents of the environment, others are synthetic chemicals. Inevitably wastes generated during the manufacturing process, the substances themselves, and perhaps their degradation products, are released into the environment. Gases (*e.g.* sulfur dioxide, carbon dioxide, and oxides of nitrogen) and particulates from combustion processes are discharged into the atmosphere. Liquids and solids containing inorganic and organic substances are discharged into water and hazardous chemical wastes are buried on land, incinerated, or dumped in sea. Fires, explosions, and tanker accidents can result in sudden unintentional but devastating pulses of pollutants into the environment, and pesticides are intentionally released at certain times of the growing season. It has to be said that there are now a number of controls regulating such discharges, but all chemicals produced by human activity enter the environment in some form or the other and there are over 100,000 in general use and a few thousand of these are produced in large volumes.

6.4 POLAR AND NON-POLAR SUBSTANCES

The molecules making up the cells of organisms can be described in terms of their polar or non-polar properties. Molecules with positive or negative charges interact with the corresponding opposite charges of water molecules and tend to dissolve in water, hence the term, hydrophilic, meaning they are water loving. For example, alcohols have polar properties that enable them to form hydrogen bonds and dissolve in water. Sugars, DNA, RNA, and the majority of proteins are hydrophilic. Non-polar molecules, on the other hand, carry little or no charge, form few or no hydrogen bonds and as a consequence are only sparingly soluble in water. They are referred to as hydrophobic (water hating) molecules and they show a strong association with other non-polar molecules, for example, they are soluble in non-polar organic solvents. Hydrocarbons are examples of hydrophobic substances, they consist of carbon and hydrogen covalently bonded together and the simplest is methane (CH_4) but many methyl (CH_3-) groups join together to form hydrocarbon chains. These form long hydrocarbon tails in fatty acids, *e.g.* stearic acid (C_{18}). Fatty acids are components of fats and lipids in cell membranes and in addition to a non-polar hydrocarbon region, they also have a hydrophilic region due to carboxyl ($-COOH$) groups. In cell membranes, the fatty acids are structurally part of the phospholipids that include glycerol and phosphate groups in their overall structure. These form the characteristic bilayer of cell membranes with the hydrophobic tails packing close together to exclude water and the hydrophilic head of phospholipids facing outwards to the external and internal environment of cells. Lipid molecules in cell membranes are described as being amphipathic (amphiphilic) in that they have a hydrophilic or polar end and a hydrophobic or non-polar end. Signalling compounds that enable cells in multicellular organisms to communicate with one another can be classed as being either hydrophilic or hydrophobic. Steroid hormones, thyroid hormones, retinoids, and vitamin D are examples of small hydrophobic signalling molecules.

Chemical contaminants and pollutants are also classified according to their polar or non-polar properties. Those carrying a charge (+ve or −ve) interact and dissolve in water, they include inorganic acids such as sulfuric and nitric acids, metals in ionic form, and certain herbicides, *e.g.* atrazine and phenols. Many of the vast number of organic chemicals produced by man are non-polar and they are therefore relatively insoluble in water. DDT and related substances, PCBs, and dioxins are all good examples of hydrophobic pollutants. The hydrocarbons generally, either as chains or ring structures and especially the aromatic ringed hydrocarbons such as benzene, toluene, naphthalene, and the polycyclic aromatic hydrocarbons (PAHs) are all non-polar compounds. Hence, crude oil released into the sea following a tanker accident does not mix with seawater, but spreads as a non-polar film over the polar seawater.

In relation to the interaction of chemical contaminants with the cells of organisms (and the environment generally), the above gives rise to an important generalization that is carried forward in this chapter. Hydrophilic substances associate with the aqueous environment of cells and bind to ligands, whereas hydrophobic substances partition to fats and lipids. Hence, non-polar chemicals are also referred to as lipophilic (fat loving) substances. Chemical pollutants with lipophilic or hydrophobic characteristics interact with the lipids of cell membranes.

The octanol-water partition coefficient (K_{ow}) is a physicochemical parameter, which indicates the lipophilicity of a substance. It measures the partitioning of a chemical between water and octanol and it is expressed as a ratio of the concentration of the chemical in octanol relative to that in water. Non-polar chemicals will partition to octanol in preference to water and so give a higher ratio. Typical values of the ratio for non-polar chemicals are in thousands and tens of thousands and so the K_{ow} is typically expressed on a logarithmic scale. Benzene, naphthalene, phenanthrene, pyrene, and benzo(*a*)pyrene have log K_{ow} values of 2.13, 3.35, 4.57, and 5.18, respectively. These four compounds progressively increase in molecular size by an addition of an aromatic ring; benzene is a single aromatic ring and pyrene is a compound of four fused aromatic rings. The progressive increase in K_{ow} is related to the increase in size of a molecule and its decrease in aqueous solubility. Chlorine substitution in organic compounds as in chlorobenzenes and PCBs results in larger K_{ow} values. This is a steric effect with additional chlorine substitution having a neutralizing effect and thereby increasing the degree of hydrophobicity. Monochlorobenzene has a log K_{ow} of 2.8 and the successive substitution of chlorine atoms around the ring increases the K_{ow} value such that hexachlorobenzene, the end member of the series has a log K_{ow} of 5.5. 2-Chlorobiphenyl has a log K_{ow} of 4.3, while that of decachlorobiphenyl PCB has a value of 8.26. The reader is referred to Schwarzenbach *et al.*[2] for a fuller explanation of K_{ow}.

6.5 TOXICITY: EXPOSURE–RESPONSE RELATIONSHIPS

6.5.1 Toxicity Tests

The hazards of chemicals in the environment are defined in terms of their toxicity. A typical toxicity test involves subjecting a specie of fish (trout, fathead minnow) or invertebrate (*Daphnia spp.*, water flea) to incremental concentrations of a chemical. The basic assumption is that a sub-population of a specie is a collection of individuals with different relative susceptibilities to a chemical; a small group is relatively sensitive to the chemical and shows the toxic effect at lower concentrations, the bulk of the group exhibit the toxicity in the middle range of concentrations and a small relatively resistant group responds to the higher range of chemical concentrations. That is, the effect of the chemical on the group conforms to a log-normal distribution and a modal concentration of the chemical can be defined that affects 50% of the population. The cumulative frequency curve (Figure 2) that has a typical sigmoidal shape, shows the increasing percentage response of the organisms, in terms of mortality, to increasing concentrations of chemical over a set period of time. Hence LC_{50} is the concentration of chemical, which is lethal to 50% of the population. The concentration of chemical in the test medium and the length of time of the test represent the exposure and so a toxicity test is an example of an exposure–response relationship. It can be of short duration (24 or 96 h) or less than the life cycle of a particular organism in which case it is said to be an acute toxicity test, or the response is measured over an extended period to give a chronic test.

Toxicity tests enable the relative toxicity of different chemicals to be compared and also the relative sensitivity of different species. As shown in Figure 2, two further pieces of information are estimated from the tests; the lowest observed effect concentration (LOEC), which can be pinpointed at the base of the cumulative curve, indicating the onset of toxicity. Below this point is the no-observable effect concentration (NOEC), which is the highest concentration of chemical that produces no significant increase in the toxic response relative to the control population.

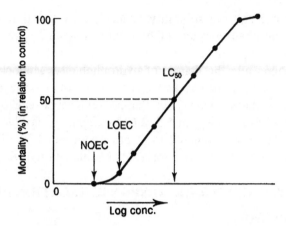

Figure 2 *Generalized sigmoidal relationship between chemical concentration and toxic response in terms of percentage mortality, demonstrating how the parameters of LC_{50}, LOEC, and NOEC are estimated*

Because chemicals discharged into the environment largely end up in aquatic systems, toxicity tests are commonly based on aquatic organisms. There is an extensive database on the acute toxicities of a variety of chemicals, although by no means on all potential contaminants. The organisms selected as test organisms are those that are easy to obtain and maintain in the laboratory but are nevertheless representative of a particular type of environment; freshwater, marine, sediments, and soils, and also of different levels of biological organization, *e.g.*, primary producers, invertebrates, and vertebrates. In the UK, toxicity tests based on the freshwater environment are commonly undertaken with the water flea (*Daphnia spp.*) and trout and in the US, the fathead minnow is the fish most frequently used.

Mortality is only one of the many possible end points of toxicity and it is the main end point of acute tests. Chronic tests typically examine the effects on the general wellbeing and survival of an organism, *e.g.* growth, reproduction, and feeding rate. The parameter used in these cases is the effective concentration or EC_{50}, the concentration that results in 50% reduction in growth or some physiological function. At the biomolecular level, the emphasis is on determining toxic mechanisms. Chronic toxicity tests are regarded as being more realistic to the ambient environment as exposures are lower and over extended periods, and the toxicity parameters that are derived from these give much lower values for NOEC.

6.5.2   Environmental Risk Assessment

ERA is a process for protecting man, his resource species, and wildlife generally from chemical pollution, and it provides a framework for assessing ongoing pollution problems as well as the impact of new chemicals. It is based on information from toxicity tests and in simple terms we say that concentrations of a chemical in the environment at or near LC_{50} or EC_{50} values are unacceptable whereas concentrations at or below the NOEC are non-hazardous and therefore acceptable. This gives rise to the hazard ratio in which the environmental exposure or concentration is divided by the NOEC with values >1 indicating environmental levels above the NOEC and a potential chemical hazard, whereas values <1 suggest that there is little or no hazard. In human risk assessment, toxicity data define an intake below which there is little or no hazard and exposure measurements and/or models provide environmental data against which to determine the risk or likelihood of a population being exposed to an intake that exceeds the tolerable value. For people living adjacent to land contaminated with chemicals, a source – pathway – receptor (human population) approach is used. Pathways through air, water, and food that link the transfer of chemicals from the source to the human population are defined and quantified. If the total intake of chemical exceeds or is likely to exceed the toxic threshold, then clearly some form of remedial action is required.

In the process of risk assessment there are many uncertainties associated with estimating NOEC and ambient exposure levels. Estimates of NOEC are dependent on the sensitivity of test species, the design of tests (acute *vs.* chronic), the end points measured, and even on the experimental conditions of the tests. Ambient concentrations of a chemical are subject to considerable variation in time and space, irrespective of whether measurements are in air, water, or soil. For a new chemical with no history of environmental contamination, predictions of environmental levels are made using models that take into account the physicochemical properties of the chemical, the proposed scale of use or production levels of the chemical and transport and degradation processes in the environment.

Reduction in the uncertainties of NOEC is achieved by gathering information on a whole range of species and toxic end points. An ideal approach is to determine the toxicity and hence the NOEC of a chemical to a host of species in, for example, freshwater systems. Plotting the sensitivity distribution in terms of NOEC, and selecting the NOEC giving protection to 95% of the species in the system provides a community-based NOEC for freshwater systems. However, this requires a considerable amount of time and effort devoted to the toxicity testing of just one chemical.

In practice, ERA is a tiered or sequential decision process, (summarized in Figure 3) in which tests of increasing complexity are applied, starting with acute tests on fish and invertebrates and initial measurements or predictions of environmental concentrations and ending with field experiments. Progress through the battery of tests depends on estimates of the hazard ratio. If the

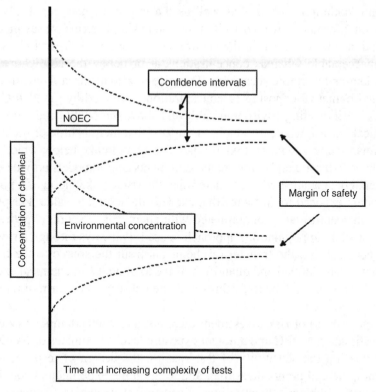

Figure 3 *Conceptual diagram of the relative uncertainties involved in estimations of envir-onmental concentration and NOEC. The diagram highlights that the increasing number and complexity of tests reduces the uncertainty in the estimates of the two parameters and that in ideal situations, the environmental concentration should be significantly below the NOEC*

ratio suggests a risk of hazardous levels in the environment, then more infor-mation is needed about the chemical to give a more precise estimate of its NOEC and likely environmental concentration. For new chemicals, the aim is to regulate their production and use patterns to ensure that environmental levels will be an acceptable margin below the hazard threshold. For existing chemical pollution problems, it is necessary to implement control measures to reduce environmental levels below those considered as hazardous. The ERA process provides a basis for setting environmental standards for chem-ical pollutants. The concept of critical loads for controlling acidification of soils and freshwater systems and critical concentrations for defining the impact of air pollution on vegetation are examples.

ERA is a far from perfect process and there are examples where it has not provided an adequate environmental protection. Tributyltin (TBT) is a case in point as initial screening tests suggested that it was not particularly toxic and not very soluble in water. As a result, it was allowed for use as an antifouling agent in paints used on the hulls of boats on the assumption that little of the compound would leach off boats into seawater and would not reach toxic levels. Reports of widespread damaging effects on oyster and dogwhelk populations that were traced to TBT led to a re-evaluation of the chemical. It was found that many gastropods and bivalves were very sensitive to TBT and showed toxic symptoms at levels found in the environment. As a result TBT was banned from use on small boats in the UK; the nature of the effects is described in more detail in a later section. Another practical problem of the ERA process as it is presently set up, is the sheer enormity of the task of screening all potentially toxic chemicals.

6.5.3 Direct Toxicity Assessment

The above described approach of ERA is very much orientated towards individual chemicals and it does not consider complex chemical mixtures typical of industrial effluents and the ambient environment where several sources may contribute to the general level of contamination. Direct toxicity assessment (DTA) examines the overall toxicity of an effluent and the waters into which an effluent is discharged (receiving waters). It is directed at ongoing inputs of chemicals into the environment and it recognizes that the receiving waters of, for example, industrial estuaries receive complex mixtures of effluents from a range of industries and that receiving waters are not pristine but contain chemicals from previous discharges. DTA examines the combined effects of chemicals and it also provides a means of linking source to exposure to ecological damage in the ambient environment.

Figure 4 is an example of a DTA for the waters around the coast of the UK. *Tisbe battagliai* was used as the test organism and acute toxicity was measured in hexane extracts of water samples from various locations. The large red circles show that waters from certain areas require little concentration to become toxic and certain samples were toxic without pre-extraction into hexane.[3]

A complementary approach for assessing pollution stress to mixtures of chemicals in the ambient environment is the measurement of scope for growth (SFG) in the common mussel, *Mytilus edulis*. It is one of the most sensitive measures of pollution-induced stress and it is based on estimates of the energy available for activity, growth, and reproduction from assimilation of food, after the energy expended in respiration and excretion have been taken into account. SFG decreases proportionately to chemical exposure and feeding rate is the component of SFG that is the most sensitive to pollutant stress.

One of the largest monitoring programmes to have been undertaken in which chemical contamination and biological effects were measured at the same time involved combined measurements of SFG and chemical contaminants in the

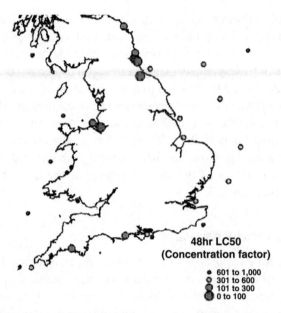

48hr LC50
(Concentration factor)

● 601 to 1,000
○ 301 to 600
● 101 to 300
● 0 to 100

Figure 4 *An example of "direct toxicity assessment", showing the acute toxicity of hexane*
extracts of coastal waters of the UK to T. battagliai; the largest circles represent sam-
ples that needed least concentration before becoming toxic and some samples were
acutely toxic in themselves
(Reproduced with permission from Ref. 3)

tissues of mussels at locations covering 1000 km in the North Sea.[4] SFG values
were higher at the northern locations than at the southern ones, which reflected
the clean water condition of the inflow of the North Atlantic into the North
Sea and the increased contamination of water associated with the urban and
industrial locations further south. Coastal regions of the Humber-Wash and
the Thames estuary recorded some of the lowest values for SFG. For several
of the contaminants detected in the mussel, experimentally derived tissue
concentration–response relationships were available and these were used as
a basis for estimating the additive contribution of the different contaminants
to the overall decline in SFG. It was found that toxic hydrocarbons (mainly
PAHs) accounted for most of the reduced SFG at the majority of sites, while
TBT made a significant contribution to the toxic effect at a number of locations.
Metal concentrations were generally below those associated with a reduction
in SFG.

6.5.4 Quantitative Structure-Activity Relationships

As part of the rationalizing process of testing the toxicity of the myriad
chemicals, it is possible to group chemicals according to their structures and

properties and relate these to their toxicity. Quantitative structure–activity relationships (QSAR) as they are known, have been established for groups or homologous series such as the chlorophenols and chlorobenzenes and the lipophilicity or K_{ow} of compounds in the series can be related to their toxicity. Linear relationships have been established between LC_{50} of members of each of the series and their K_{ow} values, showing that toxicity increases (lower LC_{50} values) as the lipophilicity of the compounds increase. QSAR are therefore useful for predicting the toxicity of chemicals with related properties. They can also give some insight into different modes of toxic action. The gradient of LC_{50} *vs.* K_{ow} for the chlorobenzenes tends towards unity and it is typical of a large number of organic hydrophobic chemicals that accumulate in the lipids of cell membranes and produce a general narcotic effect. There is a different relationship for chlorophenols and this has been taken to indicate a polar narcotic effect typical of more polar compounds that form hydrogen bond interactions with the more polar groups of lipids.

6.6 BIODYNAMICS OF CHEMICAL POLLUTANTS

So far, we have looked at the relationship between the external chemical exposure and toxic response. In this section, we start to examine the internal concentrations of chemicals in organisms from the perspective of processes of uptake and elimination, metabolism, and accumulation in the tissues and cells of organisms. These set of processes, which in this chapter are referred to as the biodynamics of chemicals, control the amount or the concentration of chemical in an organism and hence the concentration at target sites of toxicity (Figure 5). It is becoming increasingly apparent that the biodynamics of chemicals in organisms largely explain the differences in susceptibilities to chemicals found between species and even between individuals of the same species.

6.6.1 Hydrophobic Organic Chemicals – Bioconcentration, Biomagnification and Food Chain Accumulation

Hydrophobic (lipophilic) organic chemicals (HOCs)(log K_{ow} > 2) have the capacity to accumulate in the tissues of organisms to concentrations above exposure levels and in many cases the increase in concentration is by several fold. The degree of accumulation is related to the lipophilicity of the compound and therefore its K_{ow}, and it takes place in the lipid tissues of organisms and in particular the lipid component of cell membranes. It is also well established that lipophilic compounds accumulate along food chains as originally shown in the now famous incident of DDD spraying in Clear Lake in California.[5]

total or nominal concentration

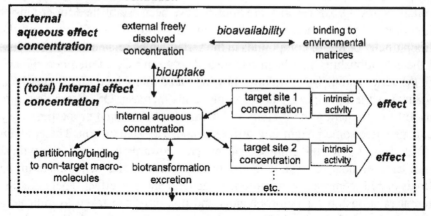

Figure 5 *The concept of the biodynamics of chemicals is highlighted by a simple representation
of an organism (box enclosed by broken lines) in which the distribution of a chemical
is shown as a function of uptake, biotransformation, and excretion and interaction
with target toxicity sites. It also highlights that the concentration of chemical available for uptake is a function of the "freely" dissolved concentration in the external
media*
(Reproduced with permission from Ref. 13)

Table 1 *Typical BCF (unitless) determined at steady state in fish for
selected chlorobenzenes and chlorophenols, together with
log K_{ow}*
(Original BCF data from A.D. Smith *et al.*, *Chemosphere*,
1990, **20**, 379)

	Log K_{ow}	BCF
1,4 dichlorobenzene	3.4	296
1,2,4 trichlorobenzene	4.1	2026
1,2,3,5 tetrachlorobenzene	4.5	4050
2,4,6 trichlorophenol	3.69	88
2,3,5,6 tetrachlorophenol	4.9	142
pentachlorophenol	5.05	216

Bioconcentration is the uptake of a chemical by an aquatic organism via the
respiratory surface (gills and/or skin) to a level in the organism that is greater
than that in the water. At steady state, the ratio of the concentration in the
aquatic organism to that in the water gives the bioconcentration factor (BCF)
and it is related to the lipophilicity of the chemical and therefore its K_{ow}. The
concentration in the organism is usually expressed on the basis of the lipid
content of the organism to take an account of the differences between organisms (lipid basis, g chem/g lipid) (Table 1).

The simplest way of explaining the process is by way of partitioning a
chemical between a polar and non-polar solvent, water and lipids, and water

and octanol. Macdonald *et al.*[6] use the term solvent switching to describe the bioconcentration process and compare it with the way that an environmental chemist extracts and concentrates a hydrophobic chemical from a large sample of water into a much smaller volume of organic solvent such as dichloromethane (DCM).

Biomagnification is the accumulation of a chemical from the diet as food containing the chemical is digested and absorbed in the gut of the organism. The chemical concentration in the organism (lipid basis, g chem/g lipid) achieves a level that exceeds that in the organism's diet and the process can again be explained by an analogy with chemical analytical procedures. Digestion and absorption of the diet in the gut of the organism can be envisaged as a sequence of solvent depletion and solvent switching steps. Solvent depletion is seen as an equivalent to evaporating the DCM solvent during a chemical extraction process and thereby concentrating the chemical in an even smaller volume of solvent.

Bioaccumulation is the sum of the two processes of bioconcentration and accumulation from the diet in conditions where a fish, for example, is exposed to the chemical in the water and its diet. It leads to the phenomenon of *food chain bioaccumulation* in which chemical concentrations in organisms (lipid weight) increase with successive levels in a food chain, causing concentrations of contaminants at the top of food chains to be greater than those in organisms at the bottom. This particularly applies to the persistent and very lipophilic chemicals, such as DDT (DDE), PCBs, dioxins, and DBDEs.

Biotransformation describes the group of enzyme-dependent processes which metabolize lipophilic substances to polar and more easily excretable substances. The capacity to metabolize such chemicals is widespread in the tissues and cells of organisms but the highest rates are associated with liver-type tissues. The system is important for the metabolism of indigenous substances such as steroid hormones and vitamins, as well as "foreign" compounds including lipophilic pollutants. The whole process is described in terms of two phases; phase I enzymatic reactions add polarity to the chemical, *e.g.* by an addition of –OH to facilitate conjugation and phase II reactions add a charged conjugate to the polar group, of which glucuronic acid is an example. Benzene, for example, is metabolized in phase I to phenol and other related products such as catechols. In the second phase, phenyl glucuronides are formed. Biotransformation is essentially a detoxification process since it acts to regulate the concentration of lipophilic chemicals in organisms. It follows that species with a capacity to biotransform a particular chemical quickly and efficiently will accumulate lower concentrations of the chemical than ones with a limited biotransformation capacity. However, metabolites produced during phase I reactions such as the hydroxylated forms are known to be very toxic and it is thought that the toxicity of many lipophilic pollutants, such as

PCBs and dioxins is the result of their hydroxylated metabolites. The group of enzymes involved in the production of hydroxylated metabolites are the monoxygenases, *e.g.* P450A1 and the sequence of events leading to their induction is becoming clearer. Important in the induction is the aryl hydrocarbon receptor (AHR) which is a member of a family of nuclear transcription factors. It is a ligand-activated transcription factor that regulates the activation of several genes that encode phase I and phase II biotransformation enzymes and it was first identified by its high-affinity binding to the environmental toxicant, 2,3,7,8 tetrachlorodibenzo-*p*-dioxin (TCDD).

Case Study 1: Bioaccumulation of Lipophilic Chemicals in Wildlife Populations

Three examples of bioaccumulation in wildlife populations are presented which consider three important influences on the accumulation process. They are: (1) the influence of food chain length on the biomagnification of toxaphene in fish from a subarctic lake in Canada; (2) the influence of migration on the magnification of PCBs, dibenzo-*p*-dioxins (PCDDs), and dibenzofurans (PCDFs) in migratory salmon in British Columbia, Canada; and (3) the influence of biotransformation on the bioaccumulation of hexabromocyclododecane (HBCD) isomers in marine mammals from waters around the coast of Western Europe.

6.6.1.1   Food Chain Length

Lake Laberge is in the subarctic region of Yukon Territory, Canada and in the early 1990s routine analyses of fish from the lake showed high concentrations of toxaphene and other lipophilic contaminants that were considered hazardous to human health. Intensive sampling of food-chain organisms, water, and sediments showed that the high levels in fish resulted from bioaccumulation along an exceptionally long food chain.[7] The source of toxaphene in the lake was traced to atmospheric inputs of low magnitude, which were similar to other lakes in the area that did not show the same degree of bioaccumulation of toxaphene. The food chains in these other lakes were shorter and more dependent on invertebrates, whereas fish such as trout in Lake Laberge had a longer-than-normal food chain that included fish.

6.6.1.2   Magnification of HOCs in Salmon during Migration

Concentrations of HOCs that included PCBs, PCDDs, and PCDFs were measured in Pacific salmon before and during migration from the Great Central Lake, British Columbia, Canada.[8] Pre-spawning migration resulted in a magnification of all classes of compound by several fold in gonad, liver, and muscle tissues. The salmon stop feeding during migration and so uptake of the HOCs ceases. It was estimated that the lipid content of their muscle tissues decreased by 84% during migration, which resulted in an increased concentration of the HOCs, essentially by depletion of the solvent (lipid).

The total HOC concentrations in the fish in terms of dioxin toxic equivalents (a summation of the relative toxicity of each compound relative to dioxin) overlapped with the LOEC derived for egg mortality in another species of salmon. There is, therefore, a potential for population level effects on current levels of HOCs in the fish and it could explain the large reductions in Pacific salmon that have been recorded since 1960s, and which coincides with high concentrations of HOCs worldwide.

Mobilization of fat reserves is a common feature of many wildlife species that undergo hibernation, migration, moulting, or experience periods of food shortages. Fat reserve depletion has been associated with increases in tissue concentrations of several hydrophobic chemicals in other populations of migrating fish, marine mammals, and polar bears.

6.6.1.3 Influence of Biotransformation on the Bioaccumulation of HBCD

In recent years, there has been a concern that brominated flame –retardants, which are widely used to improve the fire resistance of many materials have been found in wildlife. They are highly lipophilic substances and they, therefore, bioaccumulate in lipid-rich tissues of animals and because they are very resistant to environmental degradation processes, they are recognized as persistent organic pollutants. HBCD is the third most widely used brominated flame retardant in the world and toxicological studies suggest that it can disrupt thyroid function and may have developmental neurotoxicity effects. It contains three stereoisomers, of which γ-HBCD is always dominant and the α-form is only about 6% of the Σ-HBCD. A survey of HBCD concentrations in harbour porpoises and common dolphins from the North and Mediterranean Seas[9] showed that HBCD had accumulated in the marine mammals and that the highest total Σ-HBCD levels were found in harbour porpoises stranded on Irish and Scottish coasts of the Irish Sea and the northwest coast of Scotland. However, the example is interesting because it demonstrates the influence of biotransformation on HBCD bioaccumulation in the two species of sea mammals. The isomeric signature of HBCD in blubber samples was different from that in the original mixture as α-HBCD was the only form detected in the samples. Evidence was presented which shows that the change in the isomeric signature was due to the fact that these animals are able to metabolize β- and γ-HBCDs isomers but not the α-form.

6.6.2   Metal (and Metalloid) Biodynamics

It is well known that some metals (*e.g.* Cd, Hg, Pb, Cu, and Zn) are more toxic than others and that some organisms are more susceptible to metal toxicity than others but it is difficult to explain these differences using a common conceptual framework. The situation is compounded by the fact that external metal concentrations are poor predictors of toxicity as metals occur in a range of inorganic and organic species. The "free ion" form is regarded as the most available for uptake but it is not easy to measure this form directly and metal speciation is strongly influenced by water chemistry.

Case Study 2: Metal and Metalloid Biodynamics and its Role in Controlling Metal and Metalloid Exposure in Food Chains

In recent years, there has been a considerable improvement in understanding the factors that control metal biodynamics in different types of organisms. With reference to specific studies, a case study is presented that emphasizes the importance on biodynamical processes in the transfer of metal and metalloids through food chains.

Luoma and Rainbow[10] recently proposed a biologically based conceptual model for describing the basic principles of metal and metalloid bioaccumulation. It is suggested that it provides a unifying concept for metals in the same way that the partitioning model for lipophilic compounds describes the bioconcentration and biomagnification of persistent non-polar organic substances.

It has been known for some time that species have different patterns of metal bioaccumulation and these may be categorized as being either regulators or bioaccumulators. Species that regulate metal accumulation maintain tissue concentrations and these show little change over a range of metal exposures. Animals that are typical of this category have a rapid rate of excretion for some metals and therefore lose the metal as fast as they take it up. Animals that are bioaccumulators have a slower rate of excretion relative to the uptake and high concentrations are accumulated in tissues before uptake and elimination reaches a balance. Figure 6 from Luoma and Rainbow shows a good example of the two accumulation strategies. It shows Zn concentrations in the barnacle, *Balanus amphitrite* and the mussel, *Perna viridis*, collected at the same time and at the same locations in Hong Kong coastal waters and it can be seen that the barnacle bioaccumulated many times (up to 100-fold) more Zn than the mussel. Laboratory experiments established that these differences are explained by a higher feeding rate, higher Zn assimilation rate from food, and a much slower rate of Zn elimination in the barnacle than in the mussel.

Some years ago, a fragile-bone condition was diagnosed in a population of white-tailed ptarmigan (*Lagopus leucurus*) inhabiting the Animas River watershed, Silverton, Colorado, USA. The area is in the southwestern corner of the Colorado ore belt. Measurements of a range of trace metal concentrations in the food web of the ptarmigan revealed that Zn and Cd were accumulated by ptarmigan foods in sufficient quantities to pose a health hazard.[11] Of note was the finding that one particular genus of plants, the willows (*Salix sp.*), which are common food items of the ptarmigan, concentrated Cd by two orders of magnitude above the background concentrations. Cd accumulation in the willow species was responsible for an elevated Cd exposure to the ptarmigan that was associated.

Another good example is the demonstration that the nature of the food web pathway determines the degree to which selenium (Se) bioaccumulates to toxic levels in some predatory species and not in others in the San Francisco Bay area, US.[12] One food web was identified, which was based on bivalve zooplankton and another on crustacean zooplankton and the bivalve-based food web resulted in higher concentrations of Se in the top predators than the one based on crustaceans. The explanation lies in the differences in the Se biodynamics between the bivalves and crustaceans, with the bivalves having a much slower elimination rates for Se, resulting in much higher body burdens that acted as a "Se pump" into the rest of the food web and created toxic levels in the predatory fish.

An important conclusion of this case study is that the variability in trace metal bioaccumulation can be explained in terms of a biodynamic metal accumulation model which takes into account metal influx rates from water, influx rates from food, and rates of elimination from the organism. The model is developed by Luoma and Rainbow[10] using data from publications where the basic parameters were measured for a particular metal and specie, and they showed that it is possible to predict metal bioaccumulation using the biodynamic model. They compared the model predictions with measurements of metal bioaccumulation in the environment and as Figure 7 shows in terms of individual metals there is an excellent fit between concentrations predicted from the biodynamic model and measured concentrations; plotting in terms of different organisms gives a very similar linear relationship.

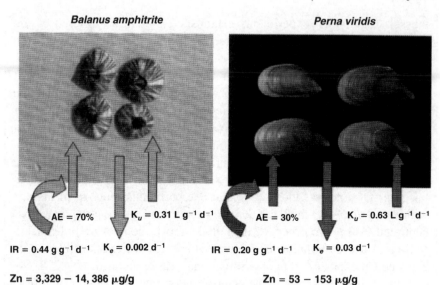

Figure 6 *Zinc concentrations ($\mu g\,g^{-1}$) in the barnacle (B. amphitrite) and mussel (P. viridis) collected at the same time and location in Hong Kong coastal waters; the barnacle is an example of a specie that bioaccumulates Zn and the mussel is of one which regulates Zn accumulation. The distinguishing features of the metal accumulator is that it has a higher ingestion rate (IR), a higher assimilation rate of Zn (AE) and a much slower Zn elimination rate constant (K_e); and K_u is the dissolved Zn influx rate constant*
(Reproduced with permission from Ref. 10)

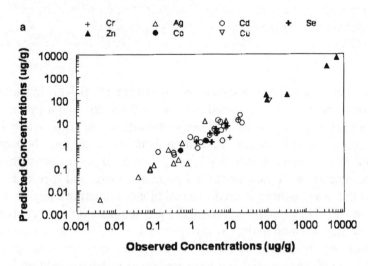

Figure 7 *Relationship between concentrations of metals predicted from a biodynamic model and metal bioaccumulated by the same species in 11 different ecosystems*
(Reproduced with permission from Ref. 10)

6.7 TOXIC MECHANISMS

A toxic effect may be observed as damage to the whole organism, specific tissues, or physiological systems. These are manifestations of interactions of a chemical at the cellular level and with specific molecular components of cells. A chemical may be said to target a receptor in the cell. In molecular bioiology, a receptor is part of the group of processes involved in cell communication in multicellular organisms. A specific protein receptor of one cell, the target cell, binds with a signalling molecule from another cell to trigger a response in the target cell. The toxic effect of a chemical can be described as a similar interaction but it is not just confined to signalling processes and it may interact with any of the components of the cells. The structure and properties of the chemical will determine which components in cells are the targets for the interaction. In broad terms, there are three receptor domains in cells and they are membranes, proteins, and genetic material (Figure 8).[13] Polarity has an important role in directing chemicals to target domains with hydrophobic chemicals partitioning to lipids and more polar chemicals diffusing through the aqueous medium of cells and interacting with polar groups of biomolecules. Van der Waals forces, hydrogen bonds, and ionic interactions are involved in binding chemicals to receptor sites such as lipids in cell membranes, protein receptors involved in cell signalling, and enzymes and these interactions form the basis of receptor-mediated toxicity. The electrophilic nature of some chemicals can lead to reactions with nucleophiles in the cell such as peptides, proteins, and DNA and the formation of irreversible covalent bonds. These types of reactions with genetic material are associated with carcinogenic and mutagenic toxicity. It should be added that chemicals can be associated with more than one toxic mechanism.

Basic cellular components and processes are conserved throughout the plant and animal kingdom and so it is believed that there are common mechanisms

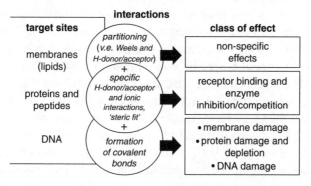

Figure 8 *Categorization of toxic mechanisms in terms of target sites, the nature of chemical interaction, and the nature of the effects*
(Reproduced with permission from Ref. 13)

of toxicity across these life forms (although specific effects on photosynthetic processes in cells would be confined to plants). It is believed that a common toxic mechanism is impaired by the same concentration of a chemical, or to put it another way, the concentration of chemical at the receptor or target site determines its effect. Furthermore, chemicals with similar structures and properties target the same mechanism and this could well explain the additive effects of chemicals.

Biodynamic processes control bioaccumulation of chemicals and therefore influence how much is available for interaction at the receptor sites. There is an underlying assumption that specific tissue or whole organism concentrations are proportional to concentrations at receptor sites and below a certain level no damaging effects occur, but above such a level toxicity is induced. From what has been said about receptor-mediated toxicity and biodynamic processes, it follows that much, if not all, the differences in sensitivity to chemicals is a consequence of different biodynamic strategies (differences in rates of elimination and biotransformation) between species. Even differences between individuals or groups of individuals of the same species (as in toxicity tests) may be explained in terms of individual variation in biodynamic processes.

6.7.1 Baseline Toxicity or Narcosis

It is a type of toxic mechanism, which describes the non-specific effects of hydrophobic chemicals on cell membranes. Partitioning of such chemicals into the lipids of membranes leads to impairment and reduced functioning of all membranes in an organism in a non-specific way and eventually it leads to the death of the organism. Partitioning to lipids is proportional to the K_{ow} and the BCF and there appears to be a critical concentration of a hydrophobic chemical in membranes that produces a lethal effect and this concentration may be attained by one chemical or a mixture of HOCs. It is not easy to measure concentrations of HOCs in membranes (the target site) but total concentrations in an organism can be used to indicate target site concentration. This has given rise to the "critical body concentration (CBC) or residues" concept in which a critical effect occurs at a specific concentration of HOC. Table 2 shows the CBCs of PAH congeners, as the CBC that causes 50% mortality to the amphipod *Diporeia spp.*, or the internal lethal concentration (ILC_{50}).[14] Also included in the table are the LC_{50}s based on the external concentration of the congeners and it can be clearly seen that the LC_{50} values vary considerably, while ILC_{50}s are similar. In addition, the ILC_{50} value for the mixture shows that the effect is additive. Normalizing the body concentration to the lipid concentration of an organism improves the variability of ILC_{50} even further and for many chlorobenzenes and many other lipophilic chemicals, the ILC_{50} is approximately $5 \, mmol \, kg^{-1}$ (wet weight) for fish. It also follows that the internal concentration

Table 2 *Lethal concentrations of some PAHs to Diporeia spp., (amphipod) in terms of external aqueous concentrations (LC_{50}) and internal concentrations (ILC_{50})*
(Taken from Ref. 14; original data from P.F. Landrum *et al.*, Chemosphere, 2003, **51**, 481)

	LC_{50} ($\mu g\,L^{-1}$)	ILC_{50} ($\mu mol\,kg^{-1}$)
Naphthalene	1266	5.8
Fluorine	542	12.3
Phenanthrene	95	7.6
Pyrene	79	6.1
Mixture	n/a	6.1

of lipophilic compounds can be estimated using a simple partitioning model that includes the K_{ow} or BCF and the lipid content of an organism.

6.7.2 Enzyme and Receptor Binding Effects

A good example of a specific toxic mechanism is the effect of organophosphate insecticides on neurotransmission across synapses or junctions of neurons or nerves with target muscle cells. These are a diverse group of compounds that include the well-known agricultural insecticides of malathion and parathion and they all have a common mechanism of toxicity. It is also a good example of a chemical affecting cell communication and in particular synaptic signalling. Organophosphates inhibit the enzyme acetylcholinesterase, which is responsible for the degradation of acetylcholine, the neurotransmitter that transmits a signal across a synapse. This results in the build up of acetylcholine at the synapse, and excessive stimulation, nervous seizure, and ultimately the death of the organism.

Endocrine disrupting chemicals (EDCs) is a broader term used for describing those chemicals whose toxic mechanism involves the disruption of the endocrine system. This is the major issue concerning chemical pollution at the present time. When it first came to attention about a decade or so ago, it was thought of in terms of a "mimicking" effect by which certain chemicals were thought to take the place of natural hormone molecules at the receptor site of the cell. It is now believed that endocrine disruption may occur in different parts of the endocrine system, which include the production sites of hormones such as the thyroid as well as interference with receptors. For example, PCBs, dioxins, and perhaps HBCD have an effect on the thyroid and as stated earlier it may be through the hydroxylated phase I metabolites, that the effects on hormone synthesis, possibly through enzyme inhibition occur.

A considerable amount of research in recent years has been directed at chemicals that have an affinity for nuclear receptor sites in cells. Most of the chemicals screened so far have binding affinities that are much less and for several of them many orders of magnitude less than that of the natural hormone–receptor system. Nonyl phenol is a good example of one of these, but it is often found in rivers in the UK at concentrations that have been shown to induce estrogenic effects in fish.[15] This leads on to the consideration of how many receptor sites need to be occupied for an environmental estrogen to have an antagonistic effect and taking account of its binding affinity, what concentration of chemical is needed in the body of an organism to achieve a critical level of receptor binding? It is important to remember that natural hormones act through specific receptor systems and that hormone-receptor binding or occupancy is needed to trigger a response in target cells.[16] For a given concentration of natural hormone or environmental estrogen at the receptor site, the affinity of the chemical for the receptor dictates how many receptor sites become occupied. Moreover, the system is more responsive at low levels of occupancy, which is commonly only up to about 10% occupancy. At higher levels, the system saturates with further receptor occupancy producing no further increases in target cell response, and saturation of response occurs below 100% receptor occupancy; at even higher levels of occupancy there is evidence that target cell response actually decreases and so the overall receptor–response relationship has been described by an upturned "U" shape. In order to understand the mimicking or antagonistic effect of environmental estrogens, it is therefore necessary to examine the interaction with natural hormones in the responsive range of occupancy of the receptor sites. This stresses the importance of determining the binding affinities of suspected EDCs, and the biodynamics of the chemicals in organisms, such that their critical concentrations at receptor sites can be established. There is growing evidence that a difference in the capacity of different species to biotransform EDCs explains much of their differences in sensitivity to estrogenic effects.

6.7.3 Metals

Metal ligand chemistry is as important inside organisms as it is in the external environment. It has been established for some time that the "free form" of a metal is the most available for uptake. In aquatic organisms and in particular, the gills of fish, ionic forms of metals bind to different types of ligands on gill surfaces. Of those that have been well characterized, one group of ligands involved in monovalent element uptake (Na^+, K^+) bind to monovalent metal cations such as Ag^+, Tl^+, and Cu^+ (Cu^{2+} undergoes reduction to Cu^+ prior to uptake). Another set of ligands involved in Ca^{2+} and Mg^{2+} uptake have been shown to bind to divalent cations such as Zn^{2+} and Cd^{2+}. Each metal cation has a specific affinity for these ligands, which has led to the development of the biotic ligand model (BLM).[17] This model predicts the affinity of metals for gill surface ligands and it has been shown that metal–ligand affinity is a determinant of a metal's toxicity, in that the greater the affinity of a metal for gill ligand the greater its acute toxicity. The BLM also helps to explain the mediating effects of trace elements such as Ca^{2+} and Mg^{2+} on metal toxicity, which is independent of any effects on the external aqueous chemistry, in that an increase of these at the gill surface provides competition and thereby reduces toxic metal–ligand binding. The toxic mechanism of metals at the gill surface seems to involve effects on the ion balance of organisms. For example, a recent study[18] showed that the effect of Ni^{2+} on *Daphnia spp.*, was to increase the loss of Mg^{2+}.

When viewed from the perspective of partitioning and reactivity, toxic mechanisms are very much in the domain of basic chemistry, and the key question becomes how do organisms regulate or control chemicals to avoid chemicals diffusing to potentially reactive sites. Once control processes have been overcome then chemical toxicity is inevitable. Therefore, the links between the biodynamics of chemicals and sites of toxic action are clearly of importance. Science is in the post-genomic era and there is no doubt that toxicity studies in conjunction with genetic characterization are going to produce some interesting insights into the regulation of control processes, the genetic basis of toxic mechanisms, and interactions between these two sets of processes. Furthermore, chemicals may react at several sites in cells and so be associated with more than one toxic mechanism. It is known for example that several EDCs in particular estrogens, as well as having effects on reproductive development are also associated with causing cancer. It is increasingly evident that endocrine disruption, particularly during the stages of rapid growth and development leads to a higher risk of cancer.[19]

6.8 EXPOSURE

Exposure is an important part of pollution studies as it links the results of controlled experiments with environmental observations. In the early stages of recognizing a pollution problem, observations relate a suspected form of pollution with an adverse effect and a consistent relationship may be established between the two; however, many factors can have an influence on the relationship and it does not demonstrate cause and effect. It is necessary to test incremental exposures to a group of organisms in controlled conditions to confirm the damaging effects of the pollution and to establish exposure–response relationships. Consistency between laboratory and field studies of exposure–response relationships goes a long way in demonstrating cause and effect. A good example is the demonstration that TBT causes IMPOSEX (imposition or development of male sex organs in females) in dogwhelk populations. Initial observations in the field highlighted the problem and showed that the prevalence of imposex was associated with marinas and with high concentrations of tin (Sn) in the tissues of the affected dogwhelks which was later shown to be TBT leached from paint on the hulls of boats. Laboratory experiments with TBT reproduced the imposex characteristics over the same range of exposures of TBT found in seawater samples close to marinas and helped to establish a much lower NOEC that led to a ban on the use of TBT on small-hulled boats in the UK.

Exposure is not an easy parameter to measure and there are many uncertainties in such measurements. Even in laboratory experiments with a controlled delivery of toxicant, variation can occur through losses of chemical to the air and container walls and different forms of chemical may be present which are not all equally available to the organisms in the test system. In the ambient environment, the situation is far more complex as chemicals interact with biogeochemical systems in air, water, and soil, which affect their distribution, their chemical form, and their persistence in the environment. Therefore, exposure to chemicals in the environment can be very different and more complex than that in laboratory experiments. It follows that measurements of exposure in the ambient environment incorporate a high level of uncertainty. We take measurements of chemicals in the environment at specific times and locations and use them for assessing exposure to species in the same environment. Such measurements need to be representative and take account of the variation in time and space of chemicals and also the different forms of the chemicals in the environment.

Exposure can arise from a single source of supply, as in the transfer of gaseous air pollutants such as sulfur dioxide (SO_2), nitrogen dioxide (NO_2), or ozone (O_3) from the atmosphere to the leaf surfaces of plants. In other cases, more than one pathway contributes to the total exposure, for example, human exposure to toxic metals such as lead and mercury and many pesticides is from water, air, and food.

General measurements of substances in air, water, or soil may not be representative of the actual amount that reaches a group of organisms. This may be a consequence of how, when, and where measurements are taken, or because a pollutant exists in more than one form, all of which may not be equally available to biota. Exposure can vary in space and time; in rivers the concentrations of pollutants vary with season, time of day, magnitude of freshwater runoff, depth of sampling, intermittent flow of industrial effluent, and hydrological factors such as tides and currents.

Human exposure to atmospheric pollutants such as airborne particulate matter and ozone is variable and therefore difficult to estimate with precision on an individual basis. Many samples are required to iron out variation due to methods of collection and analysis as well as actual spatial and time differences. In practice, exposure estimates are often based on data from a small number of monitoring stations (perhaps only one) representing a relatively large area. People move around, spending variable amounts of time inside their homes, in the workplace, in cars, and by busy roads breathing in fumes. Each person has his or her own personal exposure regime. Inside the home, other sources can affect air quality such as gas cooking and cigarette smoking and these can lead to higher concentrations of NO_2 and particulates inside the home than in the ambient air. Driving around cities in cars can also lead to higher concentrations inside the vehicle than in the outside air.

Much of the research on the effects of gaseous air pollutants on plants has been done in the controlled environment of growth chambers. The concentration of the gas in question in the inlet or outlet of the chambers has been used as an estimate of plant exposure. These values, however, can be quite different from those at the leaf surface. Design and size of growth chamber, density of plants, and air velocity are just three of many plant and environmental factors that can influence the concentration gradient between ambient air around the plants and that in contact with the leaf surfaces. These and other factors have a marked effect on the thickness of the boundary layer of air that surrounds the leaves. Gases must pass through this almost laminar flow of air by the relatively slow process of molecular diffusion to gain access to the leaf surfaces. In situations where varied responses have been reported for seemingly equivalent exposures, it is apparent that at least a part of the discrepancy can be explained by the experimental conditions and the way in which exposure was quantified.

Organic matter in soil and sediment is an important substrate for the adsorption and accumulation of lipophilic substances. Once bound to organic matter in particles lipophilic substances are less available to organisms. There is evidence that the bioavailability of these bound forms can decline with time as the compounds become more tightly sequestered or less accessible within the organic matter. A further level of complexity is that organic chemicals are susceptible to chemical or biological degradation. Some are broken down quite quickly while others, such as the larger PAHs and PCBs, are resistant to breakdown and are persistent in soil and sediment for many years. Metals exist in different forms. In soils and aquatic systems, metals are partitioned between solid and liquid phases, and within each, further partitioning or speciation occurs among specific ligands, determined by ligand concentration and the strength of each metal–ligand association. Consequently, at any one time the amount of free ion available for uptake by organisms is less, and in many instances much less, than the total concentration. As a result, organisms with different feeding habits can have very different exposures in the same ecosystem. Those foraging in sediments may be exposed to elevated levels of metals or lipophilic organic compounds whereas those swimming in the water may have a much lower exposure. The amount accumulated depends on the bioavailability of the substances in different parts of the system.

The amount or concentration of chemical in an organism or in some part of it can be related to exposure and in turn to toxicity. This was discussed earlier in respect of lipophilic organic chemicals and metals and in fact some authorities are of the view that internal concentrations of chemicals give a more realistic assessment of an integrated exposure. Epidemiological studies involving lead or mercury have used the concentration of the metal in blood as indices of exposure. For lead, there is a curvilinear relationship between intakes from air, water, and food and blood lead (PbB) concentrations.[20] In the case of methylmercury there is a good understanding of the relationship among exposure, body burden, blood concentrations, and concentrations in segments of hair. Mercury analysis of segments of hair can be used to predict both the body burden and blood concentration of mercury at the time the hair segment was laid down and it is therefore a valuable non-invasive means of screening for elevated mercury levels.[21]

In the sections below, a number of well-known forms of pollution are described in more detail. The emphasis is on what, where, and how these forms of pollution occurred, and they also highlight how understanding of pollution science has developed over the years.

6.9   HEALTH EFFECTS OF METAL POLLUTION

6.9.1   Mercury

Several major episodes of mercury poisoning in the general population have been caused by consumption of methyl- and ethylmercury- compounds. The most well known ones were in Japan and Iraq and while they occurred some time ago, they remain important as historical landmarks in the development of pollution science and for the exposure–response relationships that were developed. Methylmercury is a neurotoxin and the main clinical symptoms of poisoning reflect damage to the nervous system. The sensory, visual, and auditory functions, together with those of the brain areas, especially the cerebellum, concerned with co-ordination, are the most commonly affected. Symptoms of poisoning increase in severity in line with the increased exposure, as follows: (1) initial effects are non-specific symptoms including paraesthesia, malaise, and blurred vision; (2) in more severe cases, concentric constriction of the visual field, ataxia, dysarthria, and deafness appear more frequently; and (3) in the worst affected cases, patients may go into a coma and die. The effects in severe cases are irreversible due to destruction of neuronal cells. There is a latent period, usually of several months between the onset of exposure and the development of symptoms. The fetus is considered more sensitive and a prenatal exposure in these poisoning episodes resulted in children born with mental retardation and physical abnormalities.[21,22]

Case Study 3: Epidemics of Mercury Poisoning in the General Population; Development of Exposure–Response Relationships

6.9.1.1 The Iraqi Outbreak. This epidemic of methylmercury poisoning occurred in agricultural communities in Iraq in the winter of 1971–1972 and it is an example of an acute poisoning episode. Over 6000 people were admitted to hospitals in provinces throughout the country and over 400 people died in hospital with methylmercury poisoning. The poisonings arose from a misuse of imported seed grain treated with alkylmercury fungicide. The imported seed was intended for sowing, but in many areas the grain was ground directly into flour and used in the daily baking of homemade bread. Depending on the number of loaves consumed, individual exposure ranged from a low non-toxic intake to a prolonged toxic intake over 1–2 months. Because of the brief exposure period, epidemiological investigations began 2–3 months after it had ceased and in most cases after the onset of poisoning. This made the calculation of ingested dose and the body burden of mercury at the time of exposure to methylmercury, difficult to estimate.

Figure 9 shows both dose–effect and dose–response relationships that have been established between symptoms of poisoning and the estimated body burden at the time of cessation of ingestion of methylmercury in bread. An increased body burden of mercury is associated with an increase in the severity of symptoms experienced by patients. The dose–response curve for each effect shows the same characteristic shape, a horizontal and a sloped line, which is referred to as a 'hockey stick' line. The horizontal line represents the background or general frequency of each symptom and the sloped line shows that an increased body burden is associated with an increasing frequency of each symptom. The intersection of the two lines has been taken as the 'practical threshold' of the mercury-related response and this increases with the increasing severity of the effects; for paraesthesia it is at a body burden of about 25 mg Hg, for ataxia it is at 50 mg, for dysarthria it is at about 90 mg, for hearing loss it is at about 180 mg, and for death it is over 200 mg. These relationships do not demonstrate cause and effect and the only proof that methylmercury produced the above effects is that the effects followed a known high exposure to methylmercury, the frequency and severity of these effects increased with an increasing exposure to methylmercury, the effects are similar to those seen in other outbreaks of methylmercury poisoning, and the major signs have been reproduced in animal models.

6.9.1.2 Minamata Disease. Mass poisoning of people of Minamata in Japan in the 1950s was the first to associate severe neurological impairment with methylmercury exposure. The problem arose as a result of the local population consuming seafood contaminated with methylmercury. Mercury compounds, including methylmercury were released from industrial sources into the aquatic environment and this resulted in the accumulation of methylmercury in seafood. The outbreak of poisoning was first discovered during the 1950s and early 1960s and, by the mid-1970s, about 1000 cases (with 3000 suspects) in the Minamata area. Exposure measurements of Hg in blood and hair did not begin until several years after the outbreak of the disease, which has hampered the establishment of exposure–response relationships. By and large, symptoms of poisoning followed a very similar pattern to the Iraqi epidemic, although an important difference was in the nature of the exposure, which was lower but more prolonged. Moreover, the latent period was longer and in some isolated cases it was as long as 10 years between the initial exposure and the onset of symptoms. In other cases, it was observed that clinical symptoms worsened with time, despite reduced, or discontinued exposure. All this would appear to be related to long-term accumulation of mercury in the brain.

The exposure–response relationships suggest that blood mercury concentrations of 200–500 μg L^{-1}, hair concentrations of 50–125 μg g^{-1}, and a long term intake of 3–7 μg kg^{-1} body weight are likely to be associated with the onset of the initial symptoms of methylmercury poisoning, such as paraesthesia.

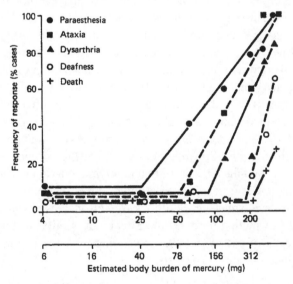

Figure 9 *The relationship between frequency of signs and symptoms of methylmercury poisoning and the estimated body burden of methylmercury (the two scales of the abscissa result from different methods of calculating body burden of the compound)* (Reproduced with permission from Bakir *et al., Science*, 1973, **181**, 230)

Clinical and epidemiological evidence indicate that prenatal life is more sensitive than adults to the toxic effects of methylmercury. Prenatal exposure of children has given rise to what is called congenital Minamata disease (CMD) and in the most severe cases the symptoms included mental retardation, primitive reflexes, cerebellar ataxia, dysarthria, and limb deformities. A similar situation was reported in the Iraqi outbreak with infants who had been prenatally exposed showing severe damage to the central nervous system. Maternal hair concentrations during pregnancy associated with these severe effects was of the order of $400 \, \mu g \, g^{-1}$. Follow-up studies with infants in Iraq found evidence of psychomotor retardation (delayed achievement of development milestones, a history of seizures, and abnormal reflexes) at maternal hair levels well below those associated with severe effects. One study has suggested that hair concentrations as low as $10 \, \mu g \, g^{-1}$ may be the threshold for neuro-developmental effects.

It is well established that the main form of mercury in fish is methylmercury and as it bioaccumulates in food chains in aquatic systems, high concentrations occur in predatory fish. This includes mercury from natural sources and in the open oceans, the majority of mercury present is believed to be from the natural biogeochemical mercury cycle. There is concern that there are a number of communities around the world that largely depend on a diet of fish and mammals and are therefore exposed to slightly elevated intakes throughout their lives. This raises the possibility that mercury accumulates to toxic levels and of particular concern is the exposure of fetuses during critical stages of their development. One such group is the Canadian Indians but because of several confounding factors, not least of which are high incidences of malnutrition and alcoholism, the assessment of health risk due to methylmercury is difficult. One study, involving 35,000 samples obtained from 350 communities, found that over two-thirds had blood mercury concentrations within normal limits ($<20 \, \mu g \, L^{-1}$) but 2.5% (over 900 individuals) had levels in excess of $100 \, \mu g \, L^{-1}$ and were considered as a group 'at risk' and in need of a close surveillance.

Three large prospective studies have specifically addressed the issue of prenatal exposure in communities dependent on a seafood diet.[23] These were in the Seychelles, Faroe Islands, and New Zealand and the emphasis in these studies was on the detection of subtle end points of neurotoxicity such as intelligence and development milestones and therefore, in defining a reference dose associated with the onset of neurological impairment. Hair concentrations in the Seychelles study ranged between 0.5 and $27 \, \mu g \, g^{-1}$ and no adverse effects were found, but in the two other studies, with similar prenatal exposures, dose-dependent neurological deficits were found. In the Faroe Island study, researchers reported that 7-year-old children that had high levels of Hg in their umbilical cords at birth showed neurological impairment and a later follow-up study suggested that these deficits were evident in children of 14 years.

These studies highlight the problems of developing exposure–response relationships in diverse communities and of the uncertainties bound up in both exposure assessment and subtle neurological measurements that could have other causes. One of the differences between the communities of the Faroe Islands and the Seychelles is that the Faroe Islanders have a unique whale-rich diet which could introduce additional lipophilic toxicants. Other work on the placental transfer of Hg suggests that the mercury level in fetal blood is 1.7 times higher than that in maternal blood, whereas it was previously assumed that the ratio was one to one. This would require revision of the safe blood level of 5.8–$3.5 \mu g L^{-1}$ and it has been estimated by the US Centres for Disease Control that some 16% of women have Hg blood levels greater than $3.5 \mu g L^{-1}$.

6.9.2   Lead

Lead pollution became a major issue in the 1970s when widespread lead contamination of the environment was extensively reported. It is acknowledged that a number of sources contributed to the environmental contamination but the overriding one was the emission of lead from cars using leaded fuel. Important achievements of health studies associated with lead pollution have been the demonstration that children are the most vulnerable to lead pollution and that damaging effects can be detected at relatively low environmental exposures.

6.9.2.1   Development of the Association between Blood-Lead Concentrations and Cognitive Development in Children.   Lead is a neurotoxin and the overt toxic effects of lead have been known for many centuries. Probably the first reported cases of lead poisoning due to environmental sources was a group of children diagnosed as having lead palsy, by clinicians at the Brisbane Hospital in Queensland, Australia, at the turn of the last century. A total of 10 cases of lead poisoning were found by health officials and it was later shown that the source of the lead was the lead-based paint which was turning to powder on the walls of homes and railings. Young children are particularly at risk from exposure to lead in the environment and according to the US Centers for Disease Control and Prevention (CDC) lead poisoning remains the most common and serious environmental disease affecting young children. They are particularly vulnerable to Pb poisoning because their playing activities can bring them into direct contact with contaminated dusts and soils, they absorb Pb more efficiently than adults and importantly, the developing nervous system of a child is particularly sensitive to the toxic effects of Pb.

In the 1970s, there were a number of sources of Pb in the environment and the main ones were leaded petrol, paint, mining operations, smelting and industrial emissions, drinking water from lead pipes, and soldered food cans. Subsequently, countries around the world have introduced programmes to reduce exposure to lead from paint, drinking water, and leaded petrol. Nevertheless, all these sources have resulted in widespread soil contamination with Pb, especially in urban areas.

Lead poisoning due to exposure to lead-based paints has affected a large number of children throughout the last century in the USA. The disease mainly affected children living in inner city areas in dilapidated buildings with surfaces of flaking and peeling lead-based paint. Children playing in the vicinity take in particles and flakes of paint by inhalation and hand-to-mouth activities and in extreme cases of children with the 'pica' habit of eating non-food items, excessive amounts of lead may be taken into the body; for example, a square centimetre of paint may contain over 1 mg of lead.

The disease in the USA was neglected for a long time and the total number of children who have suffered from lead poisoning and the number at risk from the disease emerged over several decades. The probable reasons for this time lag include the poor socio-economic status of the children in the high-risk groups and the difficulties in recognizing the disease. In 1971, the disease became fully recognized when the US Government introduced 'The Lead-Based Paint Poisoning Prevention Act'. In the decades leading up to the 1970s, there was an increasing recognition of lead as a general environmental problem, particularly in urban areas where the use of leaded-petrol and in mining and industrial areas where lead ores were extracted and processed added considerably to the total exposure of the general population.

It is important to recognize that up until the 1970s a child was diagnosed as suffering from lead poisoning only when acute encephalopathy was evident. Typically, this was characterized by a progression to intellectual dullness and reduced consciousness and eventually to seizures, coma, and, in very severe cases, death. Although unknown at the time, acute encephalopathy is associated with lead concentrations in the blood in excess of 80–$100\,\mu\text{g}\,\text{Pb}\,\text{dL}^{-1}$ of blood. At about $80\,\mu\text{g}\,\text{dL}^{-1}$ severe but not life-threatening effects of the CNS can be expected. Lead encephalopathy in children is usually accompanied by peripheral neuropathy, especially foot drop, and general weakness.

PbB concentrations have become the main tool for assessing exposure to lead as they give an integrated picture of Pb intake and exposure over the previous 1–2 days. As a result of numerous studies that have measured PbB concentrations and various indices of neurological toxicity and cognitive development, the adverse effects of elevated levels of Pb in blood have been established. The key finding has been that the adverse effect on cognitive development in children has been established at ever-decreasing PbB concentrations. At one stage, PbB concentrations in the region of $40\,\mu\text{g}\,\text{dL}^{-1}$ were associated with impaired neurobehavioural development in children.

Later, the focus became the concentration of PbB at and below $30\,\mu g\,dL^{-1}$ with a level above this being linked to behavioural and attentional problems in school children (*e.g.* disordered classroom activity, restlessness, easily distracted, not persistent, inability to follow directions, low overall functioning). Also, prenatal exposure to lead has been associated with a reduction in the mental development of infants. The effect of environmental exposure to lead on children's abilities at the age of 4 years was studied in a cohort of 537 children born between 1979 and 1982 to women living in the vicinity of a lead smelter at Port Pirie in Australia. The study indicated that an elevated PbB concentrations in early childhood affected mental development up to the age of 4 years.[24]

It is now well established that levels below $25\,\mu g\,dL^{-1}$ have effects on cognitive (reduction in IQ scores) and behavioural development in children. As a result, the US CDC defines an elevated PbB level in children under age 6 of $10\,\mu g\,dL^{-1}$. This is not a threshold but for PbB levels below $10\,\mu g\,dL^{-1}$, effects are difficult to detect because of confounding variables and lack of precision of analytical and psychological measurements. Animal studies support a causal relationship between Pb and nervous system effects and there are reports of intellectual deficits in monkeys and rats with PbB levels in the $11-15\,\mu g\,dL^{-1}$ range. However, attempts to attribute subtle deficits in development to lead exposure are difficult as many other factors (genetical, nutritional, medical, educational, and parental and social influences) can strongly influence the development of a child. For example, the nutritional status of people can affect lead absorption into their body and it is well known that children with lower intakes of dietary calcium have increased levels of lead in their blood.

Nevertheless, examples are emerging of studies showing effects of Pb on the cognitive development of children at PbB concentrations below $10\,\mu g\,dL^{-1}$. This could have important repercussions as a further refinement of what constitutes an elevated PbB concentration would vastly increase the number of children at risk from lead poisoning. Many of these are in the urban areas of large cities that have residual soil contamination from the decades of lead input from the major sources.

As a result of the programmes for reducing Pb sources to the environment, significant reductions of lead intake and blood concentrations have occurred in children. The US Food and Drug Administration (FDA) market basket surveys show that a typical daily intake of Pb for 2-year-old children dropped from $30\,\mu g\,day^{-1}$ in 1982 to about $2\,\mu g\,day^{-1}$ in 1991. The National Health and Nutrition Examination Surveys (NHANES) of PbB levels in children of 1–5 years of age showed that in the 10 years between one survey (1976–1980) and the next (1988–1991), median concentrations had decreased by 78%

from 15 to $3.6\,\mu g\,dL^{-1}$ and in 1994 the median was $2-3\,\mu g\,dL^{-1}$. However, estimates suggest that more than 2.2% of children of ages 1–5 in the US still have PbB concentrations higher than $10\,\mu g\,dL^{-1}$ with children living in cities having a higher prevalence of elevated blood lead levels. Continuing factors that enhance risk to Pb exposure, particularly during fetal life are low socio-economic status, old housing with Pb-containing paint, and poor nutrition, particularly low dietary intake of Ca, Fe, and Zn. Prenatal exposure may result from endogenous sources such as Pb in the maternal skeletal system or maternal exposures from diet and the environment.

Case Study 4: Influence of Remediation Strategies on Reducing Pb Exposure to Children at a Site with a History of Pb Contamination

A good example where intervention at a severely contaminated site has resulted in a considerable improvement in the health of the population is the Bunker Hill Superfund Site in northern Idaho in the US.[25] Uncontrolled discharges and emissions from mining and smelting activities over 100 years and extending over an area of 21 mi², had resulted in several 1000 acres of land contaminated with Pb and other metals. In the early 1970s, it came to the attention of the authorities that Pb poisoning was endemic in young children living in the area. Figure 10 provides a summary of PbB concentrations in children from three communities over the period 1970 until 2001. Clearly, closure of the smelter had the greatest impact on reducing PbB concentrations; however, subsequent remediation measures such as removing very contaminated soil and diluting with uncontaminated soil from elsewhere also resulted in a further reduction of PbB levels, such that concentrations by 2001 were much less than $10\,\mu g\,dL^{-1}$. In groups of households, links were established between soil contamination in the immediate vicinity of homes, household dust levels, lead intake, and PbB concentrations.

In assessing the risks to health from lead contaminated soil, it is necessary to establish the linkage between the source of the lead in the soil and the population living in the vicinity. This entails knowing the human exposure pathways by which the lead is transported to the population, such that intakes due to inhalation and ingestion of household dusts, soil, and food can be quantified and related to PbB concentrations. This enables identification of the groups at risk from lead poisoning and the necessary remediation action required to reduce the risk. A key aspect is the bioavailability of lead in the soil and dust, both in terms of the proportion that is available to plants used

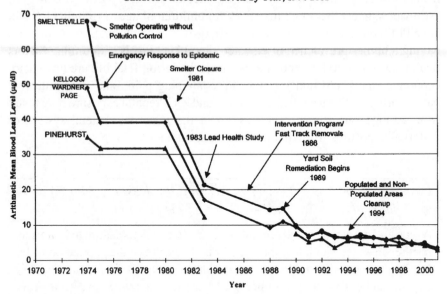

Figure 10 *Reduction in PbB concentrations over the period 1970–2001 in children from three communities at the Bunker Hill Superfund Site, Idaho, US as a result of remediation measures to reduce lead exposure*
(Reproduced with permission from Von Lindern *et al., Sci. Total Environ.,* 2003, **303**, 139)

for food and also in respect of the fraction of lead in inhaled or ingested soil and dust fractions which is available for transport across the membranes of the lung and the intestine. This all comes down to the mineralogy of lead and the soil chemistry, as the interaction between the two determines how much can be solublized and made available for uptake. For example, lead phosphates have low aqueous solubility and reduce the availability of lead in soil, and adding phosphate to soil has been shown to be very effective in reducing the PbB concentrations in swine foraging in lead contaminated soil.

6.10 HEALTH EFFECTS OF THE MAJOR AIR POLLUTANTS

Health effects of air pollution in the general population are associated with an increase in mortality and worsening of lung and heart conditions during pollution episodes. The association of smoke laden air from coal fires to increased incidences of respiratory disease in London was recognized as long ago as 1661 by Evelyn.[26] It is now acknowledged that across the world, long-term residence in cities with elevated ambient concentrations of air pollution from combustion sources is associated with an increased mortality (WHO).[31]

Case Study 5: The London Smog 1952; The Major Incident that Established the Link between Air Pollution and Health

The incident occurred in London in early December, 1952 when 3500–4000 deaths above the norm were recorded during an exceptional smog episode that lasted unremittingly for 5 days.[27,28] At the time, much of the air pollution was due to coal combustion, with numerous near-ground sources generating considerable quantities of smoke and sulfur dioxide. The fog developed across the capital as a consequence of a very stable, high-pressure zone, and an inversion layer. This prevented the dispersal of pollutants, and concentrations built up to very high levels, the particles of smoke acting as condensation nuclei, which added further to the density of the fog. The epidemic went largely unnoticed by the population of London and it was only when death certificates for the whole of the London area were later examined that the sudden upsurge in the number of deaths became apparent. Deaths were mainly confined to the elderly and those people with a history of heart and lung diseases. The central areas of London, where the fog was at its densest and most persistent, had the greatest increase in mortality, some 200% more than the average for that time of year. Estimates indicate that sulfur dioxide and smoke concentrations (48 h mean) during the London episode attained very high values and were in the region of $3.7 \, mg \, m^{-3}$ (1.3 ppm) for SO_2 and above $4.5 \, mg \, m^{-3}$ for smoke. The episode was also at the time of the Smithfield agricultural show at Earls Court and it is interesting to note that a number of cows also died during the course of the show. The London smog episode was very important in establishing the strong link between air pollution and respiratory health.

It is generally accepted that the combined effects of sulfur dioxide and smoke particles on the respiratory tract were responsible for the excess deaths and for exacerbating heart and lung disease in susceptible people. To the healthy majority of Londoners, the pollutant-laden fog was merely an inconvenience. Epidemiological studies established that the spatial and temporal trends of sulfur dioxide and smoke correlated closely with the distribution of the enhanced mortalities, but other pollutants would also have been elevated, *e.g.* carbon monoxide, sulfates, and sulfites. However, there are insufficient data on such substances to allow exposure–response relationships to be established.

Follow-up studies in other large cities concentrated on more moderate day-to-day variations of mortality and morbidity in relation to pollution levels. A WHO task group provided a summary of the salient features of many of these studies and the collated data have formed the basis for developing short- and long-term exposure–response relationships. It is important to emphasize that concentrations of sulfur dioxide and suspended particulate matter vary from place to place and from one time period to the next. Also, measurements related to levels in the outdoor environment and took little or no account of indoor exposure, which is where the elderly and the chronically sick spent most of their time. We now know that even in modern homes the indoor air quality can be as poor as that outside and that gas cooking and smoking can make indoor air quality worse than that of outside. At the time of the smogs in London, open coal fires probably further added to the poor quality of indoor air. Notwithstanding, the studies suggested that short-term exposures above $500\,\mu g\,m^{-3}$ (24 h mean) for both sulfur dioxide and smoke were associated with an increased mortality among the elderly and the chronically sick. Exposures of 250–$500\,\mu g\,m^{-3}$ (24 h mean) to both pollutants were likely to lead to worsening of the condition of patients with an existing respiratory disease and long-term exposure in places with annual-mean concentrations above $100\,\mu g\,m^{-3}$ were likely to lead to an increased prevalence of respiratory symptoms among both adults and children and an increased frequency of acute respiratory illnesses in children. Based on these relationships, and incorporating a margin of safety, WHO recommended for both pollutants that daily concentrations should be below 100–$150\,\mu g\,m^{-3}$ and annual means should remain below 40–$60\,\mu g\,m^{-3}$.

In the 1950s in America, particularly in the Los Angeles region, the problem of photochemical smog episodes and their effects on human health and plants became apparent. Photochemical episodes are a common feature of cities around world during periods of strong sunlight and the important pollutants are ozone, particles and a variety of oxygenated organic compounds, including peroxyacyl nitrates. Early epidemiological studies, based mainly on Los Angeles, failed to establish direct relationships between increased mortality rates and the frequency of smog episodes, although eye irritation (probably due to peroxyacyl nitrate compounds) and upper respiratory tract discomfort were common complaints during smog episodes. Increased breathing difficulties in heavy smokers and asthmatics and reduced performance in people indulging in physical activity are associated with periods of high oxidant (ozone) concentrations.[29] Chamber studies with short term but realistic exposures to O_3 and healthy adult subjects have consistently shown effects such as reductions in lung function (lung volume and expiratory flow rate), increases in lung reactivity to other irritants, and pulmonary inflammation.

Field studies of adults who exercise heavily for short periods have shown short-term reversible decreases in pulmonary function in association with ozone concentrations at or near the previous US National ambient air quality standard of 120 ppb. Much of the field evidence comes from children attending summer camps in North America where ambient O_3 concentrations, particularly above 120 ppb have been associated with short-term declines in the average lung function. Although, other studies have failed to reproduce this relationship, the balance of evidence points to effects of O_3 on lung function. In London, daily hospital admissions for respiratory disease over a 5-year period (1987–1992) were examined in relation to air pollution (NO_2, SO_2, O_3, and smoke) and it was found that O_3 above all the others had a small yet significant effect on admissions.[30] Such studies led in 1997 to a revised US ambient air quality standard for ozone of 80 ppb daily maximum 8 h average over 3 years.

6.10.1 The Current Situation with Respect to Air Quality and Health

It is important to emphasize that over the last two to three decades, the air quality in towns and cities across North America and Europe has improved considerably with SO_2 and particulate matter concentrations showing marked improvement. However, despite these improvements there is general agreement that current levels of exposure to air pollution, particularly in urban areas, are sufficient to cause health effects.[31] Particulate matter and ozone are thought to be the main pollutants leading to adverse health effects as they have been shown to be associated with increases in hospital admissions for cardiovascular and respiratory disease and mortality in cities in North America, Europe, and other continents.

A recent large study involving 95 large US urban communities (estimated to encompass 40% of the total US population) reported a statistically significant association between short-term changes in ambient O_3 and mortality. Of note was that a 10 ppb increase in a previous week's O_3 was associated with a 0.52% increase in daily mortality and a 0.64% increase in cardiovascular and respiratory mortality.[32]

It is believed that particulate matter poses the greatest health problem. Particulate matter in the atmosphere ranges in size from a few nanometers to several microns. Particles less than 10 μm in diameter or PM_{10} can enter the lung and are therefore important to health. The fraction less than 2.5 μm in diameter, $PM_{2.5}$, is known as the fine fraction and the coarse fraction is between $PM_{2.5}$ and PM_{10}. An ultrafine fraction of less than 0.1 μm in diameter is also recognized as a fraction, which is important for the health effects of particulate matter. Unlike the individual gaseous pollutants, particulate matter is not a single pollutant but a complex mixture of organic and inorganic substances. Over the last two decades or so, numerous epidemiological studies have reported associations between elevated PM_{10} or $PM_{2.5}$ concentrations and health effects. In the main, $PM_{2.5}$ gives a stronger association than PM_{10} and more recent studies have found very good associations with the ultrafine fraction.

Epidemiological studies have established relationships over long- and short-term exposures.

6.10.2 Effect of Short- and Long-term Exposures of Particulate Matter on Health

Important effects related to short-term exposure to particulate matter include lung inflammation responses, respiratory symptoms, adverse cardiovascular responses, increased hospital admissions, and mortality. For long-term exposures, typical effects include reduction in lung function in children, an increase in chronic obstructive pulmonary disease (COPD), and reduction in life expectancy due to cardiopulmonary mortality and probably cancer. Children, the elderly, and people with a history of cardiopulmonary diseases are more likely to suffer the adverse effects of air pollution.

The now famous six cities study monitored 8111 adults in North East and Midwest US for 14–16 years beginning in mid-1970s and found that cities with the higher concentrations of fine particles and sulfate were associated with significantly greater mortalities from all causes and particularly for respiratory and cardiovascular diseases.[33] This was followed by the larger American Cancer Society (ACS) study that involved 552,138 adults in 154 US cities over the period 1982–1989.[34] Similar findings were reported with an increased mortality from all causes and cardiopulmonary disease associated with higher concentrations of fine particles in the most polluted cities. In both these studies, the sulfate content of particulate matter was also found to be associated with the increased mortality in cities with the highest concentrations of particulate matter, but this relationship is unlikely to be causal.

Many follow-up studies in North America, Europe, and elsewhere have reported similar associations and it has led to the view that long-term residence in cities with elevated ambient concentrations of air pollution from combustion sources is associated with a reduced life expectancy arising from an increase in mortality from cardiopulmonary disease. For Europe, this has been equated with about 100,000 deaths (725,000 years of life lost) annually.

Short-term effects of particulate matter have also been established and two large multi-centre studies from the US (National Morbidity, Mortality and Air Pollution Study (NMMAPS)) and Europe (Air Pollution and Health: A European Approach, APHEA) have provided good estimates of the effects from records of daily mortality and hospital admissions. The APHEA2 mortality study covered a population of over 43 million people living in 29 European cities, for over 5 years in the early-mid-1990s. The hospital admission study involved 38 million living in eight European cities, for 3–9 years in the early-mid-1990s. The NMMAPS mortality study examined records of more than 50 million living in 20 metropolitan areas of the US between

Table 3 *Estimated effects of air pollution on daily mortality and hospital admissions from APHEA2 and NMMAPS studies*
(Taken from Ref. 31)

	Study	
	APHEA2	*NMMAPS*
Increase in total deaths per $10 \mu g \, m^{-3}$ PM_{10}	0.6%	0.5%
(95% confidence limits)	(0.4–0.8%)	(0.1–0.9%)
Increase in COPD (APHEA2; COPD + asthma) hospital admissions in persons >65 years per $10 \mu g \, m^{-3}$	1.0%	1.5%
(95% confidence limits)	(0.4–1.5%)	(1.0–1.9%)

1987 and 1994 and the hospital admission study covered 10 large metropolitan areas of the US with a combined population 1.843 million people over 65 years old. A summary of the results is shown in Table 3 where it can be seen that small but significant increases in both mortality and hospital admissions occur for every $10 \mu g \, m^{-3}$ rise in PM_{10}. In the short-term studies, it is the elderly and those with pre-existing heart and lung disease, which are the most susceptible to the effects of particulate matter on mortality and morbidity.[31]

A more detailed picture is beginning to emerge of patterns of impaired respiratory health in children within cities. In the Netherlands, a study of respiratory health of school children found that respiratory symptoms increased near motorways with high loadings of heavy vehicles and furthermore the relationship was confined to children with asthma and pre-sensitization to allergies.

Despite a considerable amount of expenditure on research around the world, the mechanisms by which particles damage human health remain uncertain. Fine particulate matter ($<2.5 \mu m$) is more hazardous than larger particles in terms of mortality and cardiovascular and respiratory end points. The characteristics that may contribute to particle toxicity are metal content, PAHs and other organic components, endotoxin, and extremely fine particles in the nanometer size range. It has been shown that particulate matter has oxidative properties, which may be involved in the production of reactive oxygen species (ROS) and oxidative stress in lung tissues. It is thought that these are responsible for production of inflammatory responses in the cardiovascular system. Quinones and certain metals such as iron (Fe) form redox active centres that produce ROS and they are both present in particulate matter. A recent study in California comparing the toxicity of particulate matter from an urban site to that from a rural location, found that ROS production was associated with the ultrafine particulate fraction and redox-cycling organic chemicals within this fraction.[35]

6.11   EFFECT OF AIR POLLUTION ON PLANTS

The main pollutants that affect vegetation are SO_2, NO_x, O_3, fluorides, and ethylene. The effects on vegetation include reduction in yields, visible leaf damage, loss of sensitive species and therefore reduction in diversity of plant communities, and changes in sensitivity to other stresses. In the UK, concentrations of SO_2 have decreased in recent decades, NO_2 concentrations are elevated but steady, and there would appear to be increased incidences of photochemical pollution producing elevated O_3 concentrations.

In the past, the main concern was over acute visible damage and impaired growth due to SO_2 (and acid aerosols) and smoke around point sources and in cities. In the early 20th century in the Leeds area, soot and sulfite deposits were associated with reduction in the yields of lettuces by up to 70% in the worst affected areas and extensive damage to evergreen privet during autumn and winter months. Certain conifer species are known to be especially sensitive to sulfur dioxide and other pollutant gases. For a number of years, attempts to establish conifer plantations in southern parts of the Pennine Hills in the North of England generally proved unsuccessful. At one stage, Scots Pine (*Pinus sylvestris*) was either absent or very sparse over a 50 km wide corridor, downwind of the major conurbations of Greater Manchester and Merseyside. The frequency of the species inversely corresponded to mean winter sulfur dioxide concentrations in the area, and no trees occurred wherever SO_2 concentrations exceeded 0.076 ppm. In the area of north of Manchester, a local genotype of the grass, *Lolium perenne* was found to have developed a tolerance to SO_2 pollution. With the decline of sulfur dioxide concentrations there has been a marked improvement in establishing conifer plantations in the area.[36]

Growth chamber experiments that subject plants to unfiltered (ambient) air and filtered (control) air have been an important feature of studies on the impact of air pollution on plants. One of the earliest studies (1950–1951) examined the effect of pollutants in the air of a Manchester suburb on the growth of perennial ryegrass, *L. perenne*. Significant reductions in growth were found in the grass exposed to the ambient air that had a mean SO_2 concentration of 0.07 ppm. Later investigations in the Sheffield area recorded reduced yields in plants exposed to much lower mean SO_2 concentrations. In other studies using more realistic open-top chamber designs, growth reductions only occurred at higher SO_2 exposures. The differences may be due to the different design of the chambers, although other factors may be involved. For instance, chronic SO_2 injury can be enhanced if plants are exposed to SO_2 during winter, when they are growing more slowly. Furthermore in these studies, only SO_2 was measured and it is now evident that other phytotoxic pollutants would have been also present, *e.g.* NO_x and O_3. The Sheffield studies coincided with some of the highest O_3 concentrations recorded in Britain.[36]

Vegetation injury caused by photochemical smog was first reported in the Los Angeles basin in 1944 and it has continued to be a chronic problem in southern California and the eastern states of the US. It was established that O_3 was the main phytotoxic agent in the smog complex. Ozone has caused widespread injury to agronomic and horticultural crops and natural and managed forest ecosystems, not only in California but also in many other states where meteorological conditions and primary pollution concentrations were favourable. Ozone has been the most economically damaging air pollutant to vegetation in the USA. Large-scale injury to tobacco crops in the eastern USA has been reported over a number of years. Extensive pine needle damage to Ponderosa and Jeffrey pines in western locations and white pine in the east were due to atmospheric oxidants (*i.e.* ozone and related pollutants). In the mixed conifer forest ecosystems in the mountains of Southern California, the dominant tree species, Ponderosa and Jeffrey pines, for many years, suffered annual mortalities of about 3%, with the result that hundreds of thousands of trees have died. In the 1970s, it was estimated that in the San Bernardino forest 46,230 acres had suffered severe ozone-type injury, 53,920 acres moderate injury, and 60,800 acres light or no injury.

Evidence that ambient ozone concentrations in Britain during the summer can inflict deleterious effects on vegetation came from a series of experiments, using open-top chambers, at a rural site near Ascot in Berkshire, some 32 km west of London. At this location SO_2 and NO_x concentrations are generally low, but episodes of high ozone concentrations have been regularly recorded. During three such incidents between 1978 and 1983, *Pisum sativum*, *Trifolium repens*, and *T. pratense*, grown in open-top chambers and receiving unfiltered air, developed visible leaf necrosis typical of O_3 damage. Concentrations of O_3 during the course of these experiments exceeded 0.1 ppm, which is considered to be the threshold above which visible leaf necrosis appears in these sensitive plants. Field observations in the surrounding areas revealed symptoms typical of O_3 injury in a variety of crop plants. It has since been realized that *P. sativum* crops grown in the UK for a number of years, frequently developed necrotic lesions typical of ozone injury without the cause being known. Field observations have established that many commercially grown crops in Europe develop visible injury, such as chlorosis or bronzing, due to ambient levels of O_3 and that these symptoms commonly develop in response to short-term episodes rather than longer-term average concentrations.[37]

The main impact of O_3 on semi-natural plant communities is through shifts in species composition, loss of biodiversity, and changes in the genetic composition. In one study over 61 plant species native to Derbyshire were exposed to the same O_3 exposure over a period of 2 weeks, also included in the study for comparative purposes was a variety of tobacco which was known to be sensitive to O_3. Whilst up to 45% of the species showed no detectable effect, over 25% showed a marked reduction in growth and with a similar sensitivity to the tobacco cultivar. Evidence of tolerance to O_3 comes from a similar experiment in which populations of the broad leaved plantain (*Plantago major*) collected from different geographical areas of the UK were exposed to O_3 and it was found that there was as much variation in sensitivity within this one species as there was between the 61 different plant species. The O_3 resistance of the populations was significantly correlated with the O_3 concentrations prevailing in the areas from which they were collected. The most resistant populations were from southern England, indicating that these populations had evolved resistance to O_3.[38,39]

Studies have shown that increasing the concentration of O_3 has a greater impact than increasing the period of exposure and levels over and above a threshold concentration give a better fit to reductions in growth. The European open-top chamber experiment (EOTC), conducted in nine countries over a 5-year period, was a major international research study on the effects of O_3 over the complete lifecycle of several crop species. Figure 11 shows the linear relationship between O_3 exposure, expressed as the accumulated exposure over a threshold of 40 ppb (AOT40), and the relative yield of spring wheat. Similar relationships have been established for tree species such as beech and Norway spruce. Beech is more sensitive to O_3 with an AOT40 value of 10,000 ppb.h associated with a 10% reduction in growth.[37]

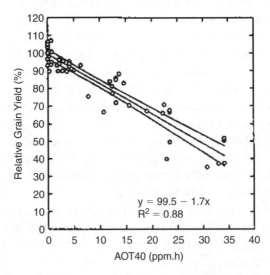

Figure 11 *Reduction of relative grain yield of spring wheat in relation to ozone exposure expressed as AOT40 for 3 months, and based on data from eight EOTC*
(Reproduced with permission from J. Fuhrer and B. Ackermann, *Critical levels for ozone*; A UN-ECE workshop report. FAC Report No. 16, Swiss Federal Research Institute for Agricultural Chemistry and Environmental Hygiene, Liebefeld-Bern)

Other effects of air pollutants include damage to the epicuticular wax layer of leaves and needles, which can lead to an increased water loss and there are reports that air pollutants can delay the onset of winter hardening. There is evidence that air pollutants acting singly and in combination (SO_3, SO_2/NO_2, and O_3) have an effect on the relative distribution of growth above and below ground, with root growth being more severely affected than shoot growth. This may be a consequence of interference with phloem translocation and assimilation in the plant. There are also reports that plants in polluted areas are more susceptible to attack by insect pests and exposure of plants to small or medium doses of SO_2 and/or NO_2 result in increases of the population growth of aphids feeding on the plants. This in turn can, result in significant increases in pest damage to plants.

The critical level concept is currently used for managing the impact of air pollution on vegetation. For each air pollutant and for a defined set of environmental conditions, a limiting value is defined above which available evidence suggests that, adverse effects are likely to occur with an increased exposure to pollutants. A limiting value is defined for different types of vegetation – crops, forests and woodlands, and semi-natural vegetation. Modifying factors such as winter stress, or nutrient deficient soils may be taken into account. Taking semi-natural vegetation as one category, the critical levels for SO_2 and NO_x are 20 and $30 \mu gm^{-3}$ respectively and that for O_3 is expressed as 3000 ppb.h (ppb O_3 above a threshold of 40 ppb accumulated during daylight hours). In the UK, it has been estimated that while the land area of semi-natural vegetation where the critical levels of SO_2 and NO_x are exceeded is in the region of only 1%, that for O_3 is of the order of 70% which highlights the continuing importance of this pollutant. The reader is referred to Bell and Treshow[40] for a more complete appraisal of the impact of air pollution on vegetation.

6.12   ECOLOGICAL EFFECTS OF ACID DEPOSITION

In the 1970s, a picture emerged of acidified lakes and streams across large parts of Europe and North America. It is now generally accepted that acid deposition with major acidic, or acidifying ions was the cause of the freshwater acidification in geologically sensitive areas. Such areas are those with slowly weathering granites and quartzite rocks that support acidic and weakly buffered podzolic and peaty-type soils. It is now known that these poorly buffered catchments were particularly sensitive to the strong acid inputs that originated from acidifying gases (SO_2 and NO_x) produced by urban and industrial centres located many miles upwind. The acidifying effect was due to the acidity of the rain and strong acid anions, in particular SO_4^{2-}, and rain more acidic than about pH 4.7 and SO_4^{2-} greater that about $20 \mu eq$ L^{-1} produced the greatest impact in the sensitive areas. Freshwater acidification is typified by a loss of acid neutralizing capacity (ANC) with SO_4^{2-} replacing HCO_3^- as the dominant anion, pH < 5, and elevated concentrations of aluminium species (Al^{n+}). The ANC indicates the acid-base status of water and is defined as the sum of base cations (Ca^{2+}, Mg^{2+}, Na^+, K^+) minus the sum of strong-acid anions (SO_4^{2-}, Cl^-, NO_3^-). Acidified waters have a negative ANC, whereas waters retaining bicarbonate buffering have a positive ANC.

Acidification has been responsible for the loss and depletion of fish populations from numerous freshwater ecosystems in parts of Norway, Sweden, UK, Canada, and USA. A considerable amount is now known about the physiological basis for the effects of acid water on aquatic organisms and there is good correspondence between experimental data and evidence from field populations. Freshwater organisms generally maintain their internal salt concentration by active uptake of ions (in particular Na^+ and Cl^-) from water against a concentration gradient. Fish mortality in acidic waters is primarily caused by a failure to regulate internal salt concentrations at gill surfaces and a characteristic symptom of acid-stressed fish is a decrease in the salt or electrolyte ion concentration of blood plasma. A study that had an important influence on pinpointing the physiological mechanism of toxicity in fish was one which followed up an extensive mortality of trout in the River Tovdal in southern Norway during spring snowmelt that resulted in a surge of acidity being discharged into the river.[41] An analysis of surviving fish showed that Na^+ and Cl^- ion concentrations of blood plasma were lower in fish from acidic reaches of the river and very low from parts of the river that recorded the most fish deaths. Field data indicated that low ion plasma concentrations and fish mortality occurred in waters of pH 5 which contrasted with laboratory experiments with purely acid solutions that could only reproduce the results in waters of pH 4. It was later shown that the dissolved concentration and chemical species of aluminium was the important factor influencing toxicity in acidified rivers and lakes. This common element in rocks and soil is much more soluble in acidic conditions, resulting in elevated concentrations in lakes and rivers of acidified areas that have been shown to be toxic in controlled experiments. The speciation of Al at different pHs is also important because the most available ionic form, Al^{3+}, predominates at around pH 5. The combined effect of acidity and Al on fish and other aquatic organism is now well established with the most toxic combination occurring at pH 5; the survival of brown trout fry is reduced at concentrations of aluminium of $250\,\mu gL^{-1}$, which is typical of acidified waters and longer-term effects on the growth of brown trout have been found at much lower concentrations. Other water chemistry constituents can influence Al toxicity, for example, increasing calcium concentrations in the water ameliorates the effect of aluminium, by competing for binding sites on gill surfaces and other constituents such as sulfates, fluorides, and soluble organic matter can form complexes with aluminium that reduces its bioavailability. All these factors help to explain the pattern of loss of fish populations in acidified areas as it was the clear water lake ecosystems, low in organic matter and low in calcium, which were the first and most badly affected and these environments were typically in mountain areas of Europe and in particular southern Norway.

The effects of acidification were not restricted to fish and various species of microorganisms, plants, and animals, have all been shown to be affected with a general tendency for a reduction in species diversity at all trophic levels. For many groups, this is accompanied by a reduction in productivity and a loss or decline of populations, *e.g.* several species of crustacea (snails and crayfish), amphibians, and fish.

The following summary of the effects of acidification on the major trophic levels in freshwater systems is based on a review of the subject by Muniz.[42]

Decomposers. Accumulation of undecomposed and partly decomposed organic matter is a characteristic of some acidified waters which indicates reduced activity of bacteria, fungi, and protozoans in the decomposition process. However, other studies have found no effects on decomposition.

Primary Producers. Reductions in the numbers of species of several phytoplanktonic algae, especially the green algae (chlorophyceae) have been reported. Dinoflagellates dominate the plankton of many acid lakes. Fewer species of periphyton occur on submerged surfaces and these habitats are commonly superseded by mass encroachments of filamentous algae (*e.g. Mougeotia, Zygogonium*, and *Spirogyra*). In many lakes in Sweden macrophyte populations of *Lobelia* declined and were succeeded by *Sphagnum* species. Synoptic surveys generally indicate that biomass and primary productivity are less in acidified lakes, although experimentally acidified lakes have shown no change and even increases in biomass.

Zooplankton. Species diversity decreases with increasing acidity, particularly below pH 5.5–5.0 and this is usually followed by a reduction in biomass. Acid sensitive species of *Daphnia* and *Cyclops* frequently disappear below pH 5.

Benthic Macroinvertebrates. Many species of mayflies, amphipods (freshwater shrimps), crayfish, snails, and clams are very sensitive to acid conditions. In Norway, snails were found to be absent in waters below pH 5.2 while mussels disappeared at pH 4.7. Recruitment failure and reduced growth are important causes of the elimination of species. Crayfish decline has been due to the effects of acidity on the eggs and juvenile stages, and the moulting stages are particularly sensitive. A similar situation has been found for freshwater shrimps, mayflies, and snails.

Fish. There are numerous examples of the loss of individual species and changes in the composition of fish communities in acidified waters. Minnows seem to be some of the most sensitive fish and population declines often start below pH 6. An approximate order of sensitivity to acid stress is as follows (beginning with the most sensitive): roach, minnow, arctic char, trout, European cisco, perch, white pike, and eels are consistently the most tolerant species. The younger life stages, particularly newly hatched or 'swim-up' fry just starting to feed, are the most sensitive stages to acid stress. As a consequence many acid-stressed populations are dominated by older fish due to recruitment failure.

Amphibians. Reproductive failure is the main effect of acid stress on amphibians. In Sweden, populations of the frog *Rana temporaria* have been lost and reproductive failure of the toad *Bufo bufo* has occurred.

Birds. Studies have shown that the breeding density of dippers in Wales (songbirds that feed almost exclusively on aquatic invertebrates) decreased as acidity has increased. This was related to the decrease in abundance of mayflies and caddis fly larvae in acidified streams. In general, some species of songbirds that breed alongside acidified lakes and streams produce smaller clutches with thinner eggshells, and breed later with fewer second clutches. Evidence also indicates that their young grow less rapidly and suffer more mortality.

Although the loss of decline of populations of animals and plants is largely attributable to changes in water chemistry, indirect effects may also cause an additional stress. These essentially arise out of changes in predator–prey relationships caused by a decline of food resources. For example, instances of increases in phytoplankton biomass may be due to a decrease in the performance of herbivores. The decline of the dipper is strongly associated with the disappearance of its prey from acidified streams.

Case Study 6: Lake Acidification; an Example of a Large-Scale Field Experiment

Experimental acidification of entire lakes has yielded very valuable information and an increased understanding of the effects of acidification on communities and ecosystem function. The Experimental Lakes Area of northwestern Ontario, Canada has been the subject of a number of large-scale field experiments, which have investigated the effect of pollution on aquatic ecosystems; they have included the effects of nutrient enrichment and synthetic estrogens (see below). Lake 223, in the Experimental Lakes Area of Canada, was acidified over an 8-year period during which time the pH was gradually reduced from 6.8 to 5.0.[43,44] One of the surprising features of the study was that primary production and decomposition showed no overall reduction. However, the composition of the phytoplankton and zooplankton communities showed distinct changes. At pH 5.9 several key organisms in the lake's food web were severely affected; opossum shrimps (*Mysis relicta*) declined from almost 7 billion to only a few animals and fathead minnows (*Pimephales promelas*) failed to reproduce. At pH 5.6 the exoskeleton of crayfish (*Oronectes virilis*) hardened more slowly after moulting and remained softer. The animals later became infected with a microsporozoan parasite.

When the acidification of Lake 223 was reversed, several biotic components recovered quickly. Fishes resumed reproduction at pHs similar to those at which it failed on acidification. The condition of lake trout improved as the small fish on which they depended for food returned. Many species of insects and crustaceans returned as the pH was raised.

Diatom communities in lakes have provided valuable information on the history of aquatic systems and in particular on their acidity. These organisms have characteristic siliceous scales that are preserved in lake sediments. Species of diatom can be classified according to their relative tolerance to acid or alkaline conditions and assemblages of different species can be used to indicate the pH of lake water. Palaeolimnological data from sediment cores have been used to construct the pH histories of lakes and in relation to recent changes they have provided evidence of the major causes. It is clear that sensitive lakes in the UK with moderate to high sulfur deposition have been acidified since the middle of the 19th century. Figure 12 shows the diatom assemblages at different depths from Lochnagar in the northeast of Scotland, and it can be seen that the onset of acidification is evident from about 1890 when the ANC had been exceeded and the pH of the lake decreased by 0.5 pH units.[45]

A key point here is that the start of acidification of the lake could be traced back to the time of the industrial revolution and this has been shown elsewhere where similar studies have been undertaken and from other records such as fish catches. For example, in southern Norway, losses of trout populations and declines in salmon populations were reported as early as the 1920s. Putting together records of water chemistry changes, in particular loss of acid neutralizing capacity, with those of fish and invertebrate populations has enabled exposure–response relationships to be established. One of the most sensitive species is the Atlantic salmon, which starts to decline when ANC drops below $20\,\mu\mathrm{eqL}^{-1}$ and this value is now widely regarded as the critical lower limit for protecting most species in freshwater systems. It should be remembered that ANC levels are negative in acidified waters and often below $-50\,\mu\mathrm{eqL}^{-1}$.

Recovery from acidification, therefore, requires a substantial increase in the ANC of surface waters which can only be achieved by decreasing the input of

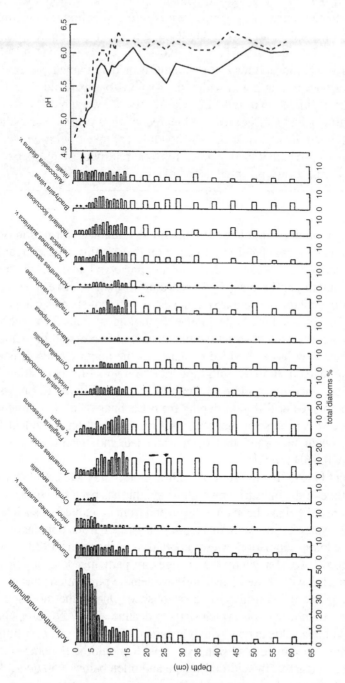

Figure 12 *Frequency distribution of diatom species in historical sediments (left) and pH reconstructions (right) from Lochnagar (Scotland); arrows show depths dated as 1900 (lower) and 1950 (upper)*
(Reproduced with permission from *J. Ecol.*, 1993, **81**, 3)

acidifying anions into catchments such that the rate of base cation production from weathering and atmospheric inputs exceeds its rate of depletion by the acid anions. This requires establishing linkages between atmospheric inputs and surface water chemistry. In Europe, emissions of acidifying gases peaked in the 1970s and by 2000, sulfur deposition had decreased by over 50%. Some of the reduction has been achieved through international agreement under the Convention on Long-Range Transport of Air Pollutants (UN economic commission for Europe), which calls for governments to implement measures to substantially reduce sulfur and nitrogen emissions. The goal is to increase ANC to above critical levels and thereby create conditions for recovery of fish and invertebrate populations. It is early days as yet, but measurements of water chemistry in Norway, Sweden, and Adirondacks (US) suggest an improvement in ANC in recent years. Modelling projections suggest that the recovery process will take several decades and this assumes that governments fully accede and implement the reductions in S and N emissions.

6.13 EFFECTS OF POLLUTANTS ON REPRODUCTION AND DEVELOPMENT: EVIDENCE OF ENDOCRINE DISRUPTION?

There are a number of instances of chemical pollution-related events in wildlife populations where changes in reproduction and development have occurred which have led to population declines and on occasions to extinctions. There is growing evidence that reproductive and developmental abnormalities observed are a consequence of effects on the endocrine system. Chemicals that are associated with disturbing normal endocrine homeostasis are referred to as endocrine disruptors or EDCs and the subject has become a major issue in chemical pollution.

The main events in wildlife populations, which are described in more detail below are

(1) Eggshell thinning in bird populations associated with an exposure to organochlorine pesticides.
(2) Reproductive and developmental problems in wildlife around the Great Lakes of North America (GLEMEDS).
(3) PCBs and reproductive impairment in marine mammals.
(4) TBT and imposex in gastropods.
(5) Abnormal vitellogenin (Vg) production in male fish populations.

The functions of all organ systems of animals are regulated by the endocrine signalling system and disturbances of this system resulting in hormone imbalances can lead to permanent damage especially if they occur during the early stages of development. There is more than one way by which EDCs interact with the endocrine system but it is believed that most of them act by binding to the nuclear receptors in cells in competition with natural hormones, in particular estrogens, and by acting in this way they are said to mimic or antagonize the action of natural hormones. The list of environmental chemicals that are known to have estrogenic activity, includes the organochlorines, DDT, PCBs, and dioxins but there are many more such as bisphenol and degradation products of detergents. It is also important to note that synthetic estrogens are found in sewage treatment effluents.

In the 1990s, several reports of reproductive and developmental problems in populations served to highlight the issue. These included reports of reproductive abnormalities in male alligators in the highly contaminated Lake Apopka (Florida, US). The contamination arose from agricultural activity, municipal runoff, and a chemical spill containing dicofol, DDT, and DDE. Male alligators had poorly developed sex organs and reduced levels of testosterone relative to estrogen hormones. Also, high incidences of intersexuality or hermaphroditic characteristics in fish populations in rivers throughout the UK were associated with discharges from sewage treatment works that were found to contain chemicals with estrogenic activity.

Looking back in the literature, it became clear that there were earlier references to the estrogenic activity of DDT. In 1968, there were reports in which DDT injected into eggs of California gulls (*Larus californious*) at similar levels to those found in wild populations, produced feminization in male birds. This corresponded to observations in gull populations in which the female:male ratio was found to be skewed to females.

Evidence from studies of human populations also suggests that some form of endocrine disruption may be occurring. There is a well-established link between women taking synthetic estrogen, diethylstilbestrol (DES) during pregnancy for preventing abortion and other complications and reproductive disorders in offspring. Epidemiological studies have reported deterioration in male reproductive health in recent decades that include reports of declines in semen quality, increased incidences of testicular cancer, and disorders of the male reproductive system. Tentative links have been made to the extensive use of DES and other estrogenic chemicals that are now commonplace in the environment.

There is, therefore, considerable concern that the widespread reproductive and developmental problems reported in humans and wildlife are related and that they are caused by chemical pollutants with estrogenic activity (or indeed with antagonists of other steroidal hormone systems). A considerable amount of research is in progress around the world which is concerned with developing assays for identifying EDCs, finding out how chemicals bind with hormone receptors, and assessing the impact on the health of humans and wildlife. In the sections below, a historical perspective is given of reproductive and development problems in wildlife populations that are associated with EDCs.

6.13.1 Case Study 7: Eggshell Thinning; an Example of a Physiological Effect in Bird Populations Associated with Organochlorine Insecticides which had a Major Influence on Recognition of the Importance of Chemical Pollution by Persistent Organic Chemicals

A wealth of literature exists on the relationship between DDT (and its metabolite DDE) and the phenomenon of eggshell thinning in various bird populations.[46,47] It was first discovered in the peregrine falcon (*Falco peregrinus*) in the UK. This particular falcon is widespread throughout Eurasia and North America, feeding almost entirely on live birds caught in flight. From about 1955 onwards, for no obvious reason, the numbers of falcons rapidly declined in southern England and subsequently, this decline spread northwards to parts of the Scottish highlands. By 1962 in the UK, some 51% of all known pre-war territories had been deserted and this figure was as high as 93% for southern England. At the time, biologists in several countries were also investigating populations of birds that had slowly declined to critical levels. It emerged that they all had residues of DDT, its metabolites in particular DDE, other chlorinated hydrocarbon insecticides, PCBs, and other chemicals in their tissues.

Ratcliffe[48] examined the eggshells of peregrine falcons and sparrow hawks in the UK from archival samples stretching back to the 19th century and samples collected during the 1950s and 1960s. Using an index of eggshell thickness (eggshell weight/(egg length $\times$ egg breadth)) it was found that during 1947 and 1948 this index decreased significantly by, on average, 19%. Eggshell thinning was also found in several American species of birds, notably the bald eagle, osprey, and peregrine falcon. In a study of over 23,000 eggshells of 25 bird species, shell thinning was found in 22 of the species and a correlation was found between the degree of thinning and DDE residues in the eggs. An experimental work with mallards and American kestrels in which captive birds were fed food dosed with DDE, produced eggs with thickness indices in proportion to DDE residues in the eggs and overlapping with DDE residues and thinning in wild populations.

It is now well accepted that DDE was the primary cause of eggshell thinning in many different kinds of birds and that some bird species were more susceptible than others to DDE-induced shell thinning. Relating eggshell thinning to low hatching rates and population decline was not straightforward for all affected species. In general, poor hatching rates and survival of chicks is increasingly evident with thinning of over 20% and this can be linked to population decline. In the case of peregrine falcon populations in the UK, the actual population decline took place at least 5 years after the onset of eggshell thinning. This coincided more closely with an upsurge in the use of dieldrin as an anti-fungicidal seed dressing.

Although it has never been substantiated, it is widely believed that the abrupt decline was due to a combination of factors, involving eggshell thinning and the later ingestion of toxic doses of dieldrin derived from a diet of contaminated pigeons.

In the years following the ban of the use of DDT, many of the species susceptible to eggshell thinning such as the peregrine falcon, guillemot, and the double-crested cormorant have undergone dramatic population increases and re-attainment of pre-DDT eggshell rigidity and thickness. A recent study of goshawk eggs (*Accipiter gentiles*) from northern Germany, which examined trends over the period of 1971–2002, showed that the eggshell thinning index had returned to the same thickness levels prior to the use of DDT.[49]

Eggshell thinning is often referred to as an example of endocrine disruption but the actual mechanism has yet to be established and it could involve a complex series of events. Shell thinning is associated with a decreased amount of calcium and the strongest evidence for an endocrine disruption mechanism comes from studies which show that DDE leads to a reduction in the synthesis of prostaglandins (PGs) by the shell gland mucosa. PGs play an important role in the control and regulation of reproduction in birds.

6.13.2 GLEMEDS

Organochlorine compounds such as PCBs, DDT, and TCDD have been associated with physiological, reproductive, developmental, behavioural, and population level problems in fish-eating birds of the Great Lakes for over 30 years.[50] Common features have been mortalities and deformities of eggs and chicks which led to the condition being called GLEMEDS (Great Lake Embryo Mortality, Edema, and Deformities Syndrome). The symptoms showed a number of similarities to chick-edema disease in poultry caused by an exposure to dioxin-contaminated food which was first recognized in the 1950s. The syndrome first came to light in common terns and herring gulls in the 1970s with observations of high incidences of egg and chick mortalities and deformities, particularly in colonies from Lake Ontario. By the mid-1970s organochlorine concentrations had declined and this corresponded with the disappearance of the gross effects in Lake Ontario herring gulls. During the same period, there were declines in bald eagle populations. Subsequent studies reported stronger associations with PCBs and dioxins, *e.g.* deformities and egg mortality in colonies of double-crested cormorants and Caspian terns were correlated with dioxin contamination and with specific congeners of PCBs in Foster's terns from Lake Michigan. Reproductive and developmental effects have been found in other wildlife populations, *e.g.* egg mortalities and deformities in snapping turtles from the Great Lakes basin, and an increased tumour incidence in the Beluga whale in the St Lawrence have been associated with levels of organochlorines and in particular PCBs. Developmental problems in human infants from regions of Lake Michigan and Lake Ontario have also been related to maternal exposure to organochlorines as a result of consumption of contaminated fish.

Between 1992 and 1994 immunological responses and related variables were measured in pre-fledgling herring gulls and Caspian terns at colonies across a broad range of organochlorine contamination (mainly PCBs), as measured in the eggs. There was a strong exposure–response relationship in both species between organochlorines and suppressed T-cell-mediated immunity. Suppression was most severe (30–45%) in colonies in Lake Ontario (1992) and Saginaw Bay (1992–1994) for both species and in western Lake Erie (1992) for herring gulls.

Although most of the evidence is through correlation, taken as a whole there is good reason to believe that the reproductive and developmental problems highlighted are due to organochlorines with PCBs and dioxins as the major problem agents. While inputs of organochlorines to the Great Lakes have largely been controlled and concentrations have declined there are still measurable levels in wildlife and concentrations in fish remain a significant hazard. The residual levels are due to the persistence of these substances, leachates from landfill sites, and inputs from the atmosphere.

6.13.3 Marine Mammals

There is a history of reproductive problems in seal populations in the Baltic and in the Dutch Wadden Sea. Between 1950 and 1970 the population in the Wadden Sea dropped from more than 3000 to less than 500 animals. The reduction in numbers is related to low breeding success. The populations from both the Baltic and Wadden were found to contain high levels of organochlorine compounds and in particular PCBs.

There have been a number of major disease outbreaks among seals and dolphins, which have been attributed to infection with known or newly recognized morbilliviruses. In 1988, the previously unrecognized phocine distemper virus caused the death of 20,000 harbour seals in north-western Europe. Other morbilliviruses have been shown to be the sources of infections in porpoises and dolphins, and mass mortalities due to this type of virus occurred among striped dolphins in the Mediterranean from 1990 to 1992. The severity and extent of these virus-related diseases has led to the speculation that pollution and in particular the bioaccumulation of organochlorines could have had damaging effects on the immune systems of the animals and as a consequence made them more vulnerable to infection.

Semi-field experiments in which groups of harbour seals were fed fish contaminated with PCBs have shown a strong link with reproductive failure. Experiments were over 2 years or more and during this time one group of harbour seals were fed a diet of fish from the contaminated Baltic Sea and another group were fed relatively uncontaminated fish from the Atlantic. The reproductive success of the group receiving the higher dose of PCBs was significantly lower than the other group with the lower dose of PCBs and the contaminated group also had lower estrogen concentrations.[51] Follow-up studies have shown more conclusively that PCBs were the main group of pollutants that were associated with the reproductive effects and immunotoxicological assays suggested that their immune responses were impaired.

Experimental studies with mink and ferret have established that PCBs are highly disruptive of reproductive processes. In the American mink (*Mustela vison*) tissue concentrations of over $50 \, mg \, kg^{-1}$ PCB were associated with reproductive failure. The otter, which is a closely related species to the mink, has experienced population declines over wide areas of Europe since the 1950s and animals from populations showing the greatest decline commonly had PCB concentrations in their tissues of over $50 \, mg \, kg^{-1}$.

PCBs are very persistent chemicals in the environment and levels have been slow to respond to the widespread ban on their production and use. In section 6.6.1, the example of the Pacific salmon highlighted that significant amounts of PCBs are still in circulation in such populations and during migration, tissue concentrations can attain levels that are associated with toxicity. Recent studies have shown that high concentrations of PCBs occur in polar bears inhabiting the Norwegian Arctic.[52] These demonstrate the long-range transport of PCBs by air and ocean currents and the influence of the cold Arctic environment in acting as a sink and a channel into the marine food web. Polar bears mainly feed on seals and they consume the blubber of these animals, which exposes them to lipophilic pollutants and especially PCBs in their prey. A key finding was that in female polar bears with offspring, concentrations of PCBs correlated with the reproductive hormone, progesterone (P_4), and while this is not a cause–effect relationship, it suggests that PCBs may affect P_4 levels and therefore act by disturbing the steroid hormone balance. It should be noted that no correlations between PCBs and either 17β-estradiol (E_2) or cortisol was found in the polar bears.

PCBs appear to have the capacity to interfere with several steps in the reproductive cycle of mammals, which includes the central hormone regulated brain-pituitary-gonadal axis. PCBs, like dioxins, have a strong affinity for the aryl-receptor, which is involved in the transcription process leading to the production of P450 enzymes and one line of evidence suggests that it is the hydroxylated products of these biotransformation reactions that act on hormone pathways to produce an imbalance in gonadal tissues, leading to reproductive damage. It has also been suggested that binding to the aryl-receptor in itself sends negative signals that result in an inhibition of hormone production. In molecular biology terminology where the stimulation of one pathway leads to an altered expression of another it is referred to as "cross talk".

6.13.4 Tributyltin and Imposex in Gastropods

Case Study 8: Imposex in Gastropods and Bivalves: an Example of Endocrine Disruption

TBT has been the cause of reproductive problems and population declines in several species of gastropods and molluscs around the coasts of Europe and North America. It is a good example of an EDC and one in which there are good linkages between the different levels of organization of molecular impacts, physiological damage, and population declines. TBT was used as an antifouling agent in paints applied to the hulls of boats and it was in the areas adjacent to large marinas that the damaging effects of TBT were largely apparent.

Sex abnormalities in neogastropod molluscs were first recorded in 1970. Many females in dogwhelk populations (*Nucella lapillus*) from Plymouth Sound, UK were found to have penis-like structures, this was followed by the reporting of male characteristics in female American mud snails from the Connecticut coast, USA. The development of male characters in females is referred to as imposex. In the 1970s, reproductive failure and major population declines of the English native oyster occurred and the introduced Pacific oyster, *Crassostrea gigas*, at certain locations along the British east coast and the French west coast, started to show poor growth and an unusual thickening of their shells. Populations of both gastropods and bivalve molluscs showing the severest abnormalities were invariably in the vicinity of yachting marinas and it was later shown that their tissues had a high tin content. In the 1980s, it was firmly established that antifouling paints containing TBT were responsible for these effects. This was confirmed by laboratory tests which reproduced imposex in gastropods and shell thickening in the bivalve molluscs at the same concentrations of TBT found near to yachting marinas and it was also shown that the tissue concentrations of TBT associated with these effects were the same in field surveys and laboratory experiments.[53]

Many other species of gastropods have been found to exhibit TBT-induced masculinization of the female and the phenomenon is worldwide in occurrence with over 100 species showing varying degrees of imposex. The consequences of imposex vary according to species and in some the abnormality does not appear to affect reproduction, while in others such as the dogwhelk, abnormality can be so severe that breeding is prevented and as a result, populations have declined drastically and even become extinct. In the dogwhelk, exposure–response relationships have been established and the onset of symptoms is associated with exposures of less than 1 ng L^{-1} TBT in seawater. Concentrations of 1–3 ngL^{-1} result in more pronounced masculinization with the penis size approaching that of males, although at this stage breeding is unaffected. However, at 5 ngL^{-1} the development of the male sex organs begins to cause blockages in the female tissues and the snail in this condition is unable to breed. The no-effect exposure with respect to imposex is very much lower than that estimated from initial screening studies using established test organisms and other end points such as mortality. In fact, in considering the environmental exposure/no effect ratio for imposex in dogwhelks, environmental concentrations in estuaries along the coast of southern England regularly exceeded levels associated with imposex.

The BCF of TBT in the dogwhelk is of the order of 30,000 and in comparison with other species it is probably near to the median value, with some exhibiting much higher factors and others' lower ones. The range of biocentration factors has been related to the ability of species to metabolize TBT. Metabolism involves successive removal of the butyl groups to give inorganic tin which can be more easily excreted, hence species with a low capacity to metabolize TBT accumulate higher concentrations and they are the most sensitive species. This is another example in which differences in the biodynamics of a chemical between species explains much if not all the variation in sensitivity to the toxin. In the case of TBT, toxicity in terms of external concentrations between species ranges over more than one order of magnitude.

The use of TBT as an antifouling agent on small vessels was banned in the UK in 1987 and affected populations have started to recover. However, it is still permitted for use on large vessels and imposex has been found in the edible whelk from the North Sea close to busy shipping routes.

At the molecular level, imposex is associated with elevated levels of testosterone and it has been shown that administration of testosterone to females produces masculine characteristics. There are some reports that TBT inhibits an aromatase enzyme involved in the metabolism of testosterone to 17β-estradiol. However, there are several hypotheses concerning the induction of imposex and a recent study demonstrated that TBT has an affinity for binding to human retinoid X receptors, which are members of the nuclear receptor family. While the endocrinology of invertebrates is not well established, it is widely acknowledged that hormone-receptor binding systems are common to species throughout the animal kingdom, and that TBT interaction with retinoid receptors is a basic step in the development of imposex.[54]

6.13.5 Case Study 9: Sexual Disruption in Fish and Amphibians – Vitellogenesis; a Biomarker Demonstrating Endocrine Disruption

In recent years, widespread induction of vitellogenesis in male and juvenile fish has been reported in Europe, UK, Japan, and USA. It is one of the most convincing biological responses associated with exposure to EDCs and the strongest responses are seen in water affected by discharges from sewage treatment works and pulp and paper mills. The main estrogenic substances involved are natural and synthetic steroidal hormones and other industrial-based chemicals that include alkylphenols, short-chain ethoxylated alkylphenols and bisphenol, and in pulp and paper mill effluents, wood derived compounds. Vg is an egg yolk precursor protein expressed only in mature female fish. It is produced in the liver and transported in the bloodstream to the oocytes. The synthesis of this protein is regulated by estrogens and during egg production it is normally elevated in blood serum of females. It is normally dormant in male fish but on exposure to estrogenic EDCs, the Vg gene is expressed in a dose dependent manner. Therefore, assays measuring Vg in plasma or Vg gene expression in male fish are specific biomarkers of exposure to estrogenic substances.

In rivers throughout the UK, elevated levels of Vg in male fish have been found wherever treated sewage discharges impact on a river. Such findings were evident using caged fish at various distances from sewage discharge points as well as in natural populations.[55] Fishes with increased Vg concentrations, frequently had enlarged livers and reduced testicular growth and in some instances this was accompanied by oocyte development in the testicular tissue. By the nature of the discharges into the river, Vg induction in male fish in the River Aire is thought to be mainly caused by nonylphenol and its ethoxylates. However, elsewhere natural and synthetic estrogens are considered to be the main estrogenic component of effluents.

Vg induction in male fish is not confined to freshwater species and extensive surveys of the European flounder in estuaries throughout the UK have reported increased levels in the major industrial estuaries relative to rural and less developed estuaries (Figure 13).[56,57] Similar trends in vitellogenesis of male fish populations have been reported across all the major continents of the world, which leads to the conclusion that endocrine disruption is widespread. However, there is only limited information on what impact these responses have on reproductive success and populations. In the Experimental Lakes Area of northwestern Ontario, Canada, a field experiment was recently set up to determine the effects of the synthetic estrogen, 17-ethynylestradiol on populations and physiology of fish, bacteria, phytoplankton, zooplankton, and insects.[58] In the ice-free period of May–September of each year ethynylestradiol was discharged into Lake 260 which achieved concentrations of $4–8\,ngL^{-1}$ in water. After 3 years, there was a dramatic decline in the fathead minnow population but no other changes were recorded in other fish species or in the lower levels of the food web. Very high Vg levels were found in male fish of fathead minnows and it is thought that this resulted in immature males and reproductive failure, although a further possibility is that the high Vg levels caused kidney damage that resulted in the death of male fish.

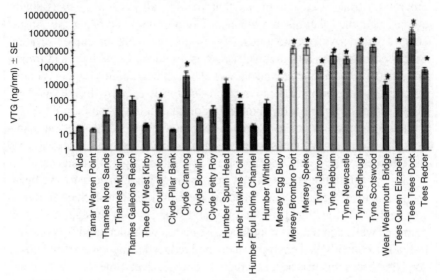

Figure 13 *Mean plasma Vg induction in male flounder from UK estuaries sampled in 1997;*
*denotes significantly different from control estuary (Alde) fish; and error bars
represent standard error of the mean (SEM)
(Reproduced with permission from Ref. 57)

A recent study has presented laboratory and field evidence for endocrine disruption in frogs caused by the commonly used herbicide, atrazine.[59] Experiments with the American leopard frog (*Rana pipiens*) showed that impaired gonad development and hermaphrodite characters were evident in frogs at atrazine concentrations well below those regularly measured in surface waters in the US. In field populations, similar characteristics were recorded at atrazine contaminated sites across the US and the investigators linked their findings to the decline in amphibians in general, suggesting that the atrazine together with EDCs are the cause of this decline.

However, it is still unclear which substances should be classed as the main endocrine disruptors and how they act at the molecular level. All parts of the endocrine system are potential targets, which include hormone receptor interactions, interference with production of the sex steroids, and effects on processes concerned with hormonal control in the hypothalamus-pituitary-gonad axis.

6.14 CONCLUSIONS

Chemical pollutants are not substances of a distinct type and all manner of physical and chemical properties are represented in this broad group of substances. It has been shown that they can be categorized in terms of their polarity, their structural properties, and their reactivity and these properties are fundamental to explaining why some types of chemicals in certain situations accumulate in organisms and produce damaging effects. Cause–effect or more specifically exposure–effect relationships form an important basis for the perception and evaluation of risk or damage from pollution. For many of the important pollutants, it is perhaps surprising to learn that these basic relationships have yet to be firmly established.

The problems stem from the complexity of the interaction of pollution with the environment and with systems within organisms. Cause–effect relationships cannot be derived from a single approach, but they are developed from field and experimental investigations and observations. In this regard, it is important to establish a consistent relationship between the measured effect and the suspected cause. The observed association should have a reasonable biological explanation. It should be possible to isolate the causal agent and to reproduce the effect under controlled conditions. The cause should normally precede the effect.

In the past, pollution studies were generally initiated in response to an obvious problem and while most of the acute problems have been identified and tackled, even today new chronic problems such as endocrine disruption are still being uncovered. This is at a time when there is an increased obligation to prevent such events by implementing chemical hazard and risk assessment procedures. As preventive pollution control strategies are implemented for existing and new forms of chemical pollution, it is important that risk assessment procedures are developed to take an account of all possible risks. There is some doubt whether or not these are sufficiently rigorous to tackle the relatively newly uncovered problem of endocrine disruption. In fact, risk assessment procedures are biased towards mortality-based toxicity and assays at the biomolecular level, which can screen for specific toxic mechanisms have yet to be incorporated into assessment procedures. The concluding remarks of the tenth report of the Royal Commission on environmental pollution are still relevant today, 'an important feature of the type of long-term environmental protection policy is that it should guard against creating situations which, though they may initially appear innocuous, have the potential for erupting disastrously – in other words, it should ensure that no new "time bombs" are set'. It is therefore, important that the next phase of research should concentrate on building under-standing from the molecular base such that key biomolecular events can be linked to whole organism damage and in turn to impacts on populations.

QUESTIONS

1. State what is meant by polar- and non-polar chemicals and give a rea-soned account of how polarity can be used as a basis for explaining the way in which toxic chemicals react with the molecules of cells.
2. The NOEC is a central feature of ERA. Describe how it is estimated, how it is used in risk assessment procedures, and critically evaluate how effect-ive it is for controlling and preventing chemical pollution.
3. The chapter introduced the concept of the "biodynamics" of pollutants; describe what is meant by this concept and using examples, discuss the extent to which it can be used for explaining differences in sensitivity between species and even between individuals of the same species to the same chemical exposure.
4. In relation to the impact of air pollution on vegetation and acidification of freshwater systems, explain the concepts of critical concentrations and loads in terms of how they are derived and applied to the management of these forms of pollution.
5. With reference to the health effects of lead and mercury, give an appraisal of the current understanding of their exposure–response relationships and provide a reasoned assessment of why they remain problem pollutants.
6. Critically assess the evidence for endocrine disruption in wildlife asso-ciated with chemical pollution.
7. The molecular basis of toxicity is the fundamental level of biological organization for explaining the ecological and health effects of chemical pollutants. Discuss the validity of this statement in the context of how the different levels of organization are believed to be related and in the light of recent advances in molecular biology.
8. Despite considerable improvements in the air quality of towns and cities throughout Europe and North America over the last 30 years, it is gen-erally believed that current levels of exposure to air pollution are sufficient to cause health effects. Critically review the epidemiological evidence that has led to this widely held view.

REFERENCES

1. M.W. Holdgate, *A perspective of Environmental Pollution*, Cambridge University Press, Cambridge, 1979.

2. R.P. Schwarzenbach, P.M. Gschwend and D.M. Imboden, *Environmental Organic Chemistry*, Wiley-Interscience Publication, New York, 1993.
3. P. Matthiessen, *Environ. Sci. Technol.*, 1998, **32**, 460A.
4. J. Widdows *et al.*, *Mar. Ecol. Prog. Ser.*, 1995, **127**, 131.
5. R. Carson, *Silent Spring*, Houghton Mifflin Co., Boston, 1962.
6. R. Macdonald, D. Mackay and B. Hickie, *Environ. Sci. Technol.*, 2002, **36**, 457A.
7. K.A. Kidd *et al.*, *Science*, 1995, **269**, 240.
8. A.M.H. Debruyn, M.C. Ikonomou and F.A.P.C. Gobas, *Environ. Sci. Technol.*, 2004, **38**, 6217.
9. B.N. Zegers *et al.*, *Environ. Sci. Technol.*, 2005, **39**, 2095.
10. S.N. Luoma and P.S. Rainbow, *Environ. Sci. Technol.*, 2005, **39**, 1921.
11. J.R. Larison *et al.*, *Nature*, 2000, **406**, 181.
12. A.R. Stewart *et al.*, *Environ. Sci. Technol.*, 2004, **38**, 4519.
13. B.I. Escher and J.L.M. Hermens, *Environ. Sci. Technol.*, 2002, **36**, 4201.
14. B.I. Escher and J.L.M. Hermens, *Environ. Sci. Technol.*, 2004, **38**, 455A.
15. C.A. Harris *et al.*, *Environ. Sci. Technol.*, 2001, **35**, 2909.
16. W.V. Welshons *et al.*, *Environ. Health Perspect.*, 2003, **111**, 994.
17. S. Niyogi and C.M. Wood, *Environ. Sci. Technol.*, 2004, **38**, 6177.
18. E.F. Pane *et al.*, *Environ. Sci. Technol.*, 2003, **37**, 4382.
19. L.S. Birnbaum and S.E. Fenton, *Environ. Health Perspect.*, 2003, **111**, 389.
20. Royal Commission on Environmental Pollution, Ninth Report, Cmnd. 8852, HMSO, London, 1983.
21. *Environmental Health Criteria, 1, Mercury*, World Health Organization, Geneva, 1976.
22. *Environmental Health Criteria, 101, Mercury*, World Health Organization, Geneva, 1990.
23. *Toxicological Effects of Methylmercury*, Committee on the Toxicological Effects of Methylmercury, National Research Council, National Academy Press, Washington, DC, 2000.
24. A.J. McMichael *et al.*, *New Eng. J. Med.*, 1988, **319**, 468.
25. Von Lindern *et al.*, *Sci. Total Environ.*, 2003, **303**, 139.
26. J. Evelyn, *The Smoke of London: Two Prophecies*, J.P. Lodge (ed), Maxwell reprint, New York (1661, reprinted), 1969.
27. Ministry of Health, *Report on Public Health and Medical and Medical Subjects No 95*, HMSO, London, 1954.
28. *Environmental Health Criteria, 8, Sulphur Oxides and Suspended Particulate Matter*, World Health Organization, Geneva, 1979.
29. *Environmental Health Criteria, 9, Photochemical Oxidants*, World Health Organization, Geneva, 1979.
30. H.R.A. Ponce de Leon *et al.*, *J. Epidemiol. Comm. Health*, 1996, **50**, S63.

31. WHO Report on a Working Group, *Health Aspects of Air Pollution with Particulate Matter, Ozone and Nitrogen Dioxide*, World Health Organization, Regional Office for Europe, 2003, EUR/03/5042688.

32. M.L. Bell *et al.*, *JAMA*, 2004, **292**, 2372.

33. D.W. Dockery *et al.*, *New Engl. J. Med.*, 1993, **329**, 1753.

34. C.A. Pope *et al.*, *Am. J. Respir. Crit. Care Med.*, 1995, **151**, 669.

35. N. Li *et al.*, *Environ. Health Perspect.*, 2003, **111**, 455.

36. J.N.B. Bell, *Gaseous Air Pollutants and Plant Metabolism*, Ch 1, M.J. Koziol and F.R. Whatley (eds), Butterworth, London, 1984.

37. United Kingdom Critical Loads Advisory Group, *Critical Levels of Air Pollutants for the United Kingdom*, Department of the Environment, London, 1996.

38. K. Reiling and A.W. Davison, *New Phytol.*, 1992, **109**, 1.

39. K. Reiling and A.W. Davison, *New Phytol.*, 1992, **122**, 699.

40. J.N.B. Bell and M. Treshow, *Air Pollution and Plant Life*, 2nd edn, Wiley, New York, 2002.

41. H. Leivestad and I.P. Muniz, *Nature*, 1976, **259**, 391.

42. I.P. Muniz, *Acid Deposition – Its Nature and Impacts*, F.T. Last and B. Watling (eds), *Proc. R. Soc. Edinburgh*, 1991, Section B, **97**, 228.

43. D.W. Schindler *et al.*, *Science*, 1985, **228**, 1395.

44. D.W. Schindler *et al.*, *Acid Deposition-Its Nature and Impacts*, F.T. Last and B. Watling (eds), *Proc. R. Soc. Edinburgh*, 1991, Section B, **97**, 193.

45. V.J. Jones *et al.*, *J. Ecol.*, 1993, **81**, 3–24.

46. F. Moriarty, *Ecotoxicology: The Study of Pollutants in Ecosystems*, Academic Press, London, 1998.

47. T.J. Peterle, *Wildlife Toxicology*, Van Nostrand Reinhold, New York, 1991.

48. D.A. Ratcliffe, *J. Appl. Ecol.*, 1970, **7**, 67.

49. W. Scharenberg and V. Looft, *Ambio*, 2004, **33**, 495.

50. M. Gilbertson, *Environ. Toxicol. Chem.*, 1997, **16**, 1771.

51. P.J.H. Reijnders, *Nature*, 1986, **324**, 456.

52. M. Haave *et al.*, *Environ. Health Perspect.*, 2003, **111**, 431.

53. P. Matthiessen and P.E. Gibbs, *Environ. Toxicol. Chem.*, 1998, **17**, 37.

54. J. Nishikawa *et al.*, *Environ. Sci. Technol.*, 2004, **38**, 6271.

55. S. Jobling *et al.*, *Environ. Sci. Technol.*, 1998, **32**, 2498.

56. Y. Allen *et al.*, *Environ. Toxicol. Chem.*, 1999, **18**, 1791.

57. Y. Allen *et al.*, *Sci. Total Environ.*, 1999, **233**, 5.

58. V.P. Palace *et al.*, *Water Qual. Res. J. Can.*, 2002, **37**, 637.

59. T. Hayes *et al.*, *Environ. Health Perspect.*, 2003, **111**, 568.

concerted action to stop development in flood plains. But in one of the most densely populated countries in the world, this is not always heeded where land for development is short. Good planning is one of the keys to good environmental management.

There are examples of 'win–win' situations where actions to improve the environment can make savings elsewhere. Waste minimisation by business is a good example of this (see Case studies 1a and 1b). Such situations are the 'utopia' which we continue to look for, but are all too rare at present.

This chapter breaks these complex interrelationships down into some of the basic principles of environmental management.

7.1.1 Environmental Management Cycle

So what is environmental management? It is helpful to think of it as a cyclical process involving evaluating the state of the environment to determine the priorities for action; deciding how these might be best achieved, taking the action and the monitoring performance to check that the action has improved the environment itself (Figure 1)[4].

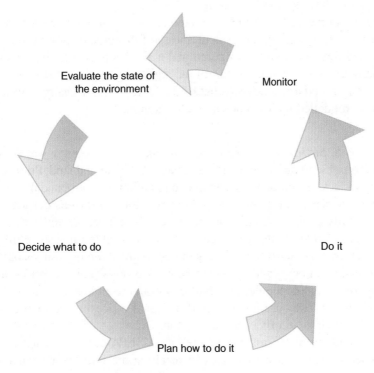

Figure 1 *The environmental management cycle*

This chapter looks at each of these components in more depth. Some aspects have already been covered in previous chapters, for example, monitoring is covered at depth in Chapter 5. This chapter shows how all these aspects are joined together for effective environmental management. It draws out the role of various players – governments, institutions, businesses and other stake-holders in achieving this aim.

7.1.2   Organisations Involved in Environmental Management

Achieving the optimum solution to complex environmental problems depends on an integrated management of environmental media and a wide range of technical specialisms. Businesses, which are the subject of regulation because their activities impact on the environment, do not like dealing with a plethora of agencies and officials.

There is no consistency from country to country as to the best way of managing the environment in an integrated way. Some countries combine the management of natural habitat within the same agency as pollution prevention, others combine regulatory activities like the protection of drinking water and aspects of public health protection, which although not part of the natural environment, have strong links to it.

Many countries in Europe have created unified pollution inspectorates but few of them link water management and the full-range of pollution control duties, and few integrate their operational activities over river basins. Most, especially in federal states, are part of the regional government structure. One of the first unified agencies was that of the United States; many other countries have modelled their agencies on their approach.

7.1.2.1   U.S. Environmental Protection Agency.   The Agency was set up in 1970 in response to public demand for cleaner water, air and land. Prior to this, there was a lack in a co-ordinated approach to pollution control and the USEPA is working to repair the environmental degradation of previous decades. This Agency leads environmental science, research, education and assessment efforts in the United States of America. It is responsible for researching and setting national standards for a variety of environmental programmes but issuing permits, monitoring and enforcement work is delegated to other bodies. It employs over 18,000 people, over half of whom are engineers, scientists and policy analysts. They are located in 10 regional offices, more than a dozen labs and in their head office in Washington, DC.

Where national standards are not met, the USEPA can issue sanctions and take other steps to assist the states in reaching the targets of environment quality.

The USEPA has a large budget, about half of which provides direct support through grants to the state environmental programmes. The funding supports high-quality research to improve the scientific basis for decisions on national environmental issues and helps the USEPA achieve its goals. Specific financial assistance is directed at targeted programmes such as the Drinking Water State Revolving Fund, the Brownfields Program and the Clean Water State Revolving fund.

With respect to environmental research, the USEPA integrates the work of various nations, private sector organisations, academics and other agencies in order to identify, understand and solve current and future environmental problems. As such, it provides strong leadership in addressing emerging environmental issues and in advancing the science and technology of risk assessment and risk management.

The USEPA works in partnership with over 10,000 industries and businesses, local governments and other bodies to tackle pollution prevention and energy conservation.

Voluntary pollution management goals are set for issues such as conserving water and energy, reducing greenhouse gases, reusing solid waste and minimising toxic emissions. The USEPA provides incentives such as access to research information and public recognition for these partnerships.[5]

7.1.2.2   Environment Agencies in the UK.   In the UK, environmental legislation has developed piecemeal, whereby successive legislation has led to new agencies being created, usually, but not always, independent of both national and local government structures. In 1996, the Environment Agency and the Scottish Environmental Protection Agency were formed, with comprehensive duties for environmental management as freestanding public bodies independent of government, known as Non-Departmental Public Bodies. They carry out environmental protection duties with a wide strategic and operational remit, ranging from national policy development and direct interface with Government and the European Community through to the management of the local field inspectorate. The agencies operate internally to environmental boundaries, based on river catchments, and are fully independent of local authorities. The agencies combine in their organisations the environmental management of the whole water cycle (integrated river basin management) and a wide range of pollution control duties (integrated pollution control). These include water resources, fisheries, flood defence, waste management and the large industrial processes that have the most potential to pollute the environment. The system in the UK is unique in Europe for its comprehensiveness of technical functions, degree of independence and method of organisation. The resultant bodies are large, employing over 11,000 people. They do not, however, have overall responsibility for biodiversity and conservation, which rests with another public body.

The UK approach maximises the possibility that regulatory decisions will be taken consistently, objectively and on the basis of sound science, independent of local political influence. But they lack accountability to the local electorate which is an issue. However, with large field inspectorates they can be responsive to local needs and can easily collaborate in local partnerships with other agencies.

7.1.2.3   Local Authorities in the UK.   In the UK, local authorities have responsibility for implementing the planning system which regulates the development and use of land in the public interest. They are responsible for developing county structure plans and development plans which define what can be located where in future. Such plans are critical in defining the pressures on the environment in future. Regional plans are developed by the Regional Development Agencies; these concentrate on economic development but must take into account social and environmental aspects.

Local authorities also have a wide range of responsibilities relevant to environmental management including waste planning, dealing with contaminated land and enforcing nuisance legislation, for example, noise pollution. They have responsibility for regulating air emissions from a range of processes including medium-sized combustion plant, waste oil burners and a number of industrial premises. They also have a duty to review and assess air quality in their areas; in a staged process related to the objectives of the National Air Quality Strategy. They also have responsibility for local transport plans and have a strong influence on this sector and hence its impact on the environment.

7.1.2.4   Central Government in the UK.   Strategies, plans and policies underpin environmental management. These are usually developed at national or international level under the auspices of central government. In the UK, overall responsibility for agriculture, land drainage and the environment lies with one government department, transport is in another and urban development and planning in another. It is inevitable that these functions need to be split but they need to work well together for effective environmental management. The UK government tries to do this through ensuring all its departments commit to sustainable development and tracks progress across the departments with a set of sustainable development indicators.

7.1.2.5   European Environment Agency.   This Agency collates, analyses and interprets environmental data and information from across the European Union. It produces periodic reports to inform the European Commission and European Parliament about the state of the environment in Europe. Its core task is to provide decision-makers with information needed for making sound and effective policies to protect the environment and support sustainable development. It does not make or enforce European policy. As such it has a very limited remit. The Agency gathers and distributes data and information through the European information and observation network (EIONET) which brings together over 300 bodies across 31 countries. The Agency employs less than 500 people. Eurostat is another European organisation that also collects statistics for the European Commission. Their remit is broader in that they collect economic and social statistics as well as some environmental statistics.

7.1.2.6 European Commission. The Commission is accountable to the European Parliament. One of its Directorates has specific responsibilities for the environment and develops specific European legislation. Member States are required to report progress to the Commission who may decide to take infraction proceedings against Member States who fail to comply. They can impose huge financial penalties for non-compliance. The European Commission sets environmental regulations in the form of European Council (EC) Directives, which must be incorporated into national legislation. They cover environmental standards, management processes and targets for emissions from specific sectors.

7.1.2.7 International Bodies. Some environmental issues require tackling on an international level and various bodies have come together under various conventions with specific aims. The United Nations Framework Convention on Climate Change resulted from the United National Conference on Environment and Development held in Rio de Janeiro in 1992 with the aim of tackling this global issue. The 1997 Kyoto Protocol to the Convention developed out of the work of this Convention, but adoption at country level can often be slow.

This conference also led to the UN Convention on Biodiversity. It promotes the restoration of degraded ecosystems and the recovery of threatened species. It requires each contracting party to develop or adapt national strategies, plans or programmes for the conservation and sustainable use of biological diversity.

There have been other Conventions, particularly relevant to atmospheric pollution. As part of the United Nations Environment Programme, the Vienna Convention to Protect the Ozone layer was signed in 1985, leading to the 1987 Montreal Protocol on substances that deplete the ozone layers.

The United Nations Economic Commission for Europe which covers all European countries, Russia, the United States and Canada has a Convention on Long Range Transboundary Air Pollution which has led to several protocols tackling different forms of pollution.

In Europe, the Paris commission (PARCOM) agreed in 1988 on surveys of riverine inputs of certain substances into the sea. This led to several 'North Sea Conferences' of Ministers of countries around the North Sea, which agreed targets for reducing the input of 36 dangerous substances to the sea in the late 1980s/1990s. These agreements were extended further at the 1998 Ministerial meeting of the Oslo and Paris Commission with more far-reaching targets for reducing pollution of the maritime area to 'near background values' by 2020.

7.2 EVALUATING THE STATE OF THE ENVIRONMENT

A sound scientific understanding of what the state of the environment is at any one time and how it is responding to the many and varied pressures placed upon it is critical to making the best management decisions. State of environment reporting can be carried out at local, national and international level. It generally follows the approach adopted by the Organisation for Economic Co-operation and Development in the early 1990s of using a 'pressure-state-response' framework.[6] This approach analyses the trends in pressures on the environment from human activities including, for example, industry, households and transport as well as analysing indicators of the state of the environment itself in terms of biodiversity, water quality, air quality and so on. It also analyses the societal and economic mechanisms in place to address the pressures and their effectiveness. This approach allows an integrated view to take place rather than a media or issue specific view. This is essential since changes in one pressure to solve one problem can give rise to other, new pressures without this inter-linkage. For example, there has been a move towards tertiary treatment in sewage-treatment works across the UK to reduce the nutrient pressure on rivers from this source. While this has led to improvements in river quality generally, the treatment processes are more energy-intensive leading to increased pressures from energy which has an impact on another media – the atmosphere. An integrated view is essential.

The 'pressure-state-response' model is widely used and has been expanded into a 'driver-pressure-state-impact-response' model by the European Environment Agency. This distinguishes between the underlying societal and economic drivers such as transport and agriculture and the actual pressures on the environment, that is, the emissions from these sources. Similarly, they distinguish between the state of the environment and the impacts of this on human health and other life forms (Figure 2).[7]

Within these frameworks, there is much information that could be used to assess each component. Policy-makers need a broad overview about what it all adds up to so there is now an emphasis of reporting on the state of the environment by using key indicators for each component.

7.2.1 Indicators

Indicators are defined as quantified information which show change over time. They have three basic functions – simplification, quantification and communication. Good indicators are scientifically sound, easily understood,

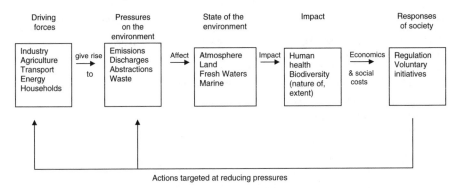

Figure 2 *The driving forces – pressures – state – impacts – responses framework*

sensitive to the change that they are intended to measure, measurable and capable of being updated regularly with readily available data and information.

The UK government produced its first set of 150 indicators of sustainable development in 1999 and has since reported annually on progress with 15 headlines to show overall progress (Table 1). The headlines have a balance of social, economic and environmental indicators.[8,9] The European Environment Agency, who is tasked with compiling and analysing information to inform policy decision of the European Commission, is itself developing a set of key indicators. It provides annual reports – Environmental Signals – to show progress in Europe with addressing environmental issues. Its reports inform the Commission of where policies are being effective and where more needs to be done to tackle the remaining and new challenges.[10]

7.2.2 Environmental Problems in Europe

The second assessment of Europe's environment in 1998 looked at the progress being made in tackling 12 key environmental problems. The indicators showed that over a five-year period, only three show progress with regard to development of policies or the state of the environment – these were acidification, stratospheric ozone depletion and technological and natural hazards. Progress in the state of the environment was unfavourable with many of the others – details are given in Table 2.[10] Some of this is due to the lag between policy development and its implementation and measuring the impact on the environment itself.

We should not underestimate the time it takes to translate a better understanding of pollution science into policy and then the realisation of actions on the ground into environmental benefits. The main barriers to progress are the complex, inter-sectoral, inter-disciplinary and international nature of both the problems and solutions. These barriers are underpinned by shortcomings in

Table 1 *The 15 headline indicators in the UK sustainable development strategy*

Maintaining high and stable levels of economic growth and development
1. Total output of the economy (GDP)
2. Total and social investment as a percentage of GDP
3. Proportion of people of working age who are in work

Social progress which recognises the needs of everyone
4. Tackling poverty and social exclusion
5. Qualifications at age 19
6. Expected years of healthy life
7. Homes judged unfit to live in
8. Level of crime

Effective protection of the environment
9. Emissions of greenhouse gases
10. Days when air pollution is moderate or higher
11. Road traffic
12. Rivers of good or fair quality
13. Population of wild birds
14. New homes built on previously developed land
15. Waste arisings and management

Table 2 *Progress in addressing environmental problems in Europe*

Key problem	Progress with policies	Improvement in state of environment
Climate change	≈	−
Stratospheric ozone depletion	+	−
Acidification	+	≈
Tropospheric ozone	≈	−
Chemicals	≈	≈
Waste	−	−
Biodiversity	≈	−
Inland waters	≈	≈
Marine and coasts	≈	−
Soils	−	−
Urban environments	≈	≈
Technological and natural hazards	+	+

Notes: +positive, ≈mixed progress, −little or unfavourable change.

institutional structures, non-implementation of commitments already made and lack of understanding of how to achieve sustainable outcomes.[11] Some of the lag is due to natural environment processes but much of it is due to the time it takes to get the required level of political will and for it to rise up the political agenda.

7.2.3 Eco-Efficiency

Eco-efficiency is a measure of the degree to which economic growth is 'decoupled' from environmental impact. This is assessed by combining various indicators. The generally accepted aim is for economic growth to continue without causing environmental degradation, although some people question whether economic growth is compatible with sustainable development. By looking at indicators in this way, the effectiveness of policies can be determined and areas for more action can be identified.

In the UK, there has been good progress in reducing the impact of industry on the environment while output has continued to grow (Figure 3).[12,13] This reflects the benefit of strong regulation on major point sources of emissions in the last 20–30 years.

The picture for transport is different. Economic growth is still strongly correlated with road transport growth and the growth in emissions of carbon dioxide. Progress has been made, however, in the reduction of other emissions such as nitrogen oxides from road transport (Figure 4).[8,13] This has largely been from the introduction of cleaner technologies but there is about a 10-year time lag for technological improvements to have an effect on reducing environmental pressures.

The degree of progress reflects awareness and the need for change. Awareness of the impact of industry dates back to the industrial revolution and much environmental legislation relates to this. Awareness of the impact of households and lifestyles is less recognised; households are often overlooked in integrated policies.

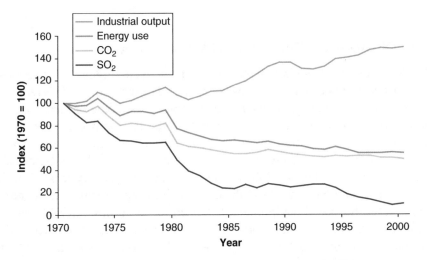

Figure 3 *Industrial output, energy use and emissions in the UK, 1970–2000*

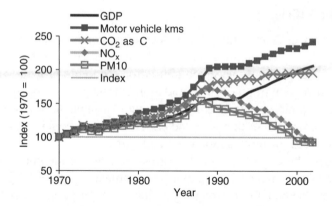

Figure 4 *Road transport and environmental impacts, 1970–2002*

Household energy use, waste generation and car ownership increased in Europe during the 1990s in line with the increase in household numbers and the rise in household expenditure. These trends are likely to continue unless progress is made in reducing the impact of household resource consumption and expenditure. In the UK, despite energy efficiency programmes aimed at households, energy use continues to rise overall due to an increase in the number of households, number of appliances used in the home and an increase in the internal temperature of many homes. These examples show how looking at indicators together can give an overview of how well we are moving towards decoupling economic growth from environmental impact. Much remains to be done to develop effective policies where the links still exist.

7.3   DECIDING AND PLANNING WHAT TO DO

The next step in the environmental management cycle (Figure 1) is to decide on what needs to be done to improve or protect the environment. This requires some consensus on aspirations, targets or objectives. Some of these may be long-term, others more immediate. Some of the decisions can be made at local level, others at a more international level and historically, human health has been a primary consideration in environment policy. Up until a decade ago, progress has depended on regulation which has largely tackled point sources of pollution. Since then, the most obvious progress in reducing pressure on the European environment has been made in those areas where an effective international framework for action has been established (such as the Vienna Convention on the protection of the ozone layer, the UNECE

Convention on Long-Range Transboundary Air Pollution and their protocols).[1] But now there are examples where these are not proving so effective – the Kyoto Protocol to address carbon dioxide emissions is proving difficult to agree and a way forward has yet to be found. Acute health problems are now rarely the result of environmental causes in the UK, and attention has shifted to the health implications of long-term exposure to environment pollution.

7.3.1 Strategies and Plans

Moving towards sustainable development requires national strategies and plans to reflect this goal. In Europe, much has been achieved through setting objectives agreed between various stakeholders and then putting in place strategies and plans for achieving them at Member State level. There are a plethora of strategies developed at national level in the UK, for example, the National Air Quality Strategy (Case study 2), National Waste Strategy (Case study 3), Sustainable Soil Strategy. These set out common objectives and targets that the country needs to work towards, but tend to look at specific sectors.

Case Study 2: The UK National Air Quality Strategy

This sets air quality standards based on recommendations from an expert panel, EC standards and World Health Organisation guidelines and provides the principle basis for assessing air quality. The standards set in 1997 were revised in 2000 to take account of new developments and to harmonise with the first EC Air Quality Daughter Directive. Short-term standards apply to acute health effects while the standards for chronic effects are based on annual mean concentration.

Some eight pollutants (such as particles, benzene and nitrogen dioxide) have standards specified in the strategy. Monitoring is carried out at specified rural and urban sites to measure compliance with the standards. Results are integrated and reported as the number of days above the standard in a given year – one of the government's headline indicators.

Under the strategy, local authorities are required to designate Air Quality Management Areas where the air quality objections are not likely to be met. For these areas, action plans must be drawn up to meet the objectives. This requires understanding the contributing sources of pollution. In many cases, it has led to local authorities introducing transport management measures to meet the objectives.[14]

Case Study 3: The UK National Waste Strategy

This recognises that all forms of waste management have an impact on the environment through emissions to the atmosphere, land and water, and may use significant quantities of resources for example, by transporting waste from one area to another. The waste strategy takes an integrated approach to waste management. It stresses the need to

- reduce the quantity of waste produced;
- recognise waste as a resource and recover value from it; and
- take an integrated approach by taking account of the collection, transport, sorting, processing and recovery or disposal of waste, involving all key players in the decision-making process and ensuring a mixture of waste-management options.

The strategy sets targets for reducing the amount of industrial and commercial waste sent to landfill and targets for recovery of municipal waste, reflecting the requirements of the European Landfill Directive. It states that the principles for decision-making should be based on the waste hierarchy, the Best Practicable Environment option, and the proximity principle, that is, that waste should generally be disposed of as near to its place of production as possible.[15]

To be effective, environmental concerns need to be integrated into strategies for specific policy areas, for example, within transport policies. There are examples of where integrated planning is beginning to take place – in coastal zone management and urban planning and development. Integrated coastal zone management seeks to manage pressures from development, recreation, fisheries, shipping and industry in a way that is environmentally sustainable, economically equitable and sensitive to local cultures. The process is interactive, proactive and adaptive bringing together various authorities and other bodies to develop an agreed work programme, concrete objectives and budgets. A similar approach is now being taken to urban planning in many European countries.

In the UK, the most integrated approach to deciding what to do lies in the preparation of development plans which provides a way of reconciling the demand for development, in all its forms, and the protection of the environment. Planning takes place at all levels – national, regional and local – and it is essential that all these plans are integrated. Because the plans take account of needs for regional development in economic terms and the need for housing, it is important that the needs of the environment are considered at the same time. Development of an appropriate form, scale and location may be necessary to secure economic and social objectives for an area and should be balanced with the needs of the environment.

It has been difficult to get the environment to feature widely in these plans which tend to be driven by economic needs. Even though the environment agencies are statutory consultees, they are only now becoming aware of the need to influence the plans early in their development if environmental concerns are to be fully integrated. But getting development in the right place with the appropriate level of infrastructure development (including sizing of sewers, sewage treatment works and incorporating sustainable drainage systems) is essential to protect the environment.

Consider the planning needed to accommodate a projected rise in the number of homes needed in England. Between 1996 and 2021, the number of homes is projected to increase by 3.8 million, 19 per cent, as a result of population change, behavioural change and greater life expectancy. The number is increasing faster than the population growth due to the fact that households are becoming smaller. The south-east of England is experiencing the greatest increase in household density – a 20 per cent growth projected from 2000. The increase in households requires more energy, water, goods and other services including increased use of transport. It is essential that this growth is planned well to minimise environmental impacts. Some of the proposed locations are where water resources are already stretched. Furthermore, sewage works are already operating to tight standards and any more flow may mean that the carrying capacity of the receiving waters is exceeded. If issues such as this are raised at the time development plans and regional spatial strategies are developed, then solutions can be factored in and problems prevented. Influencing the planning process is key to environmental management but an area which has not yet been fully grasped.

There is now a requirement for all major development plans to include a Strategic Environmental Assessment which is designed to integrate environmental considerations into policies from the outset. This reflects a European Directive transposed into UK law in 2004.

The assessment process is designed to ensure that significant environmental effects arising from proposed plans and programmes are identified, assessed, subjected to public participation and taken into account by decision-makers and monitored. The procedure sets the framework for future assessment of development projects, some of which requires detailed Environmental Impact Assessment.

The need to undertake strategic environmental assessments should improve environmental protection and promote sustainable development. Integration of environmental issues at an early stage of decision-making processes should enable 'win–win' situations to be identified, rather than mitigating or accommodating the impacts of decisions made without this consideration. But it is early days; it is important that the process is implemented within the spirit of the Directive and not seen as another 'bureaucratic' loop to get through.

The challenge rests with environmental managers to show that the process is worthwhile in making development more sustainable. There can be too much planning though, or too many types of plans. This can lead to stakeholder fatigue. The challenge ahead is for more integration.

7.3.2 Tools and Techniques

There are a host of tools and techniques to help make assessments and decisions about what to do. For example, at a more substance-specific level, life-cycle analysis is a useful tool for deciding the most sustainable approach to take. It is all too easy to decide, for example, that recycling products is a more sustainable option than incineration with energy recovery, because it is higher up the waste hierarchy (Figure 5) but this may not be the case. Recycling products often demands energy in transporting the products and recycling them. Sweden is now moving away from recycling to more energy-from-waste recovery.

Life cycle assessment and life cycle thinking are tools that can be used to assess the environmental impact of services or products throughout their life cycle ('cradle to grave'). The tools identify ways to reduce impacts and make cost savings. Life cycle analysis is the collection and evaluation of quantitative

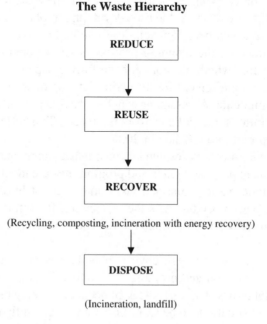

The Waste Hierarchy

REDUCE

REUSE

RECOVER

(Recycling, composting, incineration with energy recovery)

DISPOSE

(Incineration, landfill)

Figure 5 *The waste hierarchy*

data on the inputs and outputs of material, energy and waste flows associated with a product over its entire life cycle so that the environmental impacts can be determined. Life cycle thinking is more quantitative and does not require thorough data analysis. Life cycle assessments have been carried out on specific products. For example, a study in Germany found that refillable PET bottles (a type of plastic) were the best packaging for drinks, followed by cartons, refillable glass, metal cans and one-way glass bottles (many of the benefits of using plastic for recycling materials are linked to their lightweight which benefits transportation impacts). But an Austrian study found that replacing 1.5 L refillable PET drinks bottles with non-refillable PET bottles of the same size produced small but clear benefits.[16] Life cycle assessments have their place but the boundaries of the studies often have an over-riding impact on their results.

The relative merits of various environmental management options can be assessed using techniques such as the Best Practicable Environmental Option (BPEO). This is defined as "the outcome of a systematic and consultative decision-making procedure which emphasises the protection and conservation of the environment across land, air and water. The BPEO procedure establishes, from a given set of objectives, the option that provides the most benefits or the least damage to the environment as a whole, at acceptable cost, in the long term as well as in the short term".[17]

The nature of the definition and what it includes is such that general, international or even national rules are unlikely consistently to represent the BPEO. Although frameworks, such as the waste hierarchy and overarchy legislation such as the biodegradable municipal waste diversion targets in the Landfill Directive are set, the concept of BPEO means that local environmental, social and economic preferences will be important in any decision.

7.3.3 Risks and Values

Deciding what to do depends, to a large extent, on the risks and values attached to the issues. Risks can be assessed scientifically but public perception is also critical in an age where various stakeholders are involved in the decision-making process. The values attached to issues will vary from person to person and views on how much should be invested in rectifying problems will depend on value judgements about the importance of environmental expenditure against expenditure on other societal needs. Appraisal must take account of factors such as the scale of the problem and the nature of the risk it presents. For example, when there was gross pollution which had fatal impacts on human health and aquatic organisms, awareness of these risks was high. Diffuse pollution and its impacts are less well recognised, which can mean that there is less of a political driver to do something about it.

Many Directives, including the Water Framework Directive, and national legislation requires cost-benefit analysis to be undertaken. In England, there have been agreed approaches to the allocation of funding for flood-defence works for many years. This is based on a points system allocated according to three criteria – priority, urgency and economics – and then ranked on total points. The criteria take into account land use, including environmental assets of international importance, the state of repair of existing defences and hence urgency, and the economic benefit of the scheme. Urban areas tend to score more highly than rural areas due to the lives and property at risk.

The decisions on the allocation of funding to improve the environment are not as structured, although there is a requirement for cost-benefit analysis. The principle is that the cost of measures should not exceed the value of the benefits that they are designed to achieve. While it may be relatively straightforward to establish the cost of proposed improvements, it is harder to establish benefits. The problems include

- quantifying the population or resource at risk;
- knowing the effect of given levels of pollution (the 'dose response' relationship) and identifying all the effects;
- attaching monetary values to all the risks to human health, quality of life and wild life; and
- costing abatement measures when technology and markets are changing rapidly.

This leads to a high degree of uncertainty in the analysis, which creates problems for decision making.

The most widely used technique for obtaining a monetary value of environmental problems is the 'contingent valuation method', which uses surveys to assess the 'willingness to pay'. For example, a survey was done in the 1990s to evaluate how much people were willing to pay for low flows to be alleviated in the River Darent in Kent. This suggested between £3 to £6 per year amounting to £2.2 m per year when factored by the number of households in the area. This exceeded the estimated cost of the scheme so it was agreed that the scheme should be implemented.

There are difficulties with such surveys. It is not easy to determine the extent of influence of the project; people who say they may be willing to pay do not necessarily mean they would pay; the approach does not take into account other demands for resources in a limited budget. Nevertheless, such approaches add some perspective to the decision-making process.

7.3.4 Attitudes

Governments are elected representatives who make decisions on national strategies and approaches and legislation but its implementation often depends on non-elected bodies such as the environment agencies. It has become increasingly important that these bodies engage widely so that there is consensus of what to do.

The UK government carries out periodic surveys of public attitudes to the environment; face to face with about 2000 people selected to cover the whole range of socio-economic circumstances and different parts of the nation. These canvass opinions about how worried people are about issues covering local, national and global problems. Issues about which people are most worried have not changed much since the surveys in the 1980s – high on their list are chemicals put into rivers and the sea, sewage on beaches/bathing waters, radioactive waste, hazardous waste and oil spills at sea. In the last survey in 2001, concern had increased since the previous surveys over the effects of livestock farming methods. This result may have reflected the outbreak of foot and mouth in that year and BSE in previous years. The surveys also ask what environmental issues or trends would cause them the most concern in about 20 years time. Traffic congestion and fumes was the greatest concern (52 per cent of people) but global warming/climate change, air pollution and water pollution were all cited by about a quarter or a third of people in the 2001 survey.[18] But other surveys give different perceptions; the sample size, framing of questions and survey techniques can all have an impact on the results.

7.3.5 Involving Interest Groups

While the government, agencies and other regulators may have much of the overall responsibility for managing the environment, it is important that local businesses, interest groups and the public in general accept what is being proposed and the action being taken.

Consultation about proposals is a way of getting acceptance for change and expenditure but it has often been late in the process. This leads to criticism because consultees do not feel that their views will influence the outcomes. The environment agencies are now actively seeking 'stakeholder engagement' as it is called, earlier in the process, reflecting a need also recognised in the Water Framework Directive. Even so, this is not straightforward as shown by an example. The Environment Agency is developing Abstraction Management Strategies (CAMS) for some 130 catchments across England and Wales to determine how much water is available for abstraction without causing environmental damage. For each of these, a stakeholder group is set up with representatives from local government, water users, non-government organisations and other local interest groups. The group is involved throughout the development of the strategy in order to get engagement. But in reality the stakeholders are unable to influence the outcome because the strategies are based on a scientific methodology which cannot be varied. There is still much to be done to find ways of engaging with stakeholders and applying sound science in the decision-making process.

An example of where stakeholder engagement has led to environmental improvements is that of restoring flows to rivers in England. The problem lies

---

Case Study 4: Restoring Flows to the River Cherwell in Banbury

In the late 1980s, fishermen complained that low flows in a river in north Oxfordshire, England were affecting fishing potential. In certain years, the river ran completely dry due to abstractions upstream of the market town. These abstractions provide water for potable supply (after treatment) and to support a canal which links the county to the Midlands. Both abstraction licences are licences of right, issued for perpetuity over 40 years ago, meaning that the regulator is unable to vary them without paying compensation.

The regulator asked the water company, the major abstractor, to consider ways to restore flows to the river. A cost-benefit analysis was done showing that the benefit of a scheme to local amenity and users was £6 million.

They considered alternative sources of water to meet the supply needs of the towns. These involved transporting water large distances through pipes or the canal network, much more expensive than the estimated benefits. So there were only two options that could be considered further.

(i) Moving the abstraction point for the water treatment works (for potable supply) to below the town.
(ii) Transferring the effluent from the sewage-treatment works, currently entering the river below the town, upstream to provide flow in the river through the town (the water company also operates the sewage-treatment works).

Both options were costed and found to be of similar costs, about £5 million, less than the estimated benefits. Deciding on which option to adopt needed to consider other factors. Abstractions for potable supply (either with or without treatment) are traditionally above towns to reduce the risk of contamination from urban runoff. From this point of view, option 2 was preferred. But on the other hand, treated sewage-effluent is perceived to be of a lesser quality than natural river water and at times of low flow, it would be undiluted. Furthermore, in this case, the effluent contains treated effluent from a food-processing plant which is straw coloured, a potential concern aesthetically although not ecologically. In the detailed planning stage, the cost of keeping the risks to the potable supply at an acceptable level escalated, meaning that option 2 is now being implemented. Flows should be restored to the river through the town by 2007, rectifying a long-standing problem and restoring fishing potential and amenity value.

in the way in which abstraction licences were set several decades ago leading to degraded rivers in some places, too much water is abstracted at the expense of the ecology. The sites have been identified with public consultation and then various public bodies (English Nature and the Environment Agency) have worked with the water companies to find ways of rectifying the situation. The abstraction licences cannot be revoked under legislation without some form of compensation so alternative approaches have been needed. These include finding alternative sources of supply or alternative ways to abstract the same amount of water (see Case study 4). Funding for the projects required agreement of the agencies, central government and the economic regulator of the water industry, since any costs are ultimately passed on to the customer of the water companies.

7.4 TAKING ACTION

There are various approaches to taking action – regulatory, fiscal, voluntary and coercive. Each of these is addressed in turn. The traditional approach, and a cornerstone of environmental management in practice, is regulation, so more emphasis is given to this approach.

7.4.1 Regulation

7.4.1.1 History in the UK. Air and water pollution has been a concern since at least medieval times. A royal proclamation in 1306 banned the burning of coal in London because of its effect on air quality and an Act in 1383 ordered those with latrines over the Walbrook in London to pay for the cleaning of the river. However, with no adequate enforcement or means of waste disposal, together with an ever increasing population and an increase in emissions as industry developed, pollution became a growing problem.

Cholera epidemics broke out in the mid-19th century resulting in over 35,000 deaths in London alone. The impact of pollution on health was becoming clear, as was the need to keep sewage out of drinking water supplies and to keep the air clean. This led to the 1848 Public Health Act and in 1863 the Alkali, *etc.* Works Act was introduced due to the damage to health and the environment from alkali works. For example, a high incidence of infant mortality and emphysema occurred in some of the industrial cities of the times.

Despite this recognition of links between health and the environment in the mid-19th century, pollution tended to be accepted as the necessary price of industrial development. Many rivers were still grossly polluted a century later and the Great London smog of 1952 shows that air pollution was still bad (Table 3).[19]

Table 3 *Timeline of key events and legislation dealing with air pollution*

1257	Queen Eleanor leaves Nottingham because of air pollution
1285	Commission established to investigate pollution from coal burning in London
1306	The use of coal in London is prohibited
1863	The Alkali, *etc*. Works Act is introduced in response to damage to health and the environment from alkali works
1926	Public Health (Smoke Abatement) Act
1952	Smog during one week in December caused 4000 deaths in London
1956	First Clean Air Act
1961	First national survey of air pollution
1974	Control of Pollution Act
1979	Geneva convention on Long-Range Trans-boundary Air Pollution
1980	EC Directive on smoke and sulphur dioxide
1990	Integrated Pollution Control introduced and an air quality information service
1991	London traffic pollution episode costs 100–180 lives in four days
1997	Launch of UK national air quality strategy; Kyoto Protocol

New and changes in legislation since then have continued to address the problems and progress has been made in the last three decades. In general, the chemical quality of the environment has improved in the past 20 years due to investment in reducing emissions and tighter regulation. The Thames, which was devoid of oxygen through London in the 1950s, now supports over 100 species of fish, including salmon, which returned to the river in the 1980s after being absent for over a century.[20] There have been similar recoveries to the industrialised rivers of the north, such as the Tyne and the Tees, although there are still stretches in need of improvement (Table 4).

Regulation has driven reductions in emissions from industry and better sewage treatment. These improvements mean that our atmosphere and water resources have a much higher standard of protection today than in the past, and the many uses to which rivers are put – navigation, recreation and supporting wildlife and fisheries – are also protected.

But regulation has not worked so well in tackling the so-called 'diffuse sources' of pollution. These arise from activities that are dispersed over an area, as opposed to major 'point' sources of pollution. Common examples of diffuse water pollution include contaminated run-off from roads, polluted drainage from housing estates, run-off from farms polluted with surplus fertilisers and pesticides. Diffuse water pollution is widespread and is a long-term threat to the ecology of lakes, rivers and coastal waters and to the quality of groundwater and the cost of water supplies. Tackling diffuse sources of pollution is essential if we are to achieve statutory international quality standards and to make sustainable use of water resources. The main areas that need to be addressed are

Table 4 *Approach to improving river water quality in UK over past 30 years*

1970s	Many rivers anoxic, devoid of fish and suffering from sewage and industrial pollution; estuaries particularly affected
1978	River quality objectives set for all rivers. Government long-term target set to make all rivers suitable for fish
1980s	Pollution Control Act brought in. Consent standards set to ensure discharges meet needs of river. Lack of strong enforcement and investment led to slow progress, but some specific improvements in London leading to salmon found in Thames estuary again
1989	Privatisation of the water industry. Investment in sewage treatment approved so that consent standards could be met
1995	Vast improvement in water quality reported due to investment and greater awareness of pollution prevention measures
2000	Further investment plans delivering environmental benefits. Planning for the Water Framework Directive begins and the need to meet tighter objectives in future
2003	Salmon found in the Mersey estuary, first time for over a century (based on anecdotal information)

- Eutrophication (nutrient enrichment) from excess nutrients (mainly nitrate from agriculture and phosphorus from sewage-treatment works
- Nitrate contamination of groundwater
- Pollution from pesticides, solvents, mine waters and urban runoff
- Sedimentation of river habitats from disturbed soils and eroded river banks.

Diffuse air pollution includes emissions from cars and other vehicles, garden bonfires and emissions from heating systems in houses. Light and noise pollution can also be considered as types of diffuse pollution. Agriculture releases ammonia from animal waste to the atmosphere which contributes to acidification and methane from livestock contributes to global warming. Road transport is the main source of toxic air pollution in urban areas and a large source of nitrogen oxides and carbon dioxides. While improved fuel and emission standards have achieved reduction in pollution from transport, continued growth in traffic may reverse these changes and already cause exceedance of standards in many urban areas. The continued growth in air travel is also increasing diffuse sources of greenhouse gases and nitrogen oxides.

7.4.1.2 Overview of European Directives. Over the last 30 years, European policy and legislation has been an increasing driver on environment management and it is now the dominant driving force in Member States. There are well over 400 separate pieces of legislation that in some way have a bearing on environment management. They range from specifying the regulation of emissions from difference processes to the management and disposal of waste, and to the achievement of specific environmental quality standards and targets. Many of them specify the types of monitoring and reporting that needs to be done with reporting deadlines for Member States to meet. The Single European Act of 1987 set out three overall objectives:

- To preserve, protect and improve the quality of the environment
- To contribute towards protecting human health
- To ensure a prudent and rational utilisation of natural resources.

It also established five guiding principles for environmental regulation (Table 5).

While these principles are laudable aims, agreeing to what they mean in practice is often controversial. For example, in the Third North Sea Conference, 1990, participants were asked to apply the precautionary principle and take action to avoid potentially damaging impacts even when there is no scientific evidence to prove a causal link between emission and effects. The Rio Declaration on Environment and Development 1992 referred to the precautionary approach as: "where there are threats of serious or irreversible damage, lack of full scientific certainty shall not be used as a reason for postponing cost-effective measures to prevent environment degradation." But these definitions lead to many questions: how do we decide what is a serious threat? How far should we go with cost-effective measures? Hormones in beef, genetically modified organisms, and endocrine-disrupting substances are all examples of where the science is inconclusive and there is no consensus about the nature of the threat. In 2000, the European Commission attempted to provide some clarity on the precautionary principle requiring clear stakeholder involvement and the avoidance of trade disputes; there needs to be a meeting of minds between the politicians, scientists and the public if agreement is to be reached.[21,22]

Some of the Directives driving environmental improvements in Europe are given in Table 6. Many of the earliest Directives have since been amended or

Table 5 *Key principles in environmental regulation*
(From the Treaty of the European Union (Article 130r))

Sustainable	Policies should contribute to the goal of sustainable development
High level of protection at source	Environmental damage should as a priority be rectified at source rather than by treatment at the point of use
Precautionary	Where outcomes are uncertain, particularly if they are likely to be irreversible, then there should be a presumption in favour of a cautious approach
Integrated	Decisions on environmental impact should have regard to impacts and options for all media and be taken in an integrated and holistic way
Subsidiarity	Decisions should be taken at the lowest level and at the most local scale reasonable in the circumstances and recognise that, although principles of environmental protection are general, the diversity of situations applying in different regions will lead to different practices being applied

expand earlier Directives. For example, several Directives in the late 1980s specified limit values and quality objectives for specific substances listed in the Dangerous Substances Directive 76/464/EEC. The work to specify these had not been done at the time of the Directive so standards could only be specified when the science was complete. Even so, understanding of the impact of certain hazardous substances has continued to increase and many of these standards may be tightened further under the Water Framework Directive (see later). Similarly, the standards set in some of the other Directives have been

Table 6 *Example EC Directives relating to Environmental Management*

Directive number	Name	Media
75/440/EEC	Quality of surface water abstracted for drinking	Water
76/160/EEC	Quality of bathing water	Water
76/464/EEC	Pollution caused by the discharge of certain dangerous substances into the aquatic environment	Water
78/319/EEC	Hazardous Waste Directive	Waste
78/659/EEC	Quality of fresh waters needing protection or improvements in order to support fish life	Water
79/409/EEC	Directive on the conservation of wild birds	Biodiversitiy
80/68/EEC	Protection of groundwater against pollution caused by certain dangerous substances	Water
84/360/EEC	Combating of air pollution from industrial plants	Air
85/337/EEC	Environmental Impact Assessment Directive	All
89/369/EEC	Prevention of air pollution from new municipal waste incineration plants	Air
91/156/EEC	Waste Framework Directive amending 75/442/EEC	Waste
91/271/EEC	Urban waste water treatment	Water
91/441/EEC	Measures to be taken against air pollution by emissions from motor vehicles, amending Directive 70/220/EEC	Air
91/676/EEC	Protection of waters against pollution caused by nitrates from agricultural sources	Water
91/692/EEC	Standardised Reporting Directive	All
92/43/EEC	Habitat and Species Conservation Directive	Biodiversity
94/62/EEC	Packaging Directive	Waste
94/66/EC	The limitation of emissions of certain pollutants into the air from large combustion plants amending directive 88/609/EEC	Air
94/67/EC	Incineration of hazardous waste	Waste
96/62/EC	Air Quality Framework Directive	Air
96/91/EC	IPPC Directive	All
99/31/EC	Landfill Directive	Waste
2000/60/EC	Water Framework Directive	Water
2001/42/EC	Strategic Environmental Assessment Directive	All
2002/96/EC	Waste Electrical and Electronic Equipment Directive	Waste

questioned (*e.g.* bathing waters) and these could be revised in future. But often the science is not entirely conclusive and the costs and benefits of tightening standards are extensively debated. It can take many years for amendments to be agreed across Member States.

Due to the number and complexity of the Directives, there is now a move towards 'Framework' Directives. These integrate the requirements of previous Directives and allow scope for Member States to implement in the way they feel fit, following the subsidiarity principle. So EC controls are gradually taking a more co-ordinated approach to pollution rather than tackling problems separately. The Integrated Pollution Prevention and Control (IPPC) Directive is particularly wide-ranging in the number of industries it covers and takes a holistic approach by promoting energy efficiency and waste minimisation as well as preventing pollution.

Directives have to be transposed into national legislation and there is an agreed time period for this, which are often two or three years. There is then further time for many of the conditions to apply. This allows countries to prepare adequately for the new standards and requirements of the Directive, which usually require substantial planning and investment. For example, the Landfill Directive came into force in 1999, but it was not effected in the UK until 2004, some five years later.

Failing to comply with Directives is expensive. 'Infraction' as it is known, can cost tens of thousands of pounds per day. The UK has recently been challenged about its implementation of the Freshwater Fish Directive (Case study 5).

7.4.1.3 Water Framework Directive.   The Water Framework Directive is an example of a framework approach which is currently being implemented. It was agreed in 2000 after 12 years of discussion and development. The overall aim of the Directive of achieving good ecological quality has to be achieved at the latest 15 years after its date, some 27 years after recognition for the Directive; slow but steady change. The Directive will replace many of the older Directives in due course, for example, the Dangerous Substances to Water, Freshwater Fish and Nitrate Directives. The Directive requires Member States to take an integrated approach to river basin management. All waters, with a few exemptions, must meet 'Good Ecological Status' by 2015, unless Member States have negotiated an extension. The first reporting under the Directive took place in 2005 characterising the nature of river basins so that water bodies not meeting this status or at risk of not meeting it could be identified. River basin management plans must be developed by 2009 by competent authorities in conjunction with stakeholders; public participation in decision-making is very much part of this Directive. Plans must then be reviewed every six years.

Case Study 5: Freshwater Fish Directive Infraction Proceedings

The Freshwater Fish Directive is designed to protect the chemical quality of freshwater to support fish life. The Directive was introduced in 1978 and required Member States to designate rivers where the water quality should be suitable to support freshwater fisheries. Since then 20,000 km of river and canals in England and Wales have been designated as protected under the Fish Directive. Water quality at a designated site must meet chemical standards described in the Directive for 14 substances.

In January 2002, the European Commission initiated a legal action against the UK's implementation of the Directive. The main issue was insufficient designations of surface waters, particularly canals and still waters and inadequate designation to the source of rivers. The designations in the late 1970s reflected a cost-neutral position so only reaches capable of meeting the Directive were designated.

According to the EU web site, the UK designations under this Directive in 2002 were proportionally greater that most other member states of the EU but no other states have been challenged on these issues.

Further to the act, a large number of new designations were made in 2003. The impact of this was large. The water industry has been required to invest millions of pounds to ensure receiving waters comply with the requirements of these new designations. These costs have been incorporated into the 2005–2010 Asset Management Plans for the water companies, which also determine the cost for their customers.

The infraction proceedings have been a way of driving environmental improvements. So from an environmental viewpoint, reviewing the designations has been useful. But, the way in which the standards for the substances have been applied may have led to more designations than necessary, leading to costs which may not have any environmental benefit.

In this case, any investment for improvements to water quality must be based on a sound understanding of the causes and must be based on failures which are statistically significant not attributable to chance variations in quality. Risks are that the Commission pushes for a programme of designations which may include still waters and rivers not considering significant or at risk by the regulator. This will burden the regulator with a large sampling programme which may produce no benefit to the environment. Such a programme could also lead to very large costs to the water industry for discharge improvements that may not be of benefit to fish populations.

The plans must identify what needs to be done to reduce pressures on water bodies so that good status can be achieved. Co-incident with the planning, is the requirement for programmes of measures to be established with these becoming operational by 2012. The measures that can be utilised range from regulatory to voluntary mechanisms (further details later). The Directive provides a framework for addressing diffuse pollution for the first time. If this is shown to be the cause of failure to meet good status, then plans will need to be put in place to address the problems.

Even though this new Directive has been transposed in UK law, there is still much uncertainty about how it will be implemented on the ground. 'Good ecological status' has yet to be defined and a Daughter Directive, dealing with the priority and hazardous substance (to replace the former Dangerous Substances approach) is running behind schedule. The Environmental Quality Standards relating to the priority list have been debated widely; there is no common agreement about the interpretation of the underlying science, or how precaution should be approached; and much comes down to a judgement on risks and the costs and benefits of setting a certain level. There will be undoubtedly 'horse-trading' to resolve these issues and to move implementation of the Directive forward.

7.4.1.4 Air Quality Framework Directive. The Air Quality Framework Directive (96/62/EC) and its Daughter Directives aims to regulate pollutants in ambient air that pose risks to human health and requires the establishment of air quality standards for 13 substances. It requires Member States to monitor air quality and take all necessary measures, unless cost-prohibitive, to ensure that from 2012 (or other dates set in the Daughter Directives) target levels are not exceeded. Member States must draw up lists of zones where targets are exceeded and put in place measures to tackle. The UK National Air Quality Strategy relates to this (see Section 7.3.1 and later).

This Directive covers the vision of previously existing legislation and the introduction of new air quality standards for previously unregulated air pollutants, setting the timetable for the development of Daughter Directives on a range of pollutants. The list of atmospheric pollutants to be considered includes sulphur dioxide, nitrogen dioxide, particulate matter, lead and ozone – pollutants governed by already existing ambient air quality objectives – and benzene, carbon monoxide, poly-aromatic hydrocarbons, cadmium, arsenic, nickel and mercury.

An European-wide procedure for the exchange of information and data on ambient air quality in the European Community is established by the Council Decision 97/101/EC. The decision introduces a reciprocal exchange of information and data relating to the networks and stations set up in the Member States to measure air pollution and the air quality measurements taken by those stations.

The Framework Directive was followed by Daughter Directives in 1999, 2000 and 2002 which set the numerical limit values, or in the case of ozone, target values for each of the identified pollutants and a date by which they have to be met. Besides setting air quality limit and alert thresholds, the objectives of the Daughter Directives are to harmonise monitoring strategies, measuring methods, calibration and quality assessment methods to arrive at comparable measurements throughout the EU and to provide for good public information. The development of the daughter legislation is supported by expert working groups. These working groups consist of technical experts from the Commission, including their Joint Research Centre in Ispra, Member States, industry and environmental NGOs. They are supported as appropriate by the European Environment Agency, the World Health Organisation, the United Nations Economic Commission for Europe and consultants involved in cost-benefit analysis studies, among others.

The Framework Directive, as well as its Daughter Directives, requires the assessment of the ambient air quality existing in Member States on the basis of common methods and criteria; with this purpose, the Commission has prepared a Guidance report on Preliminary Assessment under EC Air Quality Directives and a Guidance on Assessment under the EU Air Quality Directives.[23]

7.4.1.5 Waste Directives. The Waste Framework Directive (75/442/EEC) as amended by 91/156/EEC established a general framework of controls for waste management that reflected the EC's general policies aimed at encouraging the recovery of waste and its management at the point of production. The Directive defines waste. Its objectives are to ensure that waste is recovered or disposed of without endangering human health or harming the environment. It requires that a public agency plans waste disposal properly, that waste disposal facilities are licensed by a competent public agency and that any exemptions from the licensing system fall within certain categories and are properly justified. It aims to harmonise landfill standards across the EC and to reduce emissions of methane (a greenhouse gas). It sets staged targets to reduce biodegradable waste to 35 per cent of its 1995 levels (by weight) by 2020 at the latest.

The Landfill Directive (99/31/EC) follows the aims and principles of the Waste Framework Directive. It requires changes to the ways in which Member States deals with waste. One major change has been the need to segregate different types of waste – hazardous from non-hazardous – in landfill sites. But, there was no agreement on certain aspects of the requirements under the Landfill Directive. For example, there is a requirement for pre-treatment of all hazardous waste but it is unclear what this actually entails. The processes are complex and involve good understanding of science (not always available) and good consultation if implementation is to be successful. There have been examples where problems have arisen because of the difficulties of implementation – 'fridge mountains', Case study 6, are an example.

> **Case Study 6: The UK's Fridge Mountain**
>
> Over two million fridges and freezers are disposed of each year in the UK. Before 2002, they were crushed and shredded for metal recovery. In 2002, some of the insulating foam became classified as hazardous waste because it contained ozone-depleting substances. These substances had to be removed before disposal. However, in 2002 there were no facilities to remove and deal with the ozone-depleting substances, which resulted in a fridge mountain. Treatment facilities are now in place to deal with fridges being disposed of as well as the backlog.[24]

The Waste Electrical and Electronics Equipment (WEEE) Directive (2002/96/EC), seeks to encourage the collection of WEEE and its reuse, recycling or recovery. The Directive places obligations on governments and, through them, producers, importers and retailers of electrical and electronic equipment. The overriding spirit is the making better use of resources and diverting waste away from landfill. To achieve this it places the burden on producers to find ways of better using discarded electrical and electronic equipment, and recovering any hazardous materials from them, a laudable aim but one which is taxing the minds of many people about how it can be put into practice. A battery-powered toy is as likely to come under the remit of this Directive as a television; new collection systems need to be sensitive to the items, for example, televisions must not be broken – the cathode ray tubes (hazardous materials) must stay intact and so on. Householders must be able to return equipment to retailers on a 'one-for-one' basis without charge. Retailers can establish alternative collection systems so long as these are not less convenient, such as operating a collection service when delivering new products, providing pre-paid envelopes for returning small items, or paying for upgrades at civic amenity sites.

The equipment collected must be treated to remove specified substances and components and there are recycling targets. This is a major challenge for producers. While collecting back their own products may be useful to them, collecting back other brands requires further investment. Computer manufacturers are worried that refurbishers may not wipe hard disks properly and power tool producers claim there is a danger that refurbished products could be unsafe. Printer cartridge re-manufacturers claim that a new 'return and recycle' service offered by one company attempts to bolster the market for its own new cartridges, although the company claims that recycling is environmentally preferable to reuse.

There is debate about the costs and the potential threat to small businesses. Product liability may also be an issue. The exact details of how it will be implemented are for each Member State. Only Greece met the EU deadline of August 2004 for transposing the Directive into national legislation. The other deadlines to be met were August 2005 for retailers to have established their in-store take-back or alternative collection systems and December 2006 for meeting the recycling and recovery targets. There are challenging times ahead.

The UK's failure to meet the deadline for transposing the Directive relates to the huge scale of the task which the government has faced. Its innovative approach and the way in which it impacts on many businesses and products should not be underestimated. The objectives of the Directive are clear: putting it into practice requires changes to our infrastructure and culture which will remain a challenge for months if we are to avoid the forming of WEEE mountains.

7.4.1.6 Regulation of Point Sources of Pollution in Practice. The basis of most regulations is for industry and businesses to get permits from a competent authority or regulator in order to operate (Figure 6). These include planning permission, abstraction licences, discharge consents, waste management licences and integrated pollution control authorisations. Businesses/individuals need to apply for permits normally with an application fee, which is then often advertised in the local paper to allow any representations to be made. The competent authority then determines whether to issue the permit or to refuse it, what conditions to attach to the permit to protect the environment and human health, and what monitoring and reporting requirements should be, if any. The exact nature of the permit depends on the relevant legislation but the principles are similar. Most permits are time-limited or subject to review after a period of years which allows for change, reflecting any better understanding, change in circumstances or new national or international requirements. By having the power to vary permits, improvements can be made as new technology comes on line (Case study 7).

Case Study 7: Rhodia Organique Fine, Avonmouth, Bristol

Rhodia manufactures organo-fluorine chemicals at Avonmouth for uses including refrigeration, medical applications and the automotive sector. The company has significantly reduced hydrofluorocarbon emissions from 192 tonnes in 1998 to 76 tonnes in 2001. The improvements were made through stricter Environment Agency authorisation coupled with the use of advanced technology. An optimisation team chaired by an Agency inspector met every two months for four years to achieve this outcome.[2]

It is then up to the regulator to ensure that the permit conditions are met. This takes the form of site audits, sampling, auditing sample results of the

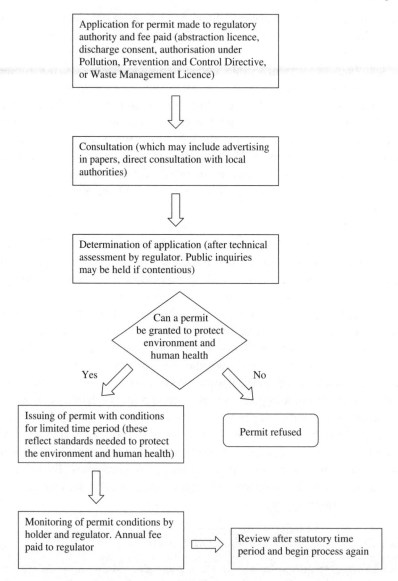

Figure 6 *Basis of regulation: the permitting process*

permit-holder and then making the results accessible on public registers. Some of these requirements are specified in relevant legislation and European Directives. When there is non-compliance with conditions, the regulator can serve enforcement notices requiring the regulated to take certain actions, serve prohibition notices where there is an imminent risk of serious environmental damage or harm to human health, suspend permits, send warning letters or cautions or prosecute in the courts.

Permitting regimes prescribe action, which society expects and which are appropriate to the level of risk. The public should have confidence in standard permits because they are perceived to be tough, transparent and consistent. But such regimes, applied in terms of 'one size fits all' are expensive to administer compared with other approaches, and they tend to be relatively inflexible. This means they are not ideal in circumstances where economic or environmental conditions change quickly.

There are a range of tools and techniques available to help environment officers set appropriate conditions on permits. All require an understanding of the state of the local environment within which the applicant wishes to operate, requiring for example, baseline data on river flows, water quality, ambient air quality conditions or whatever is applicable. Calculations are then made to ensure that any discharge to, or abstraction from, the environment complies with standards set for the receiving media. These need to take account of the variability in the nature of the receiving media and in the planned discharge or emission, having regard to the dispersion and dilution that will take place. Permit conditions are usually set to protect the environment when it is most at risk (*e.g.* low flow conditions in rivers) and to ensure environmental standards are met but many refer to 'average' conditions as well. Hence, most permits have a range of conditions which they must meet. For example, abstraction licences give an annual and peak daily allowance and most now restrict abstraction to times when river flows are above a certain rate. Discharge consents set absolute standards that can never be exceeded (maximum levels) and standards which can be exceeded by a certain number of samples depending on how many are taken during any one period (95 percentiles).

Various computer models exist to help environment officers with the tasks of setting standards and conditions. In the setting of consents to discharge to water in the UK for example, a Monte Carlo simulation technique is used that combines information on the statistical distributions of upstream river and discharge flow and quality, and calculates the appropriate emission limits to meet the downstream Environmental Quality Standard. The standard is therefore set in terms of the environmental capacity of the receiving medium rather than on the basis of what may be achieved by the process producing the emission.

An alternative approach is to use emission limits, using criteria based upon the capability of the technology. This leads to uniform emission standards for a process or a site, based upon assessments of process and pollution abatement techniques. It has largely been used for the most potentially polluting industries including power stations, oil refineries and chemical processes. This approach has the advantage of setting uniform environmental standards for any particular industry or process and is therefore favoured by industry in providing a 'level playing field'. It is also an approach that can be used without a wide knowledge of the quality and variability of the receiving medium. However, where such information is available, a sophisticated application of the emission limit approach allows local variation. In Germany, for example, uniform sectoral (*i.e.* specific industry-wide) standards applying to discharges to all media have been established in consultation with industrial representatives and independent technical associations. Uniform emission standards are derived from assessments of the state of the art in the process and pollution abatement technology for dangerous substances and non-biodegradable substances. These have become binding standards that translate directly into the process authorisations.

Uniform emission standards are relatively easy to apply. An environmental capacity-based system is harder to administer particularly if, as may be the case for air pollution control or for water pollution control to large river catchments such as the Rhine, the issue comes under the jurisdiction of a number of national regulatory agencies. However, where environmental capacity is limited, for example, in the case of small rivers with little dilution capacity, a uniform emission standard approach could easily lead to problems if account was not taken of the limited assimilative capacity of the receiving medium.

Historically, the UK favoured the approach of using quality standards to set permits whereas many other European countries favoured the emission limit approach. They have now been integrated under the IPPC Directive (96/61/EC). This Directive is built around the concept of best available technology (BAT), which describes processes or operations that are 'the most effective and advanced' of those available and which are selected "bearing in mind the likely costs and benefits and the principles of precaution and prevention". They should be capable of emission limit values designed to prevent impact on the environment, and if that is not practicable, minimise emissions and the impact on the environment as a whole. If BAT standards do not meet environmental quality standards then standards tighter than BAT must be set to ensure that the environmental quality standard is not breached. In practice BAT has been established on a Europe-wide basis in the form of standard documents drawn up by groups of process specialists.

IPPC is being implemented in a phased approach across a wide range of sectors. It includes the food and drink industry, poultry and pig farming – sectors which were not regulated under previous environmental legislation in the UK. Another facet of IPPC is the inclusion of criteria which requires good use of resources – energy and waste minimisation. This recognises the overall aim of sustainability as well as environmental protection.

Traditional regulation including IPPC has tackled point sources of pollution well. It has been less effective at dealing with diffuse pollution, the sources of which include urban and rural runoff. Designing infrastructure to deal with urban runoff appropriately (by incorporating sustainable systems) can tackle diffuse pollution from urban areas but adoption of the techniques has been limited due to lack of awareness. Similarly, diffuse pollution from rural areas can be tackled by using best practice in land management techniques. This may be brought about by financial incentives (discussed later in the chapter) rather than by regulation.

7.4.1.7 Modern Regulation. The trend towards Framework Directives has resulted from the political desire to move away from the complexity of so many Directives and so much regulation. While the benefits of regulation to improve health and the environment have been clear over the years, too many separate pieces of legislation are difficult for the regulators and regulated alike to implement and keep pace with. There is now, therefore, a move towards deregulation. In environmental management, this is manifesting itself in a move towards little or no regulation of processes which have minimal or no significant impact on the environment. For example, in the UK, the Water Act 2003, has identified a cut-off of $20\,m^3$/day for abstraction licences. All abstractions below this amount no longer require a licence because it is deemed that they will have no impact on the environment. This reduces bureaucracy and costs for both regulator and regulated.

This recognises that the approach of much regulation – 'one size fits all' – is wasteful of resources and there is now a move to more intelligent regulating, known as 'Modern Regulation'.[25] It is based on five principles set out by the UK's Better Regulation Taskforce, which are that it must be

- transparent – the rules and processes must be clear to the regulated;
- accountable – in terms of performance by regulator and regulated;
- consistent – the same approach must apply within and between sectors;
- proportionate – resources shall be allocated according to the risks involved and the scale of outcomes to be achieved; and
- targeted, or based on environmental outcomes, which must be central to the planning and approach taken.

Modern regulation aims to find the right balance – a proportionate, risk-based response that will drive environmental improvements, reward good performance, but still provide the ultimate reassurance that tough action will be taken on those who fail to meet acceptable standards. Where risks to the environment are higher, there will be more intervention and tighter regulation, where they are lower, there will be less regulation. Already some countries in Europe have reduced their burden of environmental regulation substantially; by some 25 per cent in the Netherlands.

Operator and Pollution Risk Appraisal (OPRA) forms part of the modern risk-based approach to regulation. It appraises the environmental risks and the performance of operators at sites that are regulated by the Environment Agency in the UK but the approach can be applied by others. Using OPRA helps to focus efforts where the risks are highest – and the resulting scoring system gives us unique data on company performance.

The scores have two parts – one which assesses the inherent environmental risk of processes on a site, known as the Pollution Hazard Appraisal, and the other which measures the operator's ability to manage these environmental risks, the Operator Performance Appraisal. Each score is assigned to one of five bands, from A (low hazard or well managed) to E (high hazard or poorly managed). Pollution Hazard Appraisal or Environmental Appraisal scores are unlikely to change significantly in the short term, and the score is largely beyond the control of the operator. But, the operator scores can change depending on how well the sites are managed.

Plants that score poorly under an OPRA regime will receive more attention from a regulator – in the form of inspections and general regulation. The tool therefore enables a risk-based approach to regulation: in planning resources, measuring performance and benchmarking sites with others of the same type. Compliance inspections pay particular attention to any of the attributes that scored poorly.

Charges can reflect OPRA scores as an incentive for operators to move from a lower band to a higher band. For example, the charges for holding a waste management licence in England reflect the site's OPRA score. There has indeed, been a general improvement in OPRA scores over the past three years although it is not known if this is due to the variable charging, peer pressure or media exposure.

7.4.1.8 Tradable permits. In any one environmental unit, the amount of water available for abstraction or the capacity of the environment to assimilate contamination is limited. When these conditions are reached, the unit is at its carrying capacity. Under these conditions, an approach of tradable permits is possible – it encourages the optimal use of the environmental capacity by creating a market.

Permit trading is well established in the field of water resources in arid areas. Water use permits for crops irrigation depending upon water transfer schemes incorporate various forms of rights trading, annual licence and even a rights auction to optimise the economic return from the water transfer. Well-established examples exist to the southwest USA (California, Colorado River) and in Australia (Snowy Mountain Scheme). By trading in water rights, abstractors have greater flexibility in how they manage their water needs as part of their business, through buying and selling rights in order to accommodate varying demands.

Permit trading in the UK is in its infancy. One of the problems of instituting a system of tradable permits is that it has to operate within a defined environmental unit so that the environmental impact of the trade is negligible.

Water Rights Trading is permitted under the Water Act 2003; it is too early to evaluate the extent to which this will happen.

In Europe, emissions trading scheme has been agreed in an attempt to deliver the requirements of the UN Framework Convention on Climate Change and the Kyoto Protocol. It is seen as the most cost-effective way of meeting these obligations. The scheme puts limits on carbon dioxide emissions for operators of industrial sites producing energy, cement, brick or paper, which have large combustion processes. The permits issued give each operator an allowance of carbon dioxide that they can produce. Operators are allowed to sell any surplus allowance they have if they produce less carbon dioxide than the limit set for them. Similarly, they can buy spare capacity from other operators if they are going to exceed their allowances. Setting the limits has depended on good baseline data. National Allocation Plans, which all Member States must produce, explain how the share of the national emissions cap under the Kyoto Protocol will be allocated for specific periods. There has been some difficulty in establishing these within the dates set by the Commission.

The scheme came into force in the UK in January 2005. Its effectiveness will be subject to evaluation in the coming years. By introducing the scheme, the EU hopes to use market forces to reduce the amount of carbon dioxide produced by industry and help towards achievement of the internationally agreed targets. The Environment Agency has responsibility for registering companies and monitoring compliance with the scheme in England and Wales.

7.4.1.9 Dealing with Pollution Incidents. Even with effective legislation and enforcement it is inevitable that, from time to time, pollution incidents will occur. They may arise because of a breach, accidental or malicious, of pollution control law or they may be due to unplanned circumstances, such as a road traffic accident involving a vehicle with a chemical load causing spillage that pollutes a river or aquifer. Regardless of the breach in environmental law the first requirement is to remedy the immediate situation. For this reason the environment agencies maintain an emergency response capability to attend accidents and emergencies where the environment might be placed at risk. Emergencies vary in scale and require collaborative action with other agencies but, in all cases, the objectives are to protect human life and health, protect property and protect the environment. This action will be combined with investigation of the causes of the pollution.

In the UK, pollution incidents are categorised from 1 to 4, with category 1 and 2 causing the most serious environmental impacts, where drinking water intakes may be affected or there are significant fish kills or there are potential impacts on human health.

The environment agencies have powers to prosecute people responsible for causing pollution. But the fines set by the courts do not always match the size of the crime; awareness of the seriousness of environment damage is still low, even among magistrates. During 2003, 61 companies in England and Wales were fined more than £10,000 for environmental offences, many of which had also been fined in previous years, but the average fine per prosecution for companies was only just over £8,000. The waste industry received the largest fines in total, closely followed by the water industry.

Even though fines may not be acting as a deterrent, there is good evidence that the number of pollution incidents is reducing. In 1993, there were over 7000 category 1 and 2 incidents; in 2003, less than 800. The most common sources of pollutants that caused serious harm in 2003 were the sewage and water industry (25 per cent), agriculture (13 per cent) and industry (12 per cent), although the sources of some 39 per cent were unknown. These figures reflect the spatial scale of the activities to a certain extent – there are many more consented discharges from the water companies than any other sector so more potential for things to go wrong.

Much of the reduction is due to increased awareness of businesses of the impact of their activities and the pollution prevention activities of the environment agencies. However, there are other forces in play. Since 2000, the Environment Agency has reported on the 'Spotlight on business environmental performance', which 'names and shames' the companies which have been prosecuted in the last year. It also gives credit to companies which have reduced their impacts on the environment. The reporting, which always get good media coverage, helps to create pressure on the companies to improve. There is some evidence that a poor environmental record affects share-prices – another driver for change.

7.4.2 Fiscal Measures

Environmental taxation, tax-breaks or subsidy reform, which aim to change the price signals in the market place in favour of environment-friendly approaches and products can be a powerful way to make changes. The European Environment Agency places huge emphasis on such measures. In the foreword to their Signals 2004 report they state: "Further progress in managing the environmental impact of agriculture, transport and energy … can be achieved by further increasing the use of market-based instruments to manage demand and internalise external costs, by switching … to positive subsidies and by promoting innovation".[11]

The proportion of total tax revenue derived from environmental taxes grew slowly in the decade to 1995 in Europe, but have since levelled off. Pollution and resource tax revenues have grown more than average but still contribute a very small share of overall tax burdens. Revenue from transport taxes has remained more or less constant. The overall share of total revenue from environmental taxes varies between Member States, ranging between five and ten per cent.[11] It is interesting to see the range of products to which taxes are now applied (Table 7).[26]

The Irish government, for example, introduced a tax on plastic bags (15 cents per bag) in 2002 in an attempt to curb their use and reduce litter (plastic is persistent as well as causes aesthetic pollution). Retailers are obliged under law to pass on the full amount of the levy as a charge to customers at the checkout. The levy applies to all plastic bags except those used to contain non-packaged goods, for example, fruit, vegetables and for food safety reasons, smaller bags which are used to contain fresh meat, fish and poultry. It has been very successful in reducing the number of plastic bags used, by more than 90 per cent. More than one billion plastic bags per year have been removed from circulation due to the tax.

In the UK, the landfill tax was introduced in 1996 and levied at £7 per tonne on active waste and £2 per tonne on inert waste. The purposes of the tax were to promote the 'polluter pays' principle by increasing the price of landfill to better reflect its environmental costs, and to promote a more sustainable approach to waste management by making landfill a less attractive option and to aid more recovery and recycling.

A government inquiry into its effectiveness took place in 1999. At that stage, most witnesses indicated that there had been some reduction in inert

Table 7 *Range of products or services to which environmental taxes are applied in one or more European countries*

Sector	Item/substance	Example countries
Air/energy	Carbon dioxide	Finland, Italy
	Sulphur dioxide	Denmark, France
	Nitrogen oxides	France, Italy
	Fuels	Most
	Sulphur in fuels	Norway, Sweden
Transport	Car sales and use	All
	Differential car tax	Germany, UK
Water and waste	Effluent	Most
	Dangerous waste	Finland, Germany
Noise	Aviation noise	Netherlands, Norway
Products	Tyres	Austria, Finland
(examples)	Beverage containers	Belgium, France
	Packaging	Austria, Italy
	Disposable tableware	Denmark, Iceland
	Pesticides	Belgium, Denmark
	Solvents	Denmark, Norway
	Light bulbs	Denmark
	Batteries	Italy, Belgium
	Fertilisers	Sweden

waste going to landfill (from construction and demolition waste), but little change in active waste.[27] Since then, the tax has increased annually to try to change behaviour and to encourage industry to invest in technologies to deal with waste, other than landfill. But the price is probably still too low to have any impact. In other countries, for example, Sweden and Austria, the tax on landfilling gives an economic incentive in favour of incineration. The UK government has now had to set targets for all local authorities to meet in terms of reducing the amount of biodegradable waste going to landfill. The Landfill Directive requires this; the landfill tax may have been a signal for change in 1996 but it is now through regulation that the UK hopes to meet the requirements of the Directive.

One of the counter-arguments to environmental taxes is concern about other consequences. When the landfill tax was brought in, there were concerns that fly-tipping would increase (to avoid payment). Fly-tipping does appear to have increased in the past four years but the origin is mainly domestic waste, suggesting no link to the tax, although there is evidence of large loads of inert materials (construction and demolition waste) being dumped illegally in London and other big cities. But data on fly-tipping is poor and the UK government has recently sponsored the development of 'fly-capture', a system to improve its information base. All local authorities must input data routinely. This should improve understanding of the sources and options for tackling. Initial results in March 2005 showed that there were some 75,000 fly-tips a month, 200 of which were the 'big, bag and nasty', costing local authorities 24 million pounds in clean-up over six months.

The UK is also using taxation to instigate changes in behaviour in car transport to reduce its growth. There is a differential car tax and fuel tax in favour of greener vehicles at present (linked to carbon dioxide emissions), but a fuel tax escalator, which was introduced in 1990s in an attempt to move people away from car use, had to be abandoned later. Just when the tax appeared to be biting, businesses staged a protest; the price of operating in the UK was unfavourable and not a 'level-playing field' with other countries in Europe. One of the problems with this tax was the volatility in the underlying price of oil. When this was high, the tax made it even higher and beyond what could be afforded by some businesses. It is interesting to note that despite increases in fuel duty in the 1990s, the overall cost of motoring has remained unchanged since the early 1970s because of reduction in other motoring costs. At the same time, public transport fares have continued to rise. Both these signals are going in the wrong direction of sustainable development. The example shows how hard it is to make the policy changes we need to improve the environment.

Taxing energy products is the subject of a draft EU Directive. It offers a mechanism for introducing price differentials between fuels to encourage the use of less polluting fuels. Evidence of the environmental effectiveness is limited but in Finland, differentiated excise tax on motor fuels has accounted for a 10–15 per cent reduction on carbon monoxide and hydrocarbon emissions from car traffic.[28]

Another example worth exploring in some detail is that of farm subsidies. About three-quarters of the land in England and Wales is used for agriculture so the way in which farmers operate has a large impact on the state of the environment. The European Common Agricultural Policy (CAP) is a significant economic instrument influencing agricultural practices. Its original objective was to increase agricultural productivity by promoting technical progress and making more effective use of resources. Its overall aim was to ensure that food supplies were available in the EC at reasonable prices and that rural communities could maintain a reasonable standard of living. Both these objectives and aims are consistent with sustainable development, but its implementation has led to difficulties.

To meet its aims, it set guaranteed prices for major production by intervention buying and import levies and paid export refunds to compensate traders for selling EU produce at less than world prices. Financial support from CAP since 1980 maintained the viability and increased the profitability of sheep farming, but because payment was per head, it led to greater stocking densities. This led to overgrazing in some parts of the UK affecting upland habitats and increasing the risk of soil erosion. Furthermore, the need for the use of sheep dips increased (because of stock density) leading to an increase in the incidence of pesticide pollution of headwater streams.

Similarly, there have been subsidies for various types of crops grown which have changed over the years – oilseed rape increased initially and more latterly linseed. Some of these have had unforeseen environmental consequences. For example, maize growers have used atrazine as a pesticide, leading to increased occurrence of this in fresh waters in the 1990s although this has now decreased with greater awareness of the problem. In general, the area treated with pesticides has increased from 27 million hectares in 1998 to 42 million hectares in 2002, reflecting the drive for greater productivity. The decisions made by farmers on types of crops to grow and the number of livestock or size of dairy herds was thus very much tied to quotas and economic factors, sometimes bringing marginal land into productions, without any concern for the impact on natural habitats or biodiversity. Indicators show a severe decline in farmland birds, and an increase in diffuse pollution over the years correspondingly.[29]

There have been various reforms of CAP over the years. The 1992 reforms placed a requirement on Member States to introduce measures to encourage environmentally sensitive farming. Various agri-environmental schemes were introduced in the UK and by 1999, about seven per cent of the total agricultural area in England was under some form of scheme. But, the payments associated with these were less than 3.75 per cent of all UK expenditure on grants and subsidies administered under CAP. In 2005, the Single Payment Scheme was introduced. In general, there is greater freedom to farm to the demands of the markets as subsidies are being decoupled from production. At the same time, environment-friendly farming practices will be better acknowledged and rewarded. To qualify for subsidy from the scheme, farmers must meet cross compliance standards and requirements, which are a set of rules to provide minimum levels of environmental and public health, animal and plant health and animal welfare protection. These include conditions for the protection of hedges and watercourses, soils and habitats for example.

It will be interesting to see the effectiveness of the new approach. The scheme is being phased in over seven years so it will take a while to evaluate. The level of payment will be a key element; concerns have been expressed that some farmers may not need to produce anything now – merely act as custodians of the countryside and receive payment for such. While this may meet environmental objectives, it is only one of the aims of agriculture and there is still a large population to feed – importing more and more food requires more and more transport with its consequent impact on the environment. Relationships are very complex.

7.4.3   Voluntary Measures

These relate to economic sectors agreeing to do things differently to improve the environment. Businesses may agree to adopt voluntary measures when regulatory or tax measures are being promoted and which they wish to avoid. For example, a pesticides tax was considered by the UK government with the aim of reducing pesticide use. The revenue received was to be used to offset the damage caused by pesticides. But the tax has not been introduced. Farmers have agreed to a voluntary initiative which is a programme of measures, agreed with the government, to minimise the environmental impact of pesticides. It requires farmers and pesticide advisers to follow best practice, ensure users are trained in pesticide use and for crop protection management plans to be followed.

It is early days to determine the effectiveness of this initiative but signs are good. In 2003, there was a decrease in the frequency of detection of pesticides in monitored rivers for most types of pesticides; only chlorotoluron was found more frequently than in 1998.

Perhaps one of the best examples of voluntary measures is that of the way businesses have adopted environmental management systems (EMS). This is the generic term used to show ways in which organisations set environmental objectives for themselves and is the subject of international standard ISO

14001,[30] which from 1997 has replaced the equivalent British Standard BS7750. The key reasons why a company would wish to develop an EMS are

- The requirement for regulatory compliance
- The need to establish systems to ensure that compliance is maintained
- Financial incentives
- Market place pressures
- Green credentials.

An important factor in the development of EMS has been the influence which companies have on their trading partners, the so-called supply chain pressures. Major manufacturers and traders – particularly in the automotive industry, which has been in the forefront of quality management, supermarkets and other parts of the retail trade – have been a major force, through their own environmental policies, in encouraging and often obliging their suppliers to adopt an approved EMS.

The principal formalised EMS in Europe is the European Communities EMAS (Ecomanagement and Audit Scheme). This can be achieved by meeting the ISO 14001 standard but also requires a verified public environmental policy to meet the requirements.

The take-up of EMAS within Europe has been very variable. In many countries no sites had been registered by mid-1997, more than two years after the scheme was initiated, but after eight years, the take-up is widespread, although still limited in some countries such as France, Ireland and Luxembourg. The greatest take-up is in Germany, which is generally believed to be because companies in that country believe that registration under EMAS will reduce the formal regulatory burden upon them.[31] The position of the environment agencies in the UK is that they encourage companies to establish an EMS and it is a factor considered in the company's OPRA score but this, in itself, is not a substitute for environmental regulation. Indeed, some research done by the Environment Agency suggests little difference in the performance of EMAS accredited sites to those without. Some EMA accredited sites have been prosecuted for causing pollution.

There is good evidence from many companies that investment in EMS is repaid in both tangible benefits, such as reduced operating costs, and intangibles, for example, improved safety, improved public perception and less risk of prosecution. However the achievement of EMAS or ISO 14001 accreditation and the ongoing maintenance of the necessary systems is a significant investment in time and resources, which many smaller organisations are unwilling to commit. The benefits, in reduced operating costs and reduced risk of causing pollution, of adopting environmental best practice are still important to business and bring with them an improvement to the environment. Therefore, following the adage that 'prevention is better than cure', pollution control agencies are increasingly devoting effort to education in pollution prevention and good environmental management through special campaigns, on-site advice and promotion of waste minimisation and recovery projects.

Voluntary measures include those that rely on consumer power. Various groups, organisations and individuals have raised awareness about the impact of various products on the environment, either in their production or use, which has sent strong messages to those with purchasing power, and affected their choices. This can then lead to an impact on the way in which products are manufactured including the seeking of alternatives.

An European eco-label award scheme was launched in 1992 and by 2002, 98 manufacturers, retailers and service providers were eligible to use the 'Flower' logo, with awards having been granted for over 400 products. The overall number of awards is low and concentrated in a few product groups and a few Member States. The highest number of awards exist in Denmark, France, Italy and Spain and the lowest in Austria and Luxembourg (none in either) and very few in Finland, Belgium, Ireland, the Netherlands and the UK. For example, textile products and indoor paints and varnishes account for 36 and 28 per cent of the awards respectively. But, the European Commission hopes to raise the profile of the scheme to reach between 1 and 30 per cent of the market share, depending on product type. Such targets are ambitious: currently less than 0.1 per cent of paint and varnish products carry the Flower logo.[11]

The Danish Environment Protection Agency has run an annual campaign since 2001 to promote the EU Flower and Nordic Swan eco-labels. This encourages children and consumers to understand the benefits – six other countries participate. The campaigns are considered worthwhile with an increase of 60 per cent of products to use the Flower label since 2002.

The Nordic Swan has a high level of recognition among Scandinavian consumers. In Germany, there is a Blue Angel scheme which was introduced in the 1970s and is found on over 3000 products. Other eco-labels used in the EU include the Energy Star for energy-efficient office equipment (which originates from the United States) and the logo for organic farm products.

7.4.4   Coercive Measures

Voluntary methods are often favoured by business; regulatory methods tolerated, but coercive measures are positively disliked. These measures include moves to ban or restrict certain activities or the sale of certain products that are harmful to the environment. They are generally disliked because it is often difficult and expensive to find alternative products or substitutes that are as effective as the original substances. It usually takes many years for substances to be banned – there is a need for good scientific evidence of the problem which is usually hard to obtain, and then time to develop alternatives, which need to be assessed to ensure that their impact is less than the original products.

Phthalates are an example. These are a group of chemicals used mainly in softening PVC. There has been political concern and public debate over their potential harm to human health. There is a possible risk of hormone-disrupting effects, but research is ongoing. At the end of 1999 the European Commission placed an emergency ban on the use of phthalates in soft PVC toys intended to be chewed or sucked by children under the age of three.

Changing from one product to another may lead to unforeseen problems. For example, up until the mid-1990s, sheep dips were based on organophosphates. There was concern that these were harmful to the health of farmers and a switch to synthetic pyrethroids occurred. The environmental consequences have been large. The inappropriate disposal of a small amount of these can have a devastating effect on aquatic organisms (at a rate of one teaspoonful in a swimming pool).

Another example is that of tributyl tin. While this has been found to have devastating effects on certain marine organisms, finding an adequate anti-fouling substance for use on large ships has been difficult which is why a ban in 1986 only applied to small boats. Without the anti-foulants, the ships move slower using up more energy and thus causing a greater impact on the environment in this regard. It is only now, 20 years since the original ban, agreement has been reached for its removal from all ships. Even so, some of the alternatives, which include copper-based biocides, also raise environmental concerns.

Coercive measures can have large environmental benefits, even if they are seen as unpopular initially. Lead concentrations in the atmosphere have reduced substantially since the introduction of unleaded petrol. Dog whelks are re-establishing themselves around the UK coast since the banning of tributyl tin on small boats in 1986, although there is still a problem in areas which receive large ships – hopefully to be resolved in the near future.

7.4.4.1 Diffuse Pollution: Meeting Local Air Quality Targets and Objectives. The requirements of the UK National Air Quality Strategy have been outlined previously (Section 7.3.1) and the role of the local authorities in defining Air Quality Management Areas (Section 7.1.2). It is worth considering how the authorities are working to improve local air quality to meet the objectives set; the measures taken are essentially coercive.

By June 2002, 100 authorities in England and Wales had declared an air quality management area, including 28 in London.[32] The areas range in extent from a couple of streets to much larger urban areas. Traffic is the main problem and local authorities are adopting various traffic management measures to address it. For example, Park and Ride schemes encourage people to leave their cars outside town centres to reduce the amount of traffic. These schemes have generally been successful but Oxford is an example, where air quality standards are still exceeded due to the emissions of the public transport available (buses). In the longer term, cleaner vehicle technology like liquid petroleum gas, compressed natural gas, hybrid electric vehicles and fuel cells which will greatly reduce emissions may be needed. Other strategies have included banning vehicles from certain streets all together, or creating lanes where only vehicles with passengers can pass. In some cases, coercive measures have been replaced by fiscal measures or combined with them. Congestion charging in central London is an example. This has helped to reduce pollution and congestion, but at the cost to road users.

It is not only traffic that causes objectives to be exceeded. Large industrial processes may play a part. Local authorities therefore have to liaise with the environment agencies to ensure that these sources are tackled too by direct regulation.

7.5 MONITORING AND REVIEWING

The next step in the environmental management cycle is to monitor the impact of the actions taken, which allows us to review their effectiveness. Monitoring has been considered in a previous chapter so it is not repeated here. This step then leads on to evaluating the state of environment in an integrated way – and so the environmental management cycle continues.

In conclusion, environmental management is complex. There is a range of approaches to taking actions; what is most effective may vary from time to time, from place to place and with the nature of the risks involved. This chapter gives an outline to what is involved and how sound science is needed to inform the process.

QUESTIONS

(i) Describe the environmental management cycle and how the concept helps to deliver a better environment for the future generations.

(ii) Why do we need to use indicators in environmental management?

(iii) How do European Framework Directives differ from the original approach taken by European Directives?

(iv) What is the role of national and local governments and their agencies in determining environmental improvements? Why is it important to involve the public in decision-making where possible?

(v) Explain why the Common Agriculture Policy has led to environmental problems and what is being done to address them.

(vi) Compare and contrast the benefits and disbenefits of voluntary action *vs.* environmental regulation.

REFERENCES

1. European Environment Agency, *Europe's Environment: The Second Assessment*, Elsevier, Amsterdam, 1998, 7.

2. Environment Agency, *Spotlight on Business Environmental Performances 2001 Report*, Environment Agency, Bristol, 2002, 11.

3. Environment Agency, *Spotlight on Business Environmental Performance 2002 Report*, Environment Agency, Bristol, 2003, 28.

4. Environment Agency, *Viewpoints on the Environment*, Environment Agency, Bristol, 1997, 8.

5. USEPA, About EPA, www.epa.gov/epahome/aboutepa.htm, 2005.

6. Organisation for Economic Co-operation and Development, *Environmental Performance Reviews: United Kingdom*, OECD, Paris, 1994.

7. European Environment Agency, *Europe's Environment: The Second Assessment*, Elsevier, Amsterdam, 1998, 9.

8. Department of Environment, Transport and the Regions (DETR), *Quality of Life Counts*, DETR, London, 1999.

9. Department of the Environment, Food and Rural Affairs, Headline indicators of sustainable development, www.defra.gov.uk/environment/statistics/supp/spkf09.htm, 2005.

10. European Environment Agency, *Europe's Environment: The Second Assessment*, Elsevier, Amsterdam, 1998, 16.

11. European Environment Agency, *Signals 2004*, EEA, Denmark, 2004.

12. Department of Trade and Industry, *Digest of United Kingdom Energy Statistics*, www.dti.gov.uk, 1999.

13. National Environmental Technology Centre, *National Atmospheric Emissions Inventory*, AEA on behalf of Department of Environment, Transport and the Regions, www.aeat.co.uk/netcen/airqual, 2000.

14. Department of the Environment, Transport and the Regions, Scottish Executive, National Assembly of Wales and Department of the Environment in Northern Ireland, *The Air Quality Strategy for England, Scotland, Wales and Northern Ireland*, Cm 4548, DETR, London, 2000.

15. Department of the Environment, Transport and the Regions, *Waste Strategy 2000: England and Wales*, The Stationery Office, London, 2000.

16. Agra Europe (London) Ltd, Country reports, *European Packaging and Waste Law*, Agra Europe (London) Ltd, October, 2000.

17. Royal Commission on Environment Pollution, *Best Practicable Environmental Option*, 12th Report Cm 310 HMSO, London, 1988.
18. Department of the Environment, Food and Rural Affairs, *Survey of Public Attitudes to Quality of Life and to the Environment: 2001*, www.defra.gov.uk/environment/statistics/pubatt/ch4conc.htm.
19. Environment Agency, *The State of the Environment of England and Wales: The Atmosphere*, The Stationery Office Ltd, London, 2000.
20. Environment Agency, *The Water Quality of the Tidal Thames*, The Stationery Office, London, 1997.
21. European Commission. *Communication from the Commission on the Precautionary Principle*, COM (2000) 1, Brussels, 2000.
22. European Environment Agency, *Late Lessons from Early Warnings: The Precautionary Principle 1896–2000 Environmental Issue Report No. 22*, European Environment Agency, Copenhagen, 2001.
23. Europa, *Ambient Air Quality*, www.europa.eu.int/comm/environment/air/ambient.htm, 2005.
24. Environment Agency, *Hazardous Waste, a Growing Challenge*, Environment Agency, Bristol, 2004.
25. Environment Agency, *Delivering for the Environment: A 21st Century Approach to Regulation*, Environment Agency, Bristol, 2000, 33.
26. European Environment Agency, *Europe's Environment: The Third Assessment*, European Environment Agency, Copenhagen, Denmark, 2003.
27. Environment, Transport and Regional Affairs Committee, *Thirteenth Report: The Operation of the Landfill Tax*, 1999.
28. European Environment Agency, *Environmental Signals 2002, Environmental Assessment Report No. 9*, EEA, Copenhagen, 2002.
29. Environment Agency, *The State of the Environment in England and Wales: The Land*, The Stationery Office Ltd, London, 2000.
30. Environmental Management System – Specifications with Guidelines for Use, EN ISO 14001, CEN, Brussels, 1996.
31. Europa, *EMAS Organisations and Sites, 2004*, www.europa.eu.int/comm/environment/emas/pdf/5-5articles-en.pdf/, 2005.
32. NETCEN, *The UK National Air Quality Information Archive*, AEA Technology on behalf of Defra, www.aeat.co.uk/newcen/airqual, 2002.

Subject Index